Galileo: Heretic

GALILEO HERETIC

(GALILEO ERETICO)

by Pietro Redondi

Translated
by Raymond Rosenthal

PRINCETON UNIVERSITY PRESS, PRINCETON, NEW JERSEY

Published by Princeton University Press, 41 William Street,
Princeton, New Jersey 08540

Library of Congress Cataloging in Publication Data will be found
on the last printed page of this book

ISBN 0-691-08451-3
ISBN 0-691-02426-X

First Princeton Paperback printing, 1989

Publication of this book has been aided by a grant from the
Italian Ministry of Foreign Affairs
This book has been composed in Linotron Galliard text
and Weiss display type

Clothbound editions of Princeton University Press books are printed on
acid-free paper, and binding materials are chosen for strength and
durability. Paperbacks, although satisfactory for personal collections,
are not usually suitable for library rebinding

Printed in the United States of America by Princeton
University Press, Princeton, New Jersey

CONTENTS

ILLUSTRATIONS

Facing page 150

ACKNOWLEDGMENTS

"The past, exactly because it is the *past*, forever remains inaccessible to us: it vanishes, it is no longer there; we cannot touch it, and it is only starting from its vestiges and its traces, its still-present remains—books, monuments, documents which have escaped the destructive action of time and men—that we try to reconstruct it. But objective history—the history that men make and undergo—cares very little about the history of historians; it lets things without value subsist and destroys pitilessly the most important documents. . . ."

If there is a passage of Alexandre Koyré's to which I owe my gratitude more than any other, it is this one. Even more than Koyré's great historical teachings on Galilean thought, it is this lesson, apparently banal, that has guided the passion and method with which this book has been written concerning the labile traces of another history of the Galileo case.

I am indebted to other authors of the period between the two wars for the perspectives of my research: to Lucien Febvre, for his attention to the mental universe of the religious seventeenth-century man; to Lynn Thorndike, for his profound interest in the minor figures, the marginal authors, those defeated by the progress of so-called positive science; to Delio Cantimori, for his devotion to another kind of defeated person, the witnesses of religious heresy, and for those pages of posthumous notes which have been for me more illuminating than any other work on Galileo's condemnation.

For their teaching, exchange of points of view, and suggestions, the author wishes to thank teachers and friends who are specialists in the scientific seventeenth century: Pierre Costabel, Secretary for Life of the Académie Internationale d'Histoire des Sciences; Ernest Coumet (E.H.E.S.S.); Jean-Claude Guédon (University of Montreal); Ludovico Geymonat (University of Milan); and William Shea (McGill University).

I also want to thank the directors and personnel of many ecclesiastical archives and libraries that made possible my long and complex research—in particular, His Eminence Cardinal Joseph Ratzinger, president of the Sacred Congregation for the Doctrine of the Faith (Rome) and Father Giacinto Mariani, O.S.B. Oliv.; Father Edmondo Lamalle, S.J., of the Historical Archive of the Superior General Curia of the Society of Jesus (Rome); Father Brunet, S.J. (Les Fontaines, Chantilly); Father Monachino, S.J. (Gregorian University, Rome); Father Nello Morrea of the headquarters of Clerks Regular Minor (Rome), Don Carmine Latina of the Archiepiscopal Seminary at Teano (Caserta).

I am grateful for their sensitivity and courtesy, which have given me access to original documents and which have always made up for traditional difficulties: gaps in documentation, lack of space and personnel.

That said, a historian may loyally express another feeling, with the profound freedom of one who has as his sole preoccupation the desire to understand. The privilege from which I have benefited, through a series of exceptional and casual motives, and which has allowed me to know new documents of the seventeenth century, cannot fill me with private satisfaction or moral consolation since these archives are usually closed to historians.

The archives of a great institution of the past—we speak in particular of the Holy Office—are a part of the memory of humanity. The conservators of an archive have a great responsibility; to decide to keep an archive closed to historians means to sequester and keep in the dark, in order to monopolize study and comprehension, a part of the past and of the memory of men. And this is an abuse of one's fellows undeserving of justification or comment.

The author finally wants to make it clear that he alone is responsible for any remaining inaccuracies and mistakes.

Galileo: Heretic

INTRODUCTION

> The custome of those who write histories is, to propose in the beginning, a modell of what they meane to handle. Wich I have tought fit to deferre untill this place, making it an abstract of the wich is related already, and a dessigne of that wich is to follow. Having resolved to give to the memorials, wich I had collected, some forme, wich might not exceed my power, and yet best befit the matter . . . I tought first that the forme of a Diary would best agree to the subject.*

The Inquisitor's Gaze

At a distance of three hundred years, an image can change. On the morning of June 11, 1982, I was in Rome. In a room on the first floor of the Palace of the Holy Office, I was waiting to consult a new and mysterious document on Galileo. This is a justly famous palace, framed by magnificence. The window of the room offered, with an unforgettable black-and-white effect, both a side view of St. Peter's nearby cupola and, in the foreground, in the dark panes of the Hall of General Audiences, a singular reflection of the piazza's colonnade. But, inside the room, there was another reflection with purplish tones: a painting. It was a quite large portrait of Saint Roberto Bellarmino, who for many years had been a cardinal-inquisitor within these walls. In a certain sense it was as though the master of the house had appeared to welcome me (Fig. 16).

The cardinal's minute person was depicted standing erect, his hands clasped over an open volume, dense with type, on a desk cluttered with open and closed books: the portrait of Bellarmino's fabulous erudition. There are many portraits of Bellarmino, and many that are well known. But the portrait I had before me was almost unknown. It gave one a strange sensation. In fact, the painting seemed to contain a hidden image—another portrait, identical to this and imprinted on my memory, but hidden here by something that prevented me from recognizing it at first sight. The light of the halo that crowned the cardinal's head served to clarify that sensation. No less flagrantly than a neon sign, the halo declared that the painting was a very recent work. It was a copy. Probably it went back to the twenties, the period in which Bellarmino was sensationally beatified (in 1923), after a trial unprecedented for its length, for the number of its adjournments, and for the importance of its appeals and polemics.

* *The Historie of the Councel of Trent*, written in Italian by Pietro Soave Polano (Friar Paolo Sarpi) and faithfully translated into English by Nathanael Brent, London 1620, 3rd ed. 1640, p. 707

I remembered the photograph of the original from which this copy had been made. Anyone can easily track it down. In fact, it illustrates the article entitled "Roberto Bellarmino" in the famous Treccani *Italian Encyclopedia*. The caption under that photograph (Fig. 17) is: "From a painting by Pietro da Cortona, Rome, Superior General Curia of the Society of Jesus."[1]

Do not ask me for any more because, from what I have been able to ascertain, all that exists of the original portrait is this photo and the uninformative caption that accompanies it. The rest is a mystery, which I am quite happy to hand over to my colleagues in art history.[2]

This may be too generous an attribution, owing perhaps to the fact that a very old Bellarmino, such as the man depicted in the photo of the original portrait, could have been a particularly valuable subject for a Pietro da Cortona at the beginning of his career in the Roman court. Or perhaps because, as in other youthful works of Pietro da Cortona, this painting too appears from the photo to be very "tinted," as it would have been put in the seventeenth century; that is, with a prevalence of dark areas. Or because, like all the rare portraits by Pietro da Cortona, this one too seems extremely "faithful."

As has been said, the man depicted is a very old Bellarmino, having reached through illnesses and penances the respectable threshold of eighty years. Even more than the gray of head and beard, the realistic, disproportionate dimensions of nose and ear are revelatory signs of extreme old age.

The cardinal must not have appeared very different from this to Galileo on the morning of February 26, 1616, when the scientist was summoned to the Paradiso Rooms in the Vatican Palace—Bellarmino's official residence—to be informed of Copernicus's condemnation. The exquisite regard of that private communication was proof of the cardinal's profound deference to the great scientist. In that delicate circumstance, destined later to play the crucial role which everyone knows, Galileo could appreciate his host's tact and affability.

But if we believe in the truth of that portrait, Galileo, who was "big and square," must also have felt to the bottom of his heart the penetrating and

[1] See *Enciclopedia italiana di scienze, lettere ed arti*, vol. VI, Rome 1930, p. 548 ff., concerning Fig. 17.

[2] I can add only that, going to the Superior General Curia of the Society of Jesus in Rome to do some research on certain Jesuit scientists, I did not find a trace of the original portrait. Also, at the Jesuits' central site, the Church of Gesú, all memory of that portrait has been lost. It is a mystery, too, so far as I have been told, at the other colleges and churches of the order in Rome. A curious mystery: a seventeenth-century portrait depicting Bellarmino does not drop into the void, all the more so if an indirect attribution (the caption is not very clear, but it seems to indicate a contemporary copy) to Pietro da Cortona is involved. Yet, to tell the truth, this attribution is also a mystery. Neither this nor any of Bellarmino's other portraits is in fact mentioned, even among the works falsely attributed to Cortona, in monographs devoted to the great maestro of Barberinian Rome. See G. Briganti, *Pietro da Cortona*, Florence 1962, 2nd ed. 1982; A. E. Sanchez, *Pintura italiana del XVII siglo*, Madrid 1965, p. 264.

perspicacious gaze of Cardinal Bellarmino's eyes. These were "bright eyes,"[3] an important identifying mark given us by his contemporaries, men who were accustomed to look at their fellows with the refined art of the physiognomical gaze, so as to decipher in the signs of the face the character of the spirit.[4]

In the photograph—all that is left of the original portrait—Bellarmino's short, fragile figure hangs completely on the energy of his gaze. This gaze, as in other, more youthful portraits, looks insistently into the spectator's eyes and says much about the man's determination of character which is thus emphasized.[5]

Here was a gaze worthy of the great cardinal who had navigated through all the Roman congregations, of a famous intellectual, protagonist of the principal political and religious affairs of his day, a statesman of European stature, and a cardinal who was twice a candidate for the pope's tiara. Here was a man whose name was anagrammatized in the phrase *Robur bellum arma minae*. Here was the "hammer of the heretics," as he was officially commemorated in the necrology of the Sacred College.[6] That is how his contemporaries saw Bellarmino: a fighter who had never hesitated to throw the weight of his position as cardinal into the controversy *de auxiliis* against the Dominican theologians. In short, the gaze in the original portrait is a rather frightening one, worthy of the great protagonist and the great inquisitor that Bellarmino was in his time, a period in which both politics and religion were arts that centered around the gaze.

The copy I now had before me in the Holy Office reproduced the original in all its details: the same pose, the same cardinal's cap worn with affable sim-

[3] See Father D. Bartoli, S.J., *Della vita di Roberto cardinale Bellarmino, libri quattro*, Rome 1678, p. 256.

[4] On the art of the gaze in the seventeenth century, I mention here F. Bacon, *The Essayes; or, Counsels civill and morall*, London 1597, 3rd ed. 1629, essay 22, p. 128 (with special reference to the Jesuit technique of the gaze); G. B. della Porta, *De humana physiognomia*, Vico Equense 1586; and, in a more general way, *Breviarium politicorum secundum rubricas mazarinicas*, Coloniae Agrippinae 1684, in the Italian version edited by G. Macchia, Milan 1981.

On the difference between Renaissance "acuteness of vision" and the superiority of sight in the seventeenth century, see A. Koyré, "L'apport scientifique de la Renaissance" and "Attitude esthétique et pensée scientifique," in *Études d'histoire de la pensée scientifique*, Paris 1966, 2nd ed. 1973, pp. 52 and 276.

[5] See the well-known portrait of Bellarmino by Bartolomeo Passarotti preserved at the Jesuit College of Chamartin de la Rosa (Madrid), which goes back to 1606 when Bellarmino was sixty-four. For a Bellarminian iconography, see Father A. Fiocchi, S.J., *S. Roberto Bellarmino*, Isola del Liri 1930, where the mysterious Cortona-like portrait is reproduced without explanation, but with a very conspicuous halo (p. 480).

[6] See Father G. Fuligatti, S.J., *Vita del cardinale Roberto Bellarmino*, Milan 1624 (2nd ed., Rome 1644); Father A. Eudaemon-Johannes, S.J., *De pio obitu Roberti card. Bellarmini*, Dilingae 1621; Bartoli, *Della vita di Roberto card. Bellarmino*; and the most extensive modern biographies: Father J. Brodrick, S.J., *The Life and Work of Blessed Robert Francis Cardinal Bellarmino, S.J.*, 2 vols., London 1928; and Father E. Raitz von Frentz, S.J., *Vita di San Roberto Bellarmino*, Isola del Liri 1930.

plicity by this prince of the Church, the same books, the same instruments of an indefatigable intellectual labor. And yet something was irremediably different, a detail, but a significant one: the look in his eyes.

Cardinal Bellarmino, as portrayed in this copy, no longer probes into the eyes and hearts of his spectator. He has a bemused, mildly abstracted expression. Actually his eyes do not focus on any precise point; his gaze goes beyond the painting, to wander absent-mindedly along the walls of the room in the Holy Office. What is the meaning of this change?

The look in Cardinal Bellarmino's eyes has changed because the look of the person who painted him has changed. The art of the copyist, by some secret ability, can go so far as to counterfeit an original exactly. However, this copy exhibits a different and less secret fascination and endeavor. The intention was perhaps to reproduce the original in order to camouflage it, so as to obtain an identical and at the same time unrecognizable result.

The persons who commissioned this copy are unknown, but it is entirely possible that they were motivated by the same celebratory and propagandistic rationale behind the copious Bellarminian oleography that flourished in the twenties and thirties of our own century. After the happy conclusion of the hotly contested beatification trial, tenaciously sponsored by the Jesuits for three centuries, Bellarmino obviously appeared in a new light, the light cast by the halo that crowned him.

The propaganda campaign aimed at a very rapid consolidation of that success, making use of Bellarmino's stunning career from beatified to saint to doctor of the Church in the space of a few years. This repaid his supporters for the frustrations and the political and intellectual rejections met with in the past, as well as for certain criticisms leveled at the heroism of Bellarmino's theological abilities.[7] The propaganda was therefore centered on the image of Bellarmino as the "Immaculate Doctor." Yet it is evident that in order to present that image and obtain a convincing effect of unquestionable beatitude, nourished by meekness and spiritual asceticism, it was not enough to stick a halo on a faithful portrait.[8] In order to make the now officially recognized beatitude shine through, it became necessary to retouch those eyes or depict in that gaze one of those long moments of ecstasy which, according to the first hagiographic lives, seized the old statesman (with greater frequency during his last days) with "a moderate elevation of the eyes, interrupted sighs, and slight movements of the lips."[9] So the author of the copy, though not a Guido Reni or an Andrea Pozzo, replaced the irre-

[7] On the Bellarmino trial, see Father P. Tacchi Venturi, S.J., *Il beato R. Bellarmino. Esame della nuove accuse contro la sua santità*, Rome 1923; and I. de Récalde (pseudonym of Abbot Boulin), *Un saint jésuite. La cause du vénérable Bellarmin*, Paris 1923.

[8] The mediocrity of such a crude montage is visible in the portrait above Cardinal Bellarmino's tomb in the Church of St. Ignatius.

[9] See the physiognomical analysis of Bellarmino's ecstatic poses in Fuligatti, *Vita del cardinale Roberto Bellarmino*, p. 376 (2nd ed., p. 295).

ducible gaze of the original portrait with an expression of meek spirituality and abstracted ecstasy.

The image had changed. The image of a statesman, of a great controversialist who deserved from his contemporaries the epigraph—"With force I have subdued the brains of the proud"—engraved on his tomb in the Church of Gesú, between the statues of Science and Religion, had become, for reasons of edification, that of a devout dreamer.

Questions and Proposals

Bellarmino's unexpected portrait in that room in the Holy Office is an introduction of exemplary value to the story that awaits us. I had been led there by a delicate critical problem in the history of Galileo's ideas on physics. As we shall see, that history bore some surprising analogies to the history of the portrait. In brief, this was a case of substituted theory, put forth by Galileo in the guise of an identical theoretical picture. The episode previously described—of original images that disappear, of anonymous copies that reappear, of camouflaged emotions and faces, of obvious alterations and hidden apologetic intentions—in reality offers not only a foretaste of the psychological nature of the research I have set out to describe, but also a precious methodological lesson.

One transformed detail, as we have seen, can change many things. A transformation of this kind can be owing to purely intrinsic, technical, or simply casual considerations. But the case of the portrait shows quite eloquently that completely external reasons can also prevail and act in a most subtle fashion: reasons of expediency, invisible but historically decisive, are present.

Incidentally, then, the ecstatic gaze of that portrait seems to suggest spontaneously, to anyone who is interested in Galileo, an apparently strange question. Could the same misrepresentation suffered by Cardinal Bellarmino one day also become the lot of Galileo? Centuries later, what inevitable disguise lies in store for Galileo, on that day when there might be recognized in him an orthodoxy which in his epoch, and for a long time, was denied?

The question is less idle than it might seem. Has not the Galilean vision of relations between science and Biblical revelation been officially validated since Leo XIII's 1893 encyclical *Providentissimus Deus*? Would not the recognition that Galileo was the victim of a tragic judicial error, in a trial obscurely contaminated by personal and political motives, be sufficient to rehabilitate the image that many of Galileo's contemporaries had of him? If one canceled, with a rehabilitation, the sentence of Galileo's trial, would that not also eliminate any shadow of doubt as to Galileo's doctrinal orthodoxy in his own time?

But things are not so simple. The reasons that led to the incrimination of

Galileo and the degradation of his relations with the Church were more complex than what one is allowed to see from the trial that officially led to his condemnation.

The comparison of a series of related documents, of events and problems that have deeply influenced the history of philosophical and scientific ideas, will permit us to appreciate this complexity. It already became tangible for me that morning when the document I had come to consult in the Palace of the Holy Office was placed in my hands.

This palace is an obligatory stop for anyone who wants to make an ideal pilgrimage to the key places of the history of ideas in modern Europe. Here petty and important informers, petty and important inquisitors, petty and important defendants have written many paragraphs, known and unknown, in that difficult history. The paragraph I wanted to read concerned an obscure episode in the relations between Galileo and the Holy Office, but it could furnish precious information on the aforementioned critical problem. This was a document about Galileo from the seventeenth century, duly catalogued and preserved where it was reasonable to suppose that it would be, and yet one that mysteriously remained so discreet as to be completely ignored. Unless one is inclined to believe in a principle of chance, one will find much cause for excitement in this enigma.

Indeed, emotional moments and discoveries are not lacking in the story of this retrieval. To understand its significance, we must bring back to life the logic and history in which it fits, as the destined tessera fits into the design, no matter how full of gaps, of a lost mosaic. Both stories—the story of a discovery and the discovery of the unknown story that underlies it—deserve to be told.

How is it possible to avoid a dual tonality in style? Throughout our inquiry, at a distance of three centuries, the past becomes the present, annulling the limits of time. Starting from only a few images, so as to reconstruct this past, one has to accept being engulfed in it, and at times abolish oneself.

Our goal must be to restore the plot and its motives—from among the motives and passions of that time, to reconstruct another universe of politics, society, art, and religion, not simply of science. In short, to understand from within what for us is all the more difficult to comprehend now that three centuries have passed. It is not only a matter of scientific problems and ideas, but also of men—a gallery of human cases, with their cunning shifts, their conditioning, their desires, enthusiasms and fears, their madness and their poetry. But in order to follow the proper sequence and keep things in their proper light, we must now return to the point of departure.

ONE. SUBSTITUTION OF THEORY

> And new Philosophy calls all in doubt,
> The Element of fire is quite put out;
> The Sun is lost, and th' earth, and no man's wit
> Can well direct him where to look for it.
> And freely men confess that this world's spent,
> When in the Planets and the Firmament
> They seek so many new; they see that this
> Is crumbled out again to his Atomies.*

The Matter of Light

Initially my research had begun with the problem of Galileo's ideas on the nature of light. In truth, Galileo had never formulated a general theory on this fundamental phenomenon, and he wrote or published only brief statements on this problem. Nonetheless, beginning with his first trip to Rome in the spring of 1611, he had officially reported a memorable scientific experience having to do with physical optics, one which had profoundly impressed his Aristotelian colleagues.

Galileo had brought with him to Rome, besides his telescope, a "little box" containing some fragments of a rock recently discovered by alchemists in Bologna. This, after being calcined, revealed the property of glowing in the dark. Today it is possible to identify that luminescent mineral substance as barium sulphide. But the alchemists who had artificially revealed its luminous quality gave it a much more fascinating name: solar sponge (*spongia solis*).[1]

The weak, cold glow of those mineral fragments in a dark room, after having been exposed to sunlight, demonstrated that light was a phenomenon separable from the idea of heat and from the presence of a luminous environment. That seemed fully sufficient for Galileo to illustrate to his interlocutors—who were struck speechless by this most recent Galilean discovery—that philosophical convictions about light understood as a quality of a transparent, illuminated medium were false. To separate the light: this was an audacious and unforgettable deed, an experience with a Biblical savor that called up before the eyes of witnesses—Aristotelian philosophers and liter-

* John Donne, *An Anatomy of the World: The First Anniversary*, 1611, ll. 205-212

[1] See P. Redondi, "Galilée aux prises avec les théories aristotéliciennes de la lumière," in "Matière et lumière au XVIIe siècle," *XVIIe siècle* 34 (1982), pp. 267-83.

ary men—pages from Torquato Tasso's poem *Mondo creato*, paintings by Guido Reni on the separation of light from darkness,[2] and such works of Mosaic physics as *De Naturae luce physica ex Genesi desumpta* by the Paracelsian Gerhard Dorn.[3]

If light could exist separately from an illuminated environment and in an opaque body such as a rock, this seemed to mean that light, rather than being a quality, was a quantum and therefore communicated itself by emitting invisible corpuscles. A special substance, such as that purposely calcined rock, could perhaps attract luminous corpuscles as a magnet attracts iron shavings. This crudely effective analogy and this corpuscular interpretation of the memorable Roman event of 1611 were both attributed to Galileo, although he had not officially subscribed to them.[4] Nevertheless, just a short time later—namely, during the period of the new, exciting celestial discovery of sunspots and their physical interpretation—Dom Benedetto Castelli, a direct follower of Galileo in physics, generalized quite explicitly on the basis of Galileo's corpuscular suppositions and ideas, while Galileo, in his turn, seemed to value these developments.[5]

Light revealed a mechanical action. Like heat, it could alter the structure of bodies, changing their temperature. It transmitted itself, struck objects, and was reflected by them. For Galileo and his school, incorporeal qualities were not admissible, and mechanical processes on a microscopic scale could be traced back to corpuscular-mechanical causes only. A very high velocity of propagation, perhaps instantaneous, the characteristic of penetrating many substances without this involving considerable transformation in their physical parameters, nevertheless rendered the problem of light more complex than an immediate analogy between the luminescent rock and the magnet might encourage one instinctively to assume. The nature and behavior of particles of light could not be immediately assimilated to that of microscopic corpuscles in fluids and solids.

Galileo did not discourage his pupil Father Castelli, but he reserved for himself the privilege of doubt. Critical caution was nevertheless a faculty

[2] Cf. G. Reni, *La separazione della luce dalle tenebre* (1596-1598), originally at the Palazzo Rossi in Bologna; cf. C. Garboli, ed., *L'opera completa di Guido Reni*, Milan 1971, no. 10.

[3] See G. Dorn, *De Naturae luce physica ex Genesi desumpta juxta sententiam Theophrasti Paracelsi Tractatus*, Frankfurt 1583, p. 41 ff., for the idea of the separation of light, and pp. 59 and 108 for that of light as the source of transmutation and motion.

[4] For descriptions of Galileo's experiments on light, see G. C. Lagalla *De luce et lumine*, in *De foenomenis in orbe Lunae*, Romae 1612. This last, without the appendix *De luce et lumine*, can be found in G. Galileo, *Opere*, national edition by A. Favaro et al. Florence 1890-1899 (hereafter referred to simply as *Works*), vol. III. The attribution of magnetic characteristics to the phenomenon of the Bologna rock's luminescence goes back to N. Cabeo, *Philosophia magnetica in qua Magnetis natura penitus explicatur*, Ferrariae 1629, p. 120 ff.

[5] "From the Sun spreads the light . . . flashing continually with very swift corpuscles," Father Castelli writes apropos of sunspots in May 1612 (*Works*, XI, p. 294). See Castelli's letter to Galileo on September 5, 1637, *Works*, XVII, pp. 156-69, especially p. 161.

that, in Galileo the natural philosopher, could go hand in hand with audacity when confronted by objections and obstacles. This was soon to be seen.

Some years later, after accusations were put in circulation that he supported Telesio's ideas subversive of Aristotle's physics as well as a Copernicanism opposed to the Bible, Galileo defended himself from possible suspicions about the Catholic orthodoxy of his ideas with a sensational counteroffensive. This counteroffensive actually pivots on the new hypothesis concerning the nature of light in order to sanction a new alliance between reason and faith.

In his famous letter-manifesto to Father Castelli on December 21, 1613, Galileo legitimized the profound religiosity of "necessary demonstrations" and "sense experience." In fact, it is by means of these factors that the "very acute sciences" are for mankind sources of the revelation of nature and the "laws imposed on it."

"Proceeding equally from the divine word of Holy Scripture and from nature, the former as dictated by the Holy Ghost and the latter as the observant executrix of God's orders," possible contradictions between the literal expressions of the Bible and the conclusions of natural philosophy are owing only to the figurative meaning of some passages in the Bible. Commentators must then make up for these discrepancies with a new Biblical hermeneutic enlightened by natural arguments:

> It is the office of the wise expositors to strive to find the true meanings of passages in the Bible that accord with those natural conclusions of which first the manifest meaning or the necessary demonstrations have made certain and sure.[6]

The letter to Father Castelli had a great echo. At the beginning of 1615, it earned Galileo a denunciation from the Florentine Dominican Father Lorini, addressed to Cardinal Sfondrati, prefect of the Congregation of the Index. Lorini considered the new exegesis proposed by Galileo as equivalent to "wanting to set forth Holy Scripture in his own fashion and contrary to the common interpretation of the Holy Fathers."[7]

It will help us to remember how, after the Council of Trent, respect for the principle of authority in theological and exegetical tradition had become for Rome a settled matter. If Galileo and Castelli sought inspiration for their requests in St. Augustine's *De Genesi ad litteram*, Father Lorini sought inspiration for his complaints in the *Loci theologici* of the great Tridentine and Dominican theologian Melchior Cano. On the other hand, even scientific colleagues favorable to Galileo, such as Father Christopher Grienberger, mathematician at the Collegio Romano and Clavius's successor, had ex-

[6] *Works*, V, pp. 279-88.
[7] See Father N. Lorini's denunciation of February 5, 1615, *Works*, XIX, p. 297.

pressed some reservations, since no experience or demonstration permitted one to be "certain and sure" of Copernican truth. With this state of affairs, how could one expect official Catholic exegesis to yield to the concordance between Scripture and natural philosophy upheld by Galileo?

Then Galileo entered the fray with an essay on the concordance that enhanced his new philosophy from a religious point of view: the famous letter to Monsignor Pietro Dini, theologian and apostolic referendary in Rome. Galileo spoke as a scientist, but as an inspired scientist, and presented his candidacy to an explicit exegetical role, adopting a speculative mysticism that had Augustinian accents and referring explicitly to the Neo-Platonism of Dionysius the Areopagite, a source which the new theology of St. John of the Cross rendered topical.

Galileo did not so much exhibit the "necessary demonstrations" of heliocentrism as defend the metaphysic of the sun placed at the center of the universe, of which Copernicus had spoken in the first book of his *De revolutionibus*. Thus he celebrated a true and proper "triumph of light," with apposite references to the Creation described in the book of Genesis, the Psalms, and the Prophets: "God has placed his tabernacle in the sun . . . which is as a bridegroom coming out of his chamber, and rejoiceth as a strong man to run his race." Galileo, in the footsteps of Dionysius the Areopagite, demonstrated the suggestive concordance between these poetic verses of Psalm 18 and the emanationist ideas on celestial and terrestrial light which the Bologna rock, stellar scintillation, and the sunspots had suggested to him as an alternative to the Aristotelian physics of light. "I mean this," Galileo proposes, "to be said about the radiant sun, that is, about the light and the already mentioned calorific spirit which bears heat and fecundates all corporeal substances and which, issuing from the sun's body, spreads most rapidly throughout the entire world."

What could better be wedded to the divine illuminating grace of reason than this light, this spirit of the world—"a very spiritual substance, most tenuous and swift, which, spreading itself through the universe, penetrates everywhere without opposition and heat, vivifies and renders fecund all living creatures."[8] Was the hermeneutic value for the Scriptures and for the theological mysticism associated by Galileo with the substantialist physics of light only an opportunistic, defensive justification, or was this Biblical fundamentalism a weapon to propagandize and impose on Rome the new arguments of an anti-Aristotelian philosophy? It was both: an effort at legitimation and a philosophy that bore with it new spiritual and contemplative themes.

[8] Galileo's letter to Monsignor Pietro Dini, March 23, 1615, *Works*, V, p. 289. The Galilean hermeneutic of Psalm 18 was compatible with Bellarmino's contemporary comment, *Explanatio in psalmos*, published by Lapide, Parisiis 1861, vol. I, p. 105.

In fact, the open letter to Monsignor Dini was actually addressed to the Roman theological and ecclesiastical circles that remembered with favor the liberalization of Church culture by Pope Clement VIII at the end of the century, when Francesco Patrizi and his Hermetic Neo-Platonic, anti-Aristotelian, and Copernican philosophy had entered the College of the Sapienza and regarded light as an immaterial body of which the world "partook and by which it was fecundated."[9] Patrizi's Neo-Platonic Hermeticism had been condemned, and Paul V was not Clement VIII. But neither was the philosophy of light unchanged: Galileo now disposed of experiments and demonstrations for a substantialist physics of light, and also of heat, that could be opposed to the traditional metaphysics of Aristotelian qualities and that even went beyond Hermetic metaphysics.

The presentation of this new philosophy, in the name of conciliation with certain passages of the Scriptures and in the heat of a contemplative afflatus, reawakened in a new light the need for a Christian naturalism, similar to that of the Neo-Platonic humanism of Ficino and Giovanni Pico della Mirandola, which the most cultivated and receptive Roman theological circles (including Cardinal Bellarmino himself) recognized for its religious value. Neither Cardinal Bellarmino nor Pope Paul V cast doubt on Galileo's perfect good faith in his attempt to lay at the Church's feet Copernicanism and the new philosophy of light so as to make of them new instruments of exegesis and faith. And, in fact, Galileo was not condemned for what he had written to Father Castelli.

But he was condemned in 1616 for his initiative in favor of a Biblical exegesis enriched by modern knowledge of nature. His works were condemned and placed on the Index, awaiting correction together with the work of Copernicus, and with the works printed by certain Catholic theologians, the Augustinian Diego da Zuñica and the Carmelite Father Paolo Antonio Foscarini, who posited a hermeneutic of the Bible from the heliocentric point of view. Galileo, officially warned, regarded himself as admonished; apparently at least, he never again spoke out as an exegete, and he abstained from publishing until *The Assayer*.

He continued, however, to work as a natural philosopher. In 1623, when *The Assayer* appeared, everyone realized that Galileo had, though with great critical caution, preserved and developed some rather precise ideas on the nature of light. Indeed, *The Assayer* hypothesized a corpuscular theory of light, as Galileo had already done and continued to do for the nature of heat and the structure of solids and fluids.

Though not hiding from himself all the difficulties with any corpuscular theory, Galileo now spoke out formally on the possibility of recognizing in light an "ultimate resolution to really indivisible atoms." The philosophical

[9] F. Patrizi, *Nova de Universis Philosophia*, Ferrariae, 1591, p. 10.

term "atoms," which made one think immediately of Democritus, was reserved only for light in *The Assayer*. Heat and the particles of other elements or bodies were diversely designated as "fiery particles" [*ignicoli*], "fiery minims" [*minimi ignei*], "very thin minims" [*minimi sottilissimi*], and "the smallest quanta" [*minimi quanti*].

"The Smallest Quanta" and "Non-Quantum Atoms"

The deformed and oscillating terminology employed in *The Assayer* showed an awareness and uncertainty concerning the diversity of characteristics in the structure of matter. Light was its last stage of resolution. This was an atomic stage, close to a mysterious state of transformation—"because of its, I know not whether to say its thinness, rarity, immateriality, or even another condition different from all these and unnamed,"[10] Galileo confessed in regard to the special characteristics of light, i.e. its instantaneous velocity and its universal propagation.

Since his earliest work, Galileo had been interested in the seductive, innovative perspective of atomism in physics, which the philosophy of nature and the diffusion of Lucretius's *On the Nature of Things* and Hero of Alexandria's *Pneumatica* incited him to examine.[11] In his *Discourse on Floating Bodies* (1612), Galileo had subjected to hydrostatic testing the credibility of Democritus's idea that heat is composed of fire atoms. The result, with some criticism and reservations, had been encouraging: atomism, a hypothesis born of legitimate and fertile research, was to represent the concept of quality in Aristotelian physics with the kinetic and mechanical activity of material corpuscles. It was a question of transforming that hypothesis into a theoretic program general enough in the field of physics to become an effective weapon against the Aristotelian conception of the world.

The Assayer presented this program. It offered a corpuscular theory of all the elements of nature and of all perceptible phenomena (apart from those of sound, for which an interpretation of an undulatory character was reserved). But for the rest, the world of the senses was seen as a dense movement of particles of matter.

This theory was the most generalized and advanced that Galileo had ever formulated. Naturally, too, it was as yet only a schematized theory and quite far from being logically impeccable or capable of solving the old problems

[10] G. Galileo, *Il saggiatore* [*The Assayer*], Mascardi, Rome 1623, 2nd ed. in *Works*, II, Bologna 1655-1656 (published in succession at Florence 1718, Padua 1744, Milan 1832); cf. *Works*, VI. I will quote from the version edited by L. Sosio, Milan 1965, which also offers an improved lesson in regard to the excellent restoration of the text on the part of A. Favaro and which presents in its notes the Italian translation of the *Libra*. Ibid., p. 266 (*Works*, VI, p. 351 ff.)

[11] See M. Boas, "Hero's 'Pneumatica,' A Study of Its Transmission and Influence," *Isis* 40 (1949), pp. 38-48.

of the atomist explanation of nature. But Galileo, in *The Assayer*, put his cards on the table: he did not hide those difficulties from others or from himself.

The more traditional difficulties were bound up with the fact that atoms required empty interstices in order to move. And this rendered the explanation of the internal cohesion of bodies problematical. On the other hand, if one interpreted a body as composed of a very great number of minute material particles, and if one had to admit a principle of impenetrability among these particles, then the phenomenon of condensation or contraction became difficult to explain.

The Assayer frankly recognized that one was dealing with painful matters for a physics of atoms: the phenomena of condensation and rarefaction were "among the most recondite and difficult problems in nature."[12] Other defects were less obvious. Galileo did not make a clear and definitive choice between the hypothesis of differentiated corpuscles for every natural element and that of particles of a homogeneous raw material, which would have been closer to the traditional atomic hypothesis. In reality, Galileo adopted both these ideas: for example, heat is explained in *The Assayer* as the movement of corpuscles of a substance agitated by homogeneous and all-pervasive fire atoms.

The Assayer alluringly announced that philosophy is written in the book of the universe and that the characters of this book are geometric figures. But the reader suffered the disappointment of not finding out which were the precise geometric figures for the various kinds of particles. Galileo was content to say that particles, corpuscles, and atoms were "shaped in such and such a way."[13] But what shapes? Perhaps the suggestive shapes of the Pythagorean atomism in Plato's *Timaeus*? And another question without a reply: what forces, real or fictitious, held together the atoms and the various microscopic components of the bodies?

Let us not be surprised: Galileo was in possession of a kinetic, not a dynamic. *The Assayer*'s corpuscular theory presented a kinetic model of the structure of matter. Even later, in the *Discourses*, when Galileo will try to understand the phenomenon of the cohesion of bodies, he will imagine forces of intermolecular equilibrium in reference to the empty interstices that generate forces in opposition to the void: forces of equilibrium modeled on the atomic ideas of Hero's hydrostatic theories. However, at the level of the physiological explanation for the perception of phenomena, the gaps in the theory become more embarrassing. For example, the perception of heat occurs, according to *The Assayer*, through the more or less rapid penetration of substantial particles, excited by fiery minims and capable of penetrating the

[12] *The Assayer*, p. 206 (*Works*, VI, p. 231). [13] Ibid., p. 265 (*Works*, VI, p. 351).

pores of the skin in a more or less pleasant fashion. But how to justify an identical perceptive mechanism in this way if every heated substance emits different particles? Furthermore, how was a more or less pleasant sensation connected with the velocity of corpuscular penetration?

Galileo had been asked such questions in 1619, well before publication of *The Assayer*, in an important scientific letter from the Genoan scientist Giovanni Battista Baliani. Galileo had seemed not too concerned, for he had noted in the margin of the letter: "that the minims into which wax dissolves are of a different substance from those into which iron dissolves is of little importance when it comes to generating heat in us, provided that both dissolve into very thin parts, sharp and mobile, i.e. suited to penetrating our pores."[14]

The reply confirmed the force and conviction of a corpuscular vision imbued with Democritan reminiscences. If this theory did not accommodate all the desirable logical standards, nor include the experimental and mathematical assets with which Galileo would have had to endow it, it was nevertheless tantamount to a great intuition in physical theory. Trust in the validity of the corpuscular nature of matter was profound in *The Assayer*.

A strong reason for trusting in this theory was the possibility of making the "corporeity" of light coherent and plausible. But Galileo declared that, for the moment, he did not want to face the "infinite ocean" of difficult questions associated with this "etherial substance" composed of the smallest atomic particles, perhaps the only really indivisible parts of matter. However, *The Assayer* announced that this subject would be studied more thoroughly on "another, more opportune occasion."[15]

This opportunity showed up more than ten years later in the *Discourses and Mathematical Demonstrations Concerning the Two New Sciences*, published in Leiden in 1638. During this long interval, many things had occurred: the elating preparation of the *Dialogue*, the trial, and the sentence.

In the "First Day" of the *Discourses*, Galileo studied the problem of the cohesion and resistance of solids. Thus, after so long a time, he had returned to the most problematic section of his corpuscular theory, this time with the declared intention of examining the question of internal forces. Apparently it was only a prolongation of the ideas of *The Assayer*. Galileo still seemed perfectly convinced of the corpuscular nature of matter in its different states of aggregation. He still spoke of corpuscles of light and fire. The fundamental phenomenon of thermal propagation is still explained in mechanical terms, and the prodigious energy of light is assigned to its invisible constituents.

Nevertheless, revealed here for the first time were some new hypotheses.

[14] See Galileo's handwritten postscripts to the letter to G. B. Baliani, August 8, 1619, *Works*, XII, pp. 474-78, especially p. 475.

[15] *The Assayer*, p. 267 (*Works*, VI, p. 352).

In particular, the hypothesis of *The Assayer* on the instantaneous propagation of light had been abandoned. It was replaced by the fascinating proposal of an experimental study for measuring the speed of light on earth.

It was a new and important research hypothesis, but neither decisive nor dangerous for the corpuscular theory. Even if Galileo or his pupils had achieved a real or presumed success in their chimerical attempt to measure the speed of light, the theory with which this had been studied in *The Assayer* would not have been harmed—just the contrary.[16] In essence, the corpuscular and mechanistic picture of ten years earlier seemed unchanged, having substantially the same properties, so much so that it could be developed by new arguments based on the physics of the void.

Now Galileo actually attributed the internal cohesion of solids to the presence of intercorpuscular gaps, a vacuum. In fact, for Galileo, liquids, not possessing "vacuums" among their particles, do not have an appreciable internal cohesion. In other words, the existence in solids of "vacuums" maintains internal forces of cohesion, owing to the natural tendency of matter to resist a vacuum, i.e. the traditional empirical principle of *horror vacui*.[17]

But how can this traditional principle of a physics of the continuous be reconciled with a physics of the discontinuous, "the abhorrence of a vacuum" with the seduction of a physics of vacuums? In reply to this question, Salviati and Sagredo at this point in the *Discourses* came up with a famous lesson in mathematical philosophy concerning the problem of the infinite. This, as we know, is an important page in the history of the geometry of indivisibles. Galileo arrives there by passing through the narrow door that *The Assayer* had left open: that is, the possibility, restricted then to atoms of the special material of light, of demonstrating a resolution *ad infinitum* of matter.

Galileo then used that slit as a breach through which to push a new general theory of the structure of matter. In fact, he has the participants in his dialogue assume that the empty spaces between the particles of solids are infinitesimal. As a consequence, the particles themselves are indefinitely divisible: a solid contains infinite "vacuums" and infinite particles.[18]

The explanation is offered by Salviati by means of a geometric argument that presents the key to solving the paradox of "Aristotle's wheel" in terms of infinitesimals. The solution demonstrates that a magnitude can be constituted by infinite unextended parts, as against the traditional Aristotelian rejection of this possibility.

This problem has been so often studied and repeated that we need not

[16] On the experimental method for the terrestrial measurement of the velocity of light, see *Works*, VIII, p. 87 ff. On the attempts successively carried out by the Galilean school in Tuscany, see R. Caverni, *Storia del metodo sperimentale in Italia*, 6 vols., Florence 1891-1892, vol. I, p. 41 ff.; M. L. Righini Bonelli, "La velocità della luce nella scuola galileiana," *Physis* 11 (1969), pp. 493-501.

[17] See *Works*, VIII, p. 66 ff. [18] Ibid., p. 72.

analyze it once again. But, put very briefly, it was a matter of explaining how two concentric circles that rotate together to describe a revolution can trace, as happens in reality, linear projections of equal length, despite the difference in their circumferences. In the *Discourses*, the solution of this ancient paradox had recourse to the idea of intercorpuscular voids in the line traced by the smaller circle.[19] Galileo insisted that such a conclusion also applied to surfaces and solids. Physical objects, like the geometric solids existing in Plato's heaven, would thus also be composed of an infinite number of atoms separated by an infinite number of voids. But, as one can see, this is possible because we are in Plato's heaven, in mathematical abstraction: we are speaking of atoms, but actually talking about indivisibles.

Nonetheless, the geometric argument of infinites had an advantage and an evident goal. Indeed, it offered a very strong intuitive analogy for a physics of the discontinuous that also aspired to explain the difficult problems of condensation and rarefaction.

If material bodies, too, as those in geometry, were composed of an infinite number of voids and particles, it would be permissible to suppose something that is repugnant to common sense, but not to mathematics; that is, "infinite invisible subtractions" or "unquantifiable superpositions," infinitesimal components of bodies having no extension whatsoever. "The transaction of rarefaction and condensation," a perennial source of contradictions for the atomists, could thus be resolved without incurring the difficulties linked to the impenetrability of corpuscular matter.

In addition, the changes of physical state discussed in the *Discourses* would show the advantages of this new mathematical theory of physical discontinuity. This theory could thus appear to be a refined reworking, supported by a daring and completely new mathematical intention, of the previous atomistic visions in *The Assayer*. Here too, at bottom, Galileo is speaking of "atoms," an apparently unequivocal term.

And indeed, until rather recent times, students of Galileo's ideas—from Kurd Lasswitz to Lynn Thorndike, E. J. Dijksterhuis, and Marie Boas—could suppose that Galileo had developed the same atomist point of view over the entire arc of his life, from the *Discourse on Floating Bodies* to the *Discourses*. Nevertheless, they noticed a certain laboriousness and a number of incoherences that betrayed a series of obscure waverings of thought, which earned for Galilean atomism an enduring reputation for confusion.[20] Only more recently have students of Galileo emphasized how, in reality, some-

[19] Ibid., p. 68 ff. On Aristotle's paradox of the wheel: J. E. Drabkin, "Aristotle's Wheel," *Osiris* 9 (1950), pp. 346-59; P. Costabel, "La roue d'Aristote et les critiques françaises à l'argument de Galileè," in *Galileè. Aspects de sa vie et de son oeuvre*, Paris 1968, pp. 277-88.

[20] See K. Lasswitz, *Geschichte der Atomistik vom Mittelalter bis Newton*, II, Hamburg 1892, pp. 35-55; L. Thorndike, *History of Magic and Experimental Science*, 8 vols., New York 1923-1958, vol. VIII, p. 37 ff.; M. Boas, "The Establishment of the Mechanical Philosophy," *Osiris* 10 (1952), pp. 412-541, especially p. 436 ff.; E. J. Dijksterhuis, *The Mechanization of the World Picture*, Oxford 1961.

thing substantial had changed beneath that apparently identical picture. One detail of the previous theory had been replaced by another; but this was not immediately recognizable because Galileo did not say so explicitly, and in fact continued to use the same term, "atom," to refer to what he had changed.

The conceptual substitution takes place before the reader's eyes, but it is almost imperceptible, like an adept card trick—"with your kind permission, my gentle Philosophers"—performed before the eyes of the readers of the *Discourses*. Galileo, through Salviati's mouth, speaks of points, spaces, and lines. The reader listens, fascinated by that audacious infinitesimal solution of a difficult geometric paradox—then he realizes that Salviati has begun to speak of particles instead of points, of voids instead of spaces, of bodies instead of lines, and that for every pair the terms are synonymous.

Galileo does not present justificatory explanations, nor does he disassociate himself from his previous convictions. Indeed, he does not say (either clearly or obscurely) what sort of criticisms, what problems of physics, had induced him suddenly to transfer his ideas on infinitesimals to the properties of the structure of matter. We already know the brilliant theoretical results of that methodological decision, but we will probably never know the causes that motivated it.

Now Galileo presents in the *Discourses* a mathematical theory of matter. Its constituents are "unmeasurable parts," that is, parts without extension and therefore "indivisible," lacking dimension and shape. He calls them "atoms without measure, or non-quantum atoms,"[21] but in truth they are mathematical points: we are in the world of mathematical abstraction, no longer in the material world of physics.

One could object that Galileo also spoke of indivisible atoms of light in *The Assayer*. True enough, but when a theoretical context varies, the meaning of the terms that are part of it also varies, even if their names, the words that designate them, may be the same in the two contexts.

Actually, in *The Assayer* the word "indivisible" refers to the "smallest quanta," those parts having the same nature as the continuum which contains them. In practice, in an atomist and materialist context, this was the Latin translation of the word "atom." Also, in regard to the more subtle atomic resolution of luminous material, *The Assayer* kept the notion of atomicity quite close to the materialist concept.[22] But this is no longer true here. Still, as we have said, the Galileo of *The Assayer* had played with his cards on the table, and they were not always good cards. In the *Discourses*, on the other hand, he switched the cards and suddenly broke with what there was of materialism and classic atomism in his earlier work.

For simplicity's sake, we may refer to this "variation in the semantic sig-

[21] *Works*, VIII, p. 85. [22] See *The Assayer*, p. 266 (*Works*, VI, p. 352).

nificance of theoretical terms," as a good epistemologist would put it, as a "substitution of theory." In other words, we are speaking of the substitution of *The Assayer*'s materialist metaphysical speculations by the new and, let it be said parenthetically, more powerful metaphysical speculations of a mathematical character proposed in the *Discourses*. The scientific benefit from the choice made by Galileo probably had all the justification in the world as compared to his initial atomist materialism. Unfortunately, it takes place in silence. A tacit choice: on the other hand, its methodological consequences would seem to be most eloquent, if one takes every statement literally.

Hypotheses Non Fingo?

In the private correspondence of the last decade of his life, moreover, after his sentencing, Galileo gave evidence of not wanting to hear any more talk about his original philosophical and corpuscular preferences, professing a forthright scientific phenomenonism that seemed to preclude rebuttal, but left his colleagues flabbergasted. After his condemnation, the echo of these daring hypotheses of natural philosophy and of the lively Galilean statements concerning the invisible behavior of phenomena had by no means died out among scientists and philosophers.

One of the most revealing episodes of the last period of Galileo's life was an epistolary debate with one of these philosophers, Fortunio Liceti, an Aristotelian university man "of great reputation" (as one can read in the *Dialogue*), the great teratologist, professor of philosophy and medicine at the University of Bologna. Professor Liceti had an incomparable intellectual formation in medicine, literary and archaeological erudition, and astronomy and natural philosophy. He knew a good twenty-two hypotheses on comets and all the theories of Aristotle's commentators on the nature of light.[23]

Be assured, we are not digressing. The discussion between Liceti and Galileo, which was Galileo's last public scientific intervention, actually hinged on the phenomenon of the Bologna luminescent rock from which we started.

In 1640, Professor Liceti had published not one but two books on light: one of a general character,[24] another especially devoted to the new phenomenon revealed thirty years before by Galileo and in fact entitled *Litheosphorus sive de lapide Bononiensi*. Liceti had managed to adapt that phenomenon to fit the Aristotelian theory of light and, among the many arguments accu-

[23] See F. Liceti, *De lucernis antiquorum reconditis*, Venetiis 1622 (2nd ed., Utini 1642); id., *De novis astris et cometis*, Venetiis 1622; *Pyronarchia sive de fulminum natura*, Pataviae 1634; and *De luminis natura et efficientia libri tres*, Utini 1640.

[24] Ibid.

mulated in his book, there was also that relating to the secondary lunar light which illuminates the dark part of the planet during the moon's quarter phases. Liceti thought that this was a phenomenon of luminescence like that of the Bologna rock and not the effect of the sun's rays reflected by the earth, as Galileo had been maintaining ever since his *Starry Messenger* in 1610, to the benefit of the Copernican heliocentric theory.

Professor Liceti's book also mentioned again Galileo's famous corpuscular notions about the nature of light, by now common knowledge, as we know, owing to the account of the memorable experiments with the luminescent rock performed in Rome.[25] The author sent a copy of his book to Galileo.

Immediately after reading it, Galileo dictated to his pupil Viviani a letter that had the unexpected tone of a preventive denial, disassociating himself in a quite striking manner from any sort of materialistic interpretation of light, or even of heat, which at one time could have been attributed to him. The letter, dated June 23, 1640, emphatically proclaimed that Galileo had "always been in the dark" on the question of the "essence of light" and that if he had ever been in a position to have an idea of "what fire and light were," he would have been able to understand phenomena of "remarkable force and velocity," such as the explosion of gunpowder, a phenomenon problematically evoked in the *Discourses*.

He had preferred, rather than this speculative audacity, the "truth of fact" and the study of phenomena "that in all natural effects . . . assure me of their existence, their *an sit*, whereas I gain nothing from their how, their *quomodo*." In other words, Galileo declared that he confined himself to the study of the existence of phenomena and not the manner in which they are produced.[26]

After having read Liceti's book on the Bologna rock, which book contained the corpuscular quotation universally attributed to him, Galileo hastened to send another denial, insisting again that he did not now assert materialist philosophical hypotheses about light, nor indeed had he ever done so. To support his position, he recalled having said that he was disposed to have himself imprisoned in a pitch-dark cell, living on bread and water, on the condition that he could return to the light knowing what it really was.[27]

[25] "The light is a body, a part of which is attracted by the Bolgona rock as fire is by naptha and iron by the magnet." F. Liceti, *Litheosphorus sive de lapide Bononiensi lucem* . . . , Utini 1640, p. 178.

[26] See Galileo's letter to Liceti on June 23, 1640, *Works*, XVIII, p. 208 ff. Also see Galileo's next letter to Liceti, on July 14, 1640, ibid., p. 217. In the polemic, Gassendi also intervened on the astronomical aspects; see Pierre Gassendi's letter to Liceti, August 13, 1640, ibid., pp. 228-31.

[27] See Galileo's letter to F. Liceti on August 25, 1640, *Works*, XVIII, pp. 232-37: "I am amazed that for something mentioned by the philosopher Lagalla it is attributed to me that I have regarded light as a material and corporeal thing" (p. 233).

Must we really take this hyperbole on the theme of incarceration literally, as a solemn declaration of scientific style, or would we not do better to recognize in this bitterly foreboding self-quotation the tones of a refined Galilean irony? Obviously, having been sentenced to life imprisonment by the Holy Office, Galileo no longer felt so dramatic and imperious a necessity to formulate hypotheses on this phenomenon.

Having thus challenged those who still attributed to him hypotheses stating that light was a "material and corporeal thing," Galileo defended himself from the accusation of ever having been an adversary of Peripatetic doctrines, claiming instead a "religious" fidelity to the principles of authentic Aristotelianism in philosophy:

> Against all the reasons in the world, I am accused of impugning Peripatetic doctrine, whereas I profess and am certain of observing more religiously the Peripatetic—or, to put it better, Aristotelian—teachings than many others who unworthily try to pass me off as averse to the good and Peripatetic philosophy.[28]

This profession of Aristotelian faith was amply justified at the end of the letter. Galileo had always upheld the canons of sound logic, reasoning always in terms of the rigor of argumentative deductions based on well-founded experimental premises. Even now, in fact, Galileo gave an example of his Aristotelian logic, insisting without too much reticence or hypocrisy on his Copernican point of view with respect to the phenomenon of the secondary light of the moon.

Galileo therefore had proclaimed himself to be "amazed" that corpuscular or materialist hypotheses on the moon had been attributed to him. Let us be fair and permit Professor Liceti also to be flabbergasted when receiving this letter, he who like all Aristotelian university men had secretly admired and struggled with Galileo, author of the *Discourse on Floating Bodies* and *The Assayer*, the great adversary of the physics of the continuum and the qualitative principles of the science embraced by "philosophizing" Peripatetics.

Liceti could only be delighted, even if that profession of faith sounded rather strange. But he was sensible enough not to push a defeated adversary into a corner, an adversary, moreover, who was not free to speak as he pleased, and so took Galileo at his word. Indeed, Liceti replied on September 7:

> That you, Esteemed Sir, profess not to contradict Aristotelian doctrine is good to hear and (speaking frankly) very new to me, as from your writings I would tend to gather the contrary; but it may be that on this point I am mistaken, along with many others who are of the same opin-

[28] Ibid., p. 234.

ion. I am sorry that you, Esteemed Sir, should think that I have more than once attributed to you positions that are not yours. . . .[29]

Professor Liceti had an attitude that was more than friendly, but the situation as it developed was at any rate embarrassing. Galileo could easily refute other accounts of the Bologna rock experiment of 1611. He could even shamelessly repudiate "as definitely original some thoughts of mine brought forward by Signor Mario Guiducci."[30] But he could not so easily deny what he had written, for example, in *The Assayer*. In fact, in this connection Galileo will send a clarifying comment that actually refers to *The Assayer*—not citing it, but quoting the most famous page of that book of so many years past: the famous page about the book of the universe written in mathematical characters.

Galileo will insist that he was "an admirer of such a great man as Aristotle," that he wanted to dispel all doubts about his fidelity to Aristotle's fundamental methodological canons. However, he will also explain that the world is not writ whole in Aristotle's books because it is to be discovered through experience, like a book, and deciphered with mathematics.[31]

This more specific comment was therefore clear: Galileo had, always and only, spoken of "sense experiences" and certain demonstrations. Galileo entrusted to this final statement, ideally connected with his previous work, a methodological testament of mathematical phenomenology: experience, mathematics, logic. *An sit*, not metaphysical speculations on *quomodo*,[32] as he had said from the beginning.

Hypotheses non fingo? There seems to be no doubt. Rivers of epistemological ink have been poured out concerning these final statements by Galileo, as if these were a privileged document, suggestive of some fundamental predisposition to phenomenist mathematical epistemology.

Taken literally, Galileo's words are, in fact, unequivocal. Seen with our modern eyes, accustomed to recognizing the epistemological virtues of clarity and distinctness, they could not be more eloquent. Yet, let us not forget

[29] F. Liceti's letter to Galileo, September 7, 1640, *Works*, XVIII, pp. 244-45, especially p. 245.

[30] See Galileo's letter to F. Liceti, September 15, 1640, *Works*, XVIII, pp. 247-51, especially p. 250. In 1619, Mario Guiducci had published under his name the *Discorso delle comete* [*Discourse on the Comets*], the critical edition of which (by A. Favaro in *Works*, VI) permits one to consider the fact that in great part the text had been directly edited by Galileo. "It may be that I had not had a part in it," Galileo nonetheless feels the need to say, "although I believe it is an honor that such concepts are thought to be mine" (ibid.).

[31] Letter to Liceti in preceding note, p. 248; letter to Liceti in January 1641, ibid., pp. 293-95, especially p. 295.

[32] See Galileo's letter to F. Liceti, June 23, 1640: "And here I would not like it to be said that I did not ascertain the truth of the matter, because it is experience that shows me how this happens; which, I might say, in all the effects of nature observed by me assures me of the *an sit*, but offers me no gain whatsoever regarding the *quomodo*" (*Works*, XVIII, p. 208). For the use of this terminology in the political and moral tracts of the period, see A. G. Brignole Sale, *Tacito abburrattato*, Venice 1646.

that to the virtues of clarity and distinctness, the philosophers and scientists of the seventeenth century—who were in fact called "virtuosi"—added another intellectual requisite for the expression of their ideas: the virtue of "honest dissimulation," a virtue whose methodological rules have been rigorously developed in works of historical and political methodology by Traiano Boccalini, Virgilio Malvezzi, and later by Torquato Accetto.

Let us keep this in mind, for the expressions used by Galileo in his correspondence with Liceti—the *an sit* instead of the *quomodo*—which Galileo used in speaking of scientific methodology, belonged at that time to the current language of historical and political methodology. Should one limit oneself to describing the effects, or go back to the causes? There was a great debate on this question. The writers of political, moral, and historical treatises in those days had made fashionable again the Latin formulas of scholastic philosophy, discussing whether one should reveal the secrets of history and men with "the political spyglass" or instead confine oneself to describing the facts pure and simple.

The common solution consisted in recommending "the expediency, indeed the necessity for a wise and free man, not to simulate nonexistent virtues (always an abject thing), but to dissimulate." That is what Torquato Accetto wrote about the art of pretence, in a work that appeared in 1641: for the love of truth, remain silent about a part of one's thought.

In 1635, as if to remind us that on him was imposed the necessity of "dissimulating in the face of unjust power," as Accetto would have put it, Galileo had written from Arcetri to his friend Nicolas Fabri de Peiresc that, despite his "very religious and holy mind," of which he had given proof in all his works, he was compelled to leave his own motives in the dark:

> Not only is it advisable for me to succumb and remain silent about the attacks that have rained down on me in such great numbers, even on natural subjects, in order to suppress the doctrine and publicize my ignorance, but it is also advisable that I swallow the sneers, sarcasms, and insults.[33]

That reticence was an exercise in wisdom and piety, imposed by "slanders, frauds, trickeries, and deceits" which under "the simulated mask of religion" had not only ruined him, but continually besieged him. Not only was the *Dialogue* (condemned in the 1633 sentence) prohibited, but other works "now in print for many years" were also obscurely condemned—"there being an express order to all inquisitors not to allow any of my works to be reprinted."

It is permissible to think that to many of Galileo's contemporaries who

[33] Galileo's letter to N. Fabri de Peiresc, February 21, 1635, *Works*, XVI, p. 215 ff.

did not have our refined instruments of epistemological discrimination, his profession of Aristotelian faith, which today can be called mathematical phenomenonism, must have appeared to be a disconcerting artifice of the intellectual virtue of dissimulation to oneself and for others. But there can be no room for doubt, even for us, that when he denied that the opinions published under the name of Guiducci were attributable to him, Galileo was dissimulating. To deny having advanced, privately and publicly, corpuscular hypotheses on light and heat, of having opposed a materialistic metaphysic to the Aristotelian metaphysic; to conceal the past with declarations of mathematicism and experimentalism free from any hint of a cognitive intent concerning the real constitution of phenomena—all this must understandably appear to be a denial and a prudence so dissimulated as to remain absolutely silent about its own ideas.

It is appropriate here to add that the correspondence with Liceti was destined to be spread about and discussed in Italy, and that even before denying Liceti's philosophical allegations, Galileo had expressed his concern and distrust. The distrust was directed at his interlocutor, who "has always adorned my name with specious titles, but then on the other hand obscured my thoughts"; the concern had to do with the experience, perhaps tragic, of previous "scientific disputes" where "the censures inflicted on the propositions and opinions of the beleaguered party, with the intention of showing them to be false and erroneous, are certainly not lacking in seriousness and hurt."[34] As for the seriousness of the latter consequences, we must remember that Galileo was in the position of a sentenced man not authorized to speak in any way about the reasons for his sentence.

From these different points of view, the virtue of caution and dissimulation could be more than admirable. What might surprise us is that Galileo did not also apply this virtue to some other old convictions of his. In this correspondence, and in other, later letters, Galileo could permit himself to defend less scrupulously, barely within the limits of the most elementary norms of prudence, the Copernican system, even though his sentence hinged precisely on this defense. Using the minimum of dissimulation necessary in this very delicate problem, Galileo had remained faithful to his old astronomical convictions. Not so in physics. The old convictions of Galilean physics were abandoned without explanation, not even a scientific one.

Even today, the silent abandonment of corpuscular physics which *The As-*

[34] Letter to Liceti cited in note 32, above. Galileo's letter of March 31, 1640, concerning the moonlight, was to appear in a revised version, with Galileo's consent, in F. Liceti, *De lunae suboscura luce*, Utini 1642. See L. Geymonat, *Galileo Galilei*, Turin 1957, 2nd ed. 1969; and A. C. Crombie, "The Primary Properties and Secondary Qualities in Galileo Galilei's Natural Philosophy," in *Saggi su Galilei*, vol. II, Florence 1972, pp. 71-90. We should emphasize the epistemological interest of these final statements of Galileo, recognizing their importance for a comprehension of the role of logic (Geymonat) and the phenomenistic character (Crombie) of Galileo's scientific methodology.

sayer had sketched, and which is viewed by historians as foreshadowing fertile mechanistic insights, is a complex critical problem. It is as though certain limits and obstacles confronted that confident natural philosophy pregnant with atomist ideas, persuading Galileo eventually to disassociate himself from it in favor of a radical alternative: a purely mathematical theory of the structure of matter, its methodology-of-choice being a prudent mathematical phenomenonism.

The historians who have dealt with this problem have reached, through a purely internal examination of Galileo's ideas and works, some rigorous and convincing justifications for this epistemological conversion (but not an exact reconstruction of his conscious and necessary steps, since these are not documented). We ask, as does one of the most recent studies of early-seventeenth-century atomism, why this "really surprising" step?[35]

By now, everyone sees in the contradictions of an atomist physics such as that of *The Assayer* the implicit possibility of abandoning it. But this reasonable observation by itself cannot explain Galileo's silence on the issue that has evoked such problematic fascination.

In reality, the substitution of theory and the meaning of the terms that characterize the atomist language of Galileo's physics have received different interpretations at the hands of recent historians. Some have confined themselves to pointing out the profound modifications of that intellectual history. Others have highlighted the *terminus ad quem*, that is, the final mathematical version, denying that there had ever been in Galileo another, thought-out form of atomist physics. Still others underscore the continuity of latent anticipations of infinitesimal thought, going back even to his original atomist bent in physics.[36] Evidently and inevitably, it was the final outcome of the affair—the mathematical solution proposed for atomism in the *Discourses*—that directed, with the wisdom of hindsight that historians must possess, all these attempts at internal logical justification.

When my research into Galileo's physics of light came to a halt, faced by the need to understand but also to reconstruct this problem, instead of looking at the end of the affair, I asked myself whether the reasons for the abandonment of *The Assayer*'s ideas could be considered in another manner. I asked myself whether those reasons could be sought, apart from the well-

[35] J. Henry, "Thomas Harriot and Atomism: A Reappraisal," *History of Science* 20 (1982), pp. 267-96, especially p. 281.

[36] See respectively W. Shea, "Galileo's Atomic Hypothesis," *Ambix* 17 (1970), pp. 13-27; H. E. Le Grand, "Galileo's Matter Theory," in *New Perspectives on Galileo*, edited by R. E. Butts and J. C. Pitts, Dordrecht 1978, pp. 197-208; U. Baldini, "La struttura della materia nel pensiero di Galileo," *De Homine* 56-58 (1976). Greatly revelatory of the still controversial state of the critical problem concerning the substitution of atomistic theory in Galileo are the questions raised on this aspect by A. M. Smith, "Galileo's Theory of Indivisibles: Revolution or Compromise?" *Journal of the History of Ideas* 37 (1976), pp. 571-88.

known difficulty inherent in those ideas, in the circumstances that surrounded them, relating to *The Assayer*—namely, the reactions that the book had produced.

Could not Galileo's final methodological propositions have a value of truth beyond language? Could they not be dependent on a reality other than that which they enunciated? These were the questions I asked myself. Indeed, if one looks closely, many more obscure points are clustered around *The Assayer*. Thus, as often happens during research, from one problem another was born, and a new enigma awaited me.

TWO. COMETS: PRESAGE OF MISFORTUNE

> Doubt thou the stars are fire,
> Doubt that the sun doth move,
> Doubt truth to be a liar
> . . . .
> this brave o'erhanging firmament
> this majestical roof fretted with
> golden fire—why, it appears
> no other thing to me but a foul and
> pestilent congregation of vapours.*

A Brilliant Success, a Dark Remorse

In 1623, when *The Assayer* was published in Rome, it was welcomed as an unquestionable success. Galileo's new book was a very polemical piece of writing as regards the most prestigious institution of Catholic culture: the Collegio Romano. And this obviously magnified his provocation to the point of making it scandalous.

The Assayer had the form of an essay of controversy, the most passionate and popular essayistic form in the philosophical and theological culture of his time. It had a scientific content, in good part accessible, and a refined and modern literary style, with long and fascinating digressions and literary quotations that delighted the humanists and the rhetoricians in the literary academies. With unerring skill this book brought together all the requisite qualities for it to be read and to deserve the acclamation of that particular audience which can turn a book into a literary and intellectual sensation. And so it was. *The Assayer* crowned Galileo's head with the approbation of a cultured audience composed of the "curious," the "virtuosi," and the "innovators" and, more important yet, the official favor of the highest ecclesiastical authorities. Not even the most fervent and optimistic Roman supporters of *The Assayer*'s publication had initially dared predict a triumph on such a scale.

Galileo came to Rome soon afterward to reap the benefits, and found all the solidarity, encouragement, and prestige that one could possibly get. It was the highest moment in Galileo's parabola as a public figure and the determining moment in his destiny. In fact, it was on the crest of that wave of

*W. Shakespeare, *Hamlet*, fol. 1, London 1623, Act II, Scene 2, ll. 116-18, 297-99.

triumph over his adversaries and that approbation from the highest levels of the Church hierarchy that Galileo, influenced perhaps too positively, gauged the possibility of reopening his strategic campaign on the Copernican front.

A great deal has been written about *The Assayer*. Rivers of ink have been poured out concerning its most famous pages. But the historians have mainly studied the facts leading up to it—that is to say, the difficult, indeed the *most* difficult polemic ever engaged in by Galileo, at his initiative, against Father Grassi, a Jesuit mathematician at the Collegio Romano. This was a polemic, on the nature and movement of comets, which *The Assayer* closed victoriously perhaps more because of its literary verve, its irony, its murderous wordplay, the poetry of its allegories, and its boundless intellectual passion than by the irresistible force of rational disputation. But never mind why.

Still left to be told, however, is the story of the fortune and misfortune of *The Assayer*: by whom, where, and how it was read, praised, or blamed. Indeed, this story is very important because *The Assayer*, besides being a book by Galileo and the outcome of a polemic, was in a certain sense a work of collaboration and an intellectual manifesto. Behind *The Assayer* there is not only Galileo, but a group of scientific and literary Roman intellectuals who incited, revised, corrected, and published it: the Roman nucleus of Federico Cesi's Accademia dei Lincei [Academy of Lynceans].

The Accademia dei Lincei was a private institution, animated by a lively desire to do research and in competition with the institutions of official culture. Its exponents, "extremely avid for new philosophy," were searching for a legitimation of their polemic against traditional "Scholastic" thought. In papal Rome, moreover, as under any dictatorship, their privileged personal relations with the ecclesiastical power had a significance and political value which rendered their aspirations realistic and their desire for innovative transformations credible. *The Assayer* was presented in Rome as the official manifesto of their intentions and as their effort at polemical legitimation vis-à-vis the crushing force of institutions which based their power on tradition and authority.

But a history of the welcome given *The Assayer* is at the same time difficult, because if one wanted to track down its readers it would not suffice to look for them in the convent libraries and the university classrooms; that is, in the places of an easily traceable scientific community. On the contrary, an inquiry of this kind could prove disappointing because the official scientific culture did not at all greet *The Assayer* with the sort of fervent enthusiasm evoked ten years earlier by the announcement of Galileo's astronomical discoveries.

The document that is most revealing of the embarrassment and perplexity aroused by *The Assayer*'s astronomical arguments among contemporary astronomers and mathematicians, even those who were admirers of the author of *Starry Messenger*, is magnificently given us by Kepler's review of Galileo's new book, which appeared under the title "Spicilegium ex Trutinatore Galilaei" as an appendix to the *Tychonis Brahei Dani Hyperaspistes adversus Scipionis Claramontii anti-Tychonem*, published in Frankfurt in 1625. *The Assayer* had brought to the notice of Kepler, then in Vienna, Galileo's rejection of Tycho Brahe's theories on the comet. The quotation of Kepler's opinions on phenomena of luminous reflection, from his books on optics (in connection with which Kepler, if only as an example, had mentioned comets) had been prominent.[1]

The great friendliness that he felt for Galileo did not prevent Kepler—"Professor Astronomiae Tychonicae"—from underlining this and other misapprehensions in *The Assayer*, correcting the mistaken attributions used by Galileo in support of his polemic, and stigmatizing Galileo's doubts about Tycho Brahe. Those doubts had put Tycho's work on the same plane with the astronomical errors of Scipione Chiaramonti, the Pisan astronomer and philosopher discredited by Kepler and incautiously quoted by Galileo in *The Assayer* for the good of his cause. At the same time, Kepler kept his distance from Galileo's antagonist, for and against whom *The Assayer* had set off a dialogue among the deaf, meaningless as far as astronomy was concerned.

Kepler nonetheless tried to justify Galileo, urging the readers of *The Assayer* to keep in mind the book's defensive and polemical motives and not to judge solely on scientific grounds the incomprehensions, obscurities, and verbal excesses that might make Galileo look like an envious anti-Tychonian and a misinformed and obscurantist astronomer. Kepler said that he justified the work's counterproductive result on the astronomical plane because Galileo had to defend himself at all costs: his adversary seemed to want to attack him "on a subject matter of great importance regarding faith in the dogmas." Kepler, like everyone, finally recognized that the originality and importance of *The Assayer* did not lie in its occasional astronomical controversy, but in its new "argumentation and experiments of uncommon diligence, such that I hope"—concluded Kepler—"that it receives very well deserved praise and success among the lovers of Philosophy."[2]

[1] See J. Kepler, "Spicilegium ex Trutinatore Galilaei," in *Tychonis Brahei Dani Hyperaspistes* . . . , Frankfurt 1625, p. 191. This deals with the effect of refraction, similar to that of the comet, produced on the surface of a glass globe filled with water. In reality it has been F. Sizi, in *Dianoia astronomica, optica et physica*, Venetiis 1612 (*Works*, III), who used this example, given by Kepler in *Astronomia pars optica* (1609), to support his thesis on the purely optical nature of the Medicean planets discovered by Galileo's spyglass. It should be noted that now Galileo was adopting the same strategy against Tycho's and Father Grassi's theory of comets. Kepler protested, reminding Galileo that he had never thought of denying the comets' nature as celestial bodies, nor had he ever spoken of them as optical effects.

[2] Ibid.

Kepler's justified good wishes had already been realized, independently of the weaknesses and inexact astronomical arguments inherent in the polemic on comets. But to keep the record straight, we must once more describe this controversy, which has become famous in official historiography.

Galileo had written *The Assayer* following an astronomical debate that had occurred four years before, when three comets had appeared in the European sky, between the end of 1618 and the start of the following year. In 1577, Tycho Brahe had already been able to study the problem of the position and motion of those celestial bodies. The narrowness of their parallaxes, much inferior to those of the moon, and their observed velocity had led Tycho Brahe to place the comets beyond the moon, among the celestial bodies revolving around the sun. But their motion seemed to disprove the possibility of a circular orbit.

In 1619 Father Orazio Grassi, at the height of the great excitement of public opinion over the appearance of these comets, had again proposed Tycho Brahe's conclusions, brought up-to-date by new observations. Galileo began a polemic on the positions, motions, and nature of the comets, using arguments that were absolutely unequal to those of Tycho Brahe or his epigone. But there was a hidden motive: to make sure that the comets would not discredit Copernicus. Galileo had not seen these comets. One of his arthritis attacks had prevented him from making observations even on the very brilliant comet of 1619. In any event, comets were not involved in the Copernican system which he defended. Still, Galileo had many observations and new ideas in favor of Copernicanism and the earth's motion; but he could not talk about them, for in 1616 Cardinal Bellarmino had notified him to abstain from doing so in the future. Thus, the only thing he could do was defend Copernicanism from its possible opponents: destroy the non-Copernican astronomies and cosmologies. In other words, not being able to demonstrate Copernicanism, Galileo's sole choice was to eliminate possible falsifications of it.

A celestial body endowed with noncircular motion, an unforeseen and highly dangerous hypothesis, was a threat to the Copernican system. So, without observations and without calculations, Galileo, first indirectly and then directly through intermediaries, took a position in the debate, maintaining an alternative to Tycho Brahe's and Father Grassi's very modern and sophisticated theory on comets. Which among the various, traditionally available theories on comets should be adopted? There was a very seductive one, attributed to Democritus and Anaxagoras, which explained comets as stellar agglomerations.

But the 1616 injunction conditioned the possibility of choice, and Galileo did not want to run risks over what, after all, was a secondary question in astronomy. Although (as revealed in a private note in the margin of the Baliani letter, which we have already quoted) Galileo liked that hypothesis, officially he discarded it, as he discarded other traditional explanations.

Since the lack of parallaxes was the strongest argument for placing comets at a great height in the sky, Galileo came up with an ingenious suggestion. He proposed to deny the physical reality of comets. They were not celestial bodies, but luminous appearances like rainbows or the sun's reflection on the sea at sunset: subjective optical meteors, "apparent simulations."

Parallax is valid when one deals with "real and permanent" luminous objects such as the moon, which when seen from one point appears to be close to the horizon, while from another appears to be at the zenith. But does not the "very bright streak" that lies on the sea at twilight always seem directed toward the sun, wherever the observer may be, with the effect of annulling the parallax? The problem, if any, resided in the thing, since the tail of the comet invariably appeared to be opposed to the sun. Galileo proposed to explain the comet as a luminous reflection on the atmospheric exhalations raised beyond the cone of terrestrial shadow, like an aurora borealis. The diminution of the size and speed of the phenomenon became plausible when one admitted that these vapors moved with a rectilinear motion, in a radial direction with respect to the surface of the earth: a solution identical to the one Galileo had proposed in 1604 for the new star.

Convinced that the comet was very low, and concerned to keep it far away from Copernicus's sky, Galileo thus proposed a theory on comets that was nothing but an optical variation on the explanation offered by Aristotle in the *Meteors*—with one great difference. While Aristotle hypothesized a conflagration of terrestrial vapors in motion around the earth, for Galileo the comet's light was a pure reflection on those vapors, without any thermal aspect caused by movement. In any case, it was still an idea taken from *Meteors* that seemed to be able to save Copernicus from being proved false by Tycho Brahe.

As one can see, *The Assayer* did not bring to astronomy either observations or original theses, but merely brilliant polemical arguments, all the more brilliant insofar as they were paradoxical. They were intended to discredit with their impeccable, provocative logic the certitude of the opponent's argument—Tycho Brahe's patient and rigorous observations as well as those of his anti-Copernican followers—rather than establish in astronomy a new theory about comets.

These were not the only qualities which led to *The Assayer*'s being read and valued highly by those who perhaps until that moment had not read any of Galileo's books and who did not have enough mathematical knowledge to understand the astronomical terms of the polemic. Above all, its chief attraction was the pleasure of its style. *The Assayer* was truly a "page turner." In fact, it was written as a controversial comment: every paragraph of the opponent's book was discussed separately from the others, often provoking long digressions that formed chapters in themselves. In these digressions, literary and philosophical in nature, Galileo gave proof of possessing to a high degree the qualities of a man of letters. In *The Assayer* were quotations from poets and historians, pages of literary prose, and pages of biting satire. There were subtleties and brilliant metaphors that made the book agree perfectly with the taste of the literary avant-garde and with the common struggle of the nature investigators and literary virtuosi. Therefore, one should look for the welcome accorded *The Assayer* not only in the academic and scientific world, but also in that of the literary academies, the political and cultural circles uneasy about the past.

Not by chance was the book's success decreed officially by the appreciation with which one literary man and great patron of the arts, endowed with a liberal and innovative political vision, Pope Urban VIII, enjoyed the most fascinating literary pages and the most pungent ironies in *The Assayer*. Probably, these were the only pages that he, like the majority of its readers, had read.

It is probably the literary, more than the scientific, nature of *The Assayer*'s success that has dissuaded historians of science from analyzing more closely the problem of the reactions aroused by those few paragraphs which were generally read and discussed. Having reached this point, monographs on Galileo, usually, imitate their protagonist and, like Galileo, once the success is recorded, switch their thought rapidly to the new adventure: the *Dialogue*.

It is appropriate, though, to dwell on this matter, because over that apparently unclouded success hangs the shadow of an obscure remorse, which has remained a perfect mystery. In the *Racconto istorico della vita di Galileo Galilei* [The Historic Tale of the Life of Galileo Galilei] (1717, posthumously), Galileo's first historian, his direct pupil and biographer Vincenzo Viviani, while describing the polemic over the comets that was at the heart of *The Assayer*, ended his account of that victorious dispute with an omen to which he does not give us the key.

In fact, Viviani writes, that polemic was the cause of "all the misfortunes to which, from that hour until his last days, Signor Galileo was subjected via relentless persecution for his every deed and word."[3] But that was not all. Having presented Galileo's opponent in that dispute in the sinister light of envy and pretended infallibility, Viviani added that the polemic had led to "slanders and refutations from his enemies and opponents, who then kept

[3] V. Viviani, "Racconto istorico della vita di Galileo Galilei" (1634), in S. Salvini, *Fasti consolari dell'Accademia fiorentina*, Florence 1717 (post.); critical edition by A. Favaro in *Works*, XIX, pp. 597-632. I quote from the next edition, that by F. Flora, Milan 1954, p. 47 (*Works*, XIX, p. 616).

him almost always in distress, [and] rendered him even more reluctant to perfect and publish the most marvelous doctrine of his principal work."[4]

What slanders? What refutations had induced Galileo not to "perfect" his doctrines, but to resort to the prudent virtue of dissimulation after *The Assayer* and because of it?

Did Viviani perhaps mean to say that Galileo would have done better not to give us "the very lofty concepts and singular speculations"[5] of *The Assayer*, to have abstained from that polemic? Historians have not found any serious justification for such a grave imputation and so severe a remorse. *The Assayer* had rendered the adversary ridiculous and had propelled Galileo into the pontiff's paternal embrace. If ever there was a book of Galileo's which in Rome had a brilliant success, it was *The Assayer*: it even earned Galileo an official license in the form of a papal brief as a "devout son" of the Church.

The polemic had been severe, but once won it was never mentioned again. Not even in the darkest moments of the subsequent Galilean story did the polemic over comets come up again, and no one ever thought of putting *The Assayer* on the Index, even though some would have been pleased if it had been.

Historians today have practically expunged this presage of misfortune concerning the comets of 1618. Viviani's "Tale" overflowed with naive apologetic fervor; it was after all, from beginning to end, a eulogy written in view of the imminent republication of Galileo's *Works*, in the hopes of seeing the master rehabilitated. This remorse was legendary.

At most, that presage is explained in a general sort of way, as one of the numerous proofs of the Jesuits' hostility to Galileo. This explanation, more than legitimate, is perfectly obvious. The Jesuits had seen pitilessly derided one of their most able scientists, Father Orazio Grassi, who had not even been the first to set off the cometary polemic with Galileo, a polemic in which he had all the weight of astronomy on his side. It was obvious—more than obvious—that the Jesuits also had every reason to see in Galileo one of their innumerable detractors, competitors, and adversaries. When Galileo was tried by the Holy Office in 1633, the rumor spread throughout Europe—in Rome, Venice, and Paris—that it was the Jesuits who had denounced Galileo: authoritative personalities such as Naudé, Descartes, and Grozio testified to this common opinion, subscribing to it in their letters.[6] But each and every one of these suppositions referred to the attacks launched

[4] Ibid., p. 48 (*Works*, XIX, p. 617). [5] Ibid.

[6] See G. Naudé's letter to Gassendi, April 1633, *Works*, XV, p. 88; Descartes' letter to Father Mersenne, *Works*, XVI, p. 526 ("Je me suis laissé dire que les Jesuites avaient aidé à la condemnation de Galilée"); and Grozio's letter to Vossius, ibid., p. 266 ("Ieusuitorum in impsum idio"). If not because of the Jesuits' direct work, what basis was Galileo condemned in Rome? On this question and on other testimonies concerning the activities and obscure motives of the Society of Jesus, see A. Favaro, "Oppositori di Galileo. Il padre Cristoforo Scheiner," in *Atti del Regio Istituto Veneto di Scienze, Lettere et*

against the *Dialogue*; there is no longer any trace of the comet debate, which seemed over and forgotten.

Only one piece of testimony seemed to support Viviani's disturbing judgment; and it was a qualified testimony, since it came from a Jesuit scientist in the Collegio Romano, Father Christopher Grienberger, who had been Father Orazio Grassi's professor of mathematics and astronomy: "If Galileo had known how to keep the affection of the fathers of this college," Father Grienberger stated after Galileo's sentence, "he would live in glory before the world and none of his misfortunes would have occurred, and he would have been able to write at his pleasure about any subject, even, I say, about the movements of the earth."[7]

But this statement, too, does not allow us to understand Viviani's feeling of remorse. Lacking any historically ascertainable proof of such an unspecified and reasonable hostility on the part of the Jesuits, apologists for the Society have demanded evidence of the "hateful insinuations of Galileo's partisans at that time."[8] But, apart from the quoted inferences and opinions, modern historians could show no evidence supporting the unfortunate presage of the comets reported by Viviani.

A predecessor of theirs, the great historian of science Montucla, in his famous *Histoire des mathématiques*, had no such doubts: "in any event," he wrote in connection with Galileo's incrimination, "there have been lively controversies on questions of hydrostatics, on comets, etc., with a certain Father Orazio Grassi, a Jesuit, and it is asserted that this fine father contributed quite a lot to inciting the inquisitors."[9] Montucla's information is sensational, without precedent and without sequel, but unfortunately also without proof. Living in the eighteenth century, Montucla consequently had the inestimable advantage of being two hundred years closer to Galileo than we are. Thus, Montucla may have been able to gather information, memories, and still fresh comments on the Galileo affair, echoes that to us are irremediably lost. His revelation squared exactly with Viviani's remorse, but it was not confirmed; Jesuit authors, then, could rightly denounce Montucla for an Enlightenment regurgitation of the "odious insinuations" against their seventeenth-century brother.[10] Montucla's denunciation as

---

Arti 78 (1918-1919), p. 90 ff. Favaro, however, was satisfied that purely personal reasons could explain the hostility for a persecution of Galileo.

[7] See Galileo's letter to Elia Diodati, July 25, 1634, *Works*, XVI, p. 117.

[8] Father von Hartmann Grisar, S.J., *Galileistudien. Historisch theologische Unterschungen über die Urtheile der romischen Congregationem im Galilei-Process*, New York and Cincinnati 1882, p. 329.

[9] J. E. Montucla, *Histoire des mathématiques*, Paris year VIII, vol. IV, bk. V, p. 294.

[10] "This statement would have need of proof," Montucla was reproached by an angry P. C. Sommervogel ("H. Grassi," in *Bibliothèque de la Compagnie de Jésus*, vol. III, Brussels and Paris 1892, cols. 1684-86, especially col. 1686). In our century, after Antonio Favaro's researches, an opinion of innocence has been tacitly formed in favor of Father Grassi, as well as a tendency on the part of historians to lay

well as Viviani's remorse had been forgotten by Galileo's historians. To a man, they have presented us with a completely unsolved mystery surrounding *The Assayer*.

Operation Sarsi

How was *The Assayer* born? How had it become the manifesto of an active intellectual and political group in Rome? What was contained in it to make it so? The notoriety of the circumstances that induced Galileo to write this book allows us to answer the first question simply by giving, in rapid progression, the scenario of a series of events which preceded and accompanied the publication of that work.

May 1612. Barely a year after his public appearance as a philosopher (with the experience of the Bolognan rock), Galileo confirms his propensity for natural philosophy in a polemic with Aristotelian tradition, publishing the *Discourse on Floating Bodies*. The book earns him sensational and controversial fame. It deals with the problem of floating and hydrostatic principles, starting with a discussion of the flotation of ice. It introduces strictly philosophical problems on the nature of heat and cold. Galileo opposes the qualitative physics of the Aristotelians, critically reviving Democritus's notion of "fiery atoms" moving among the particles of liquids.[11]

Tuscan university Aristotelians unite against Galileo's position. He is, however, supported by a cardinal destined for future greatness—Maffeo Barberini.

June 23, 1612. Cardinal Roberto Bellarmino, with the cordial courtesy for which he is known, thanks Galileo for the gift of a copy of the *Discourse on Floating Bodies*, promising to read it as soon as possible.

August 1612. Among the numerous arguments against the atomist declarations of Galileo's new book stands out that of Giorgio Coresio, the Pisan university professor. Dom Benedetto Castelli replies with a series of refutations. The polemic drags on, but is not recorded. It must be pointed out that Galileo's supporters as well as his detractors agree on the legitimacy of connecting his ideas with those of Democritus.[12]

1613. The Academy of Lynceans—which since 1611 has boasted of Galileo as its associate and prestigious standard-bearer—publishes in Rome the *History and Mathematical Demonstrations concerning Sunspots*. The Academy,

---

the most serious suspicions at Father Scheiner's door (see S. Drake, *Galileo at Work*, Chicago and London 1978, p. 467). In any case, however, proof does not exist for either tendency.

[11] See *Works*, IV, p. 48. The *Discourse* had two editions in 1612. See S. Drake, *Galileo Studies*, Ann Arbor 1970, p. 159 ff.; W. Shea *Galileo's Intellectual Revolution: Middle Period, 1610-1632*, London and Basingstoke 1972 (2nd ed., New York 1977), pp. 14-48.

[12] See G. Coresio, *Operetta intorno al galleggiare de' corpi solidi* (Florence 1612), *Works*, IV; and Father B. Castelli, *Errori di Giorgio Coresio* (1612), ibid., p. 281 ff.

a private, secular institution, thus assumes full responsibility for a book in which Galileo is for the first time in a direct polemic with an authoritative Jesuit scientist, Father Christopher Scheiner. The controversy, originating from a question of priority in the discovery of sunspots, swells under Galileo's pen and the aegis of the Lynceans, attaining the shrill tones of a cultural dispute. Galileo, in the third letter, hurls a great challenge: "in science, the authority embodied in the opinion of thousands is not worth a spark of reason in one man."[13] That spark is nourished by "love of the Divine Artificer . . . the source of light and truth."

Galileo denounces Aristotelian physics, accusing it of pure nominalism, and appropriates for the first time the slogan of "the book of nature" counterposed to the books of Aristotle and his commentators, as if "nature had not written this great book of the world to be read by others besides Aristotle."[14] Aristotle's texts are described as the prison of reason. Thus, hostilities with official Jesuit philosophy begin.

Is not the signal indeed given by that provocative appeal for a direct reading of the book of nature? That image which to our modern ears sounds like a poetic metaphor had in fact a less innocent sound to Roman ears, accustomed to picking up echoes from beyond the Alps. Was not that appeal perhaps a slap at the Scholastic Catholic tradition and a hand treacherously extended to the heretics, who had made a precept of the direct reading of that book, as of the Bible? "Nature stands before our eyes like a beautiful book, in which all created things are like letters expressing the invisible thoughts of God"—is this not perhaps a fundamental article, Number II, in the *Confessio belgica* (1561) of the Reformed Dutch? Could it be that Galileo, by making the reading of that book into an irreverent anti-Aristotelian jibe, dissimulates the ideals of an irenic Christianity similar to those of his friend Kepler?

It is problems like these that count most for the Jesuits. Silencing Galileo with the proper tact and authority will soon be carried out by the greatest Jesuit: the fragile Bellarmino.

February 24, 1616. A year has passed since the Holy Office opened pro-

[13] *Works*, V, p. 200.

[14] *Istoria e dimostrazioni intorno alle macchie solari e loro accidenti* (Rome 1613), *Works*, V, p. 190. For the *Confessio belgica*, see G. de Brès, *Confession de foy* [Leiden 1562], p. 2, art. II, inspired by St. Paul (Romans 1:20). In the writings called *Juvenilia*, Galileo used the expression apropos of the "book of nature," inspired by Biblical quotations: "And the heavens shall be rolled together as a scroll" (Isaiah 34:6) or "and the heavens departed as a scroll when it is rolled together" (Revelation 6:14). See *Works*, I, p. 64. On the striking presence of the theme of the book of nature, which was immediately appreciated by everyone in the intellectual and religious world of the Reformation, see R. Hooyakaas, *Religion and the Rise of Modern Science*, Edinburgh and London 1972, p. 105 ff. and p. 114. On the various literary and philosophical traditions of the book of nature, the encyclopedism and mathemetical symbology, see E. Garin, *La cultura filosofica del Rinascimento*, Florence 1961, pp. 451-55; and P. Rossi, *Clavis universalis. Arti mnemoniche e logica combinatoria da Lullo a Leibniz*, Milan and Naples 1960 (2nd ed., Bologna 1983).

ceedings against Galileo, denounced, as we said, by Father Lorini for having dared adapt the Bible to heliocentrism. Now, while Galileo is in Rome attempting to avert a condemnation of Copernicanism, the preliminary investigation has reached its climax. With unanimity of opinion, the theological experts of the Holy Office judge the heliocentric doctrine to be philosophically foolish and absurd, formally heretical, and the doctrine of the earth's movement to be erroneous *de fide*, inasmuch as the first contradicts the Scriptures and the second does not conform to them.

February 26, 1616. The condemnation of Copernicus's theory is officially communicated to Galileo by Cardinal Bellarmino at the Vatican Palace. At present the Father Commissary of the Holy Office, Bellarmino, a dilettante astronomer, devoted admirer of scientists and of that "fine mathematician"[15] which Galileo is in his eyes, manages with this official but personal admonition to block the judicial procedure opened against him by Father Lorini's Florentine denunciation—that Galileo tried to apply to the Bible the idea "that the earth moves and the sky stands still, which tramples on all of Aristotle's philosophy (which has been of such service to Scholastic theology)."[16]

What did Bellarmino say to Galileo? Did he simply inform him of Copernicus's being put on the Index, as shown by the indisputable certificate later released by the cardinal at Galileo's request to counter the slanderous rumors of an abjuration,[17] or was Galileo formally warned not "to defend or hold" the condemned theory in any manner, as shown by the unusual, very famous record of the meeting which without signatures and without legal validity was yet of great importance at the 1633 trial?[18] Now, for one hundred years, historians have still been unable to reach an agreement on this point. In fact, the two documents are incompatible and contradictory. They have in common only an official juridical purpose, not the aim of having us understand how matters actually stood. Yet, the private setting of that admonition, the cordiality of relations between the two men, and the results of the event lead one to think that the cardinal might well have told Galileo how things really stood.

The Holy Office did not concern itself with astronomy, but with questions of Catholic orthodoxy. It had been the denunciation of Galileo's exe-

[15] See Cardinal Bellarmino's letter to Father Clavius and to other members of the mathematical community of the Collegio Romano, April 19, 1612, *Works*, XI, p. 87 ff.; and the unanimous reply in favor of the celestial discoveries announced by *Sidereus nuncius* [*Starry Messenger*], April 23, 1611, ibid., p. 92. From this official initiative of Bellarmino comes Galileo's public recognition on the part of the mathematical component of the Collegio, which culminates in the celebration of *laurea honoris causa*, bestowed on Galileo on the evening of May 13, 1611, in the Great Hall of the Collegio Romano.

[16] *Works*, XIX, p. 297 ff.

[17] See Bellarmino's testimonial to Galileo on May 26, 1616, *Works*, XIX, p. 348.

[18] See *Works*, XIX, p. 321 ff.

getical and unifying initiatives that brought Copernicus before the qualificators of the Holy Office. Only now did the contraposition of the astronomical problem of the earth's motion and the geocentric expressions of the Bible pose a dramatic conflict. The effort to bend the Biblical interpretation to the exigencies of that theory, renewed by the Carmelite theologian Foscarini, compelled the Holy Office to intervene.

But let us be clear. Cardinal Bellarmino, friend of the late lamented astronomer Father Clavius since his school days at the Collegio Romano, knows very well that the Copernican description of planetary movements could be perfectly plausible—were it not false. In an affable letter to Father Foscarini, who had published a theory of the concordance between heliocentrism and the Scriptures, the cardinal ironically let it be understood that he was waiting for a demonstration of heliocentrism.[19] But Bellarmino felt at peace with his conscience in declaring, along with the theologians of the Holy Office, that the doctrine of the earth's motion was "foolish in philosophy." Clavius, in his famous *Commentarius* on Sacrobosco's *Sphaera*, has in fact shown that, granted Aristotle's laws of physics, "the earth, insofar as it is the greatest weight, tends naturally to that natural point which is the center"[20] of the universe. Vice versa, if *ab absurdum* the earth were not immobile, but moved with a straight or circular motion, all the physical consequences deriving from this would run against experience. Thus, although Bellarmino thought, erroneously, that the Copernican hypothesis was simpler than the Ptolemaic, he knew that heliocentric truth could not be established without the prior demonstration of the physical impossibility of Aristotle's cosmology. The accord of the Biblical dictum with Aristotelian geocentricism was only a further binding argument. And who more than Bellarmino, the great inquisitor of Bruno's trial, had the far-sighted Christian prudence to prevent the new Copernican philosophical fantasies and the new scandals from having free rein?

Bellarmino had nothing to teach Galileo in astronomy and physics. But

[19] See Cardinal Bellarmino's letter to Father Foscarini (Scarini), April 12, 1615, *Works*, XII, p. 171 ff. As is known, Father Foscarini had published a *Lettera sopra l'opinione dei Pitagorici e del Copernico . . . nella quale si accordano e si appaciano i luoghi della Sacra Scrittura e le proposizioni teologiche che giammai potessero addursi contro tale opinione*, Naples 1615 (see *Opere di Galileo*, ed. E. Alberi, 1852-1856, V, pp. 455-94). As for the book of the Augustinian theologian Diego da Zuñica, *Commentaria in Job*, Toledo 1584 (2nd ed., Rome 1591), it had already been criticized by Father J. Pineda, S.J., in *Commentaria in Job*, Matriti 1597-1601. See H. Grisar, *Galileistudien*, p. 263 ff.

[20] See C. Clavius, *In Sphaeram Joannis de Sacri Bosco Commentarius*, Venetiis 1591, p. 195 ff. In 1612, the Collegio Romano had accepted the possibility of local celestial corruption, a possibility adopted by Father Scheiner and also known to Cardinal Bellarmino, as becomes clear from the correspondence with F. Cesi (1618) which we shall deal with later. No concession, however, was granted to the hypothesis of the earth's motion: in the succeeding edition of the *Sphaera* (in *Opera mathematica*, 5 vols., Moguntiae 1611-1612, tome III, p. 301), Father Clavius, while looking toward a new astronomical system and appreciating Galileo's celestial discoveries, remained quite aware of the absurdity of Copernicanism in the physics of motion.

nobody could teach him the necessity of keeping Biblical exegesis within the Scholastic tradition. Now the passages of Scripture relating to the problem of the sun's motion were not so numerous or so important, nor for that matter had any council ever stipulated geocentrism as an article of faith. But it was a matter of principle. If one were to admit the possibility of interpreting the Scriptures with new natural reasons, the danger might arise that this reform would also extend to other, more fundamental points of faith and traditional interpretation.

March 3, 1616. With the decree of the Congregation of the Index, controlled by Cardinal Bellarmino, the February 24 unanimous opinion of the theological qualificators is ratified (without officially committing the Holy Office to a sentence). Copernicus's *De revolutionibus* and the commentary on the Book of Job by the theologian Diego da Zuñica are suspended, pending corrections of any and all references to the real (absolute) character of the condemned doctrine. However, the book of the theologian Foscarini and every exegetical text *idem docentes* are condemned.

On the wave of the most recent exigencies of controversy, Rome has thus opted for an all-out defense of the Aristotelian-Scholastic cosmology and the literal significance of the Bible, restricting its interpretation to the accredited theological tradition. Thus, the Augustinian instances supporting the notion of a "figurative meaning" in Biblical exegesis are disavowed. And, indeed, how many heresies have been committed in the name of St. Augustine!

September 1616. "A new philosophy, consonant with the doctrine of the saints, to be introduced into Christian schools in place of the authority of the Peripatetic sect and the pagan philosophers." Thus Brother Tommaso Campanella insists at the start of the manuscript *Apology for Galileo*, which now is being handed about in Cesi circles, presenting an untimely challenge—after the battle—to the Roman decisions with its proposals for a renewed Scriptural hermeneutic and its Augustinian theses on nature as God's book in correspondence to the Bible. The new philosophy is the very ancient one of Pythagoras, Hebraic in origin; the Mosaic astronomy and physics are more orthodox than those of the pagan Aristotle.

Campanella, in perpetual imprisonment, is condemned as a heretic several times for his Telesian, atomist, and anti-Aristotelian ideas. Galileo could not have hit upon a more ill-starred and compromising champion. The fact is that neither now nor later will Galileo want to compromise himself by answering the generous and catastrophic initiatives taken by Campanella in his defense. But in 1622, when the *Apology* will be published in Frankfurt by the Lutheran Tobias Adami, and when the names of Galileo, Cardinal Cusano, Foscarini, Bruno, Kepler, Patrizi, Telesio, and Hill are associated by the writer of the preface in the irenic light of a new philosophy that does not know religious frontiers, then Galileo will truly be publicly compromised.

May 23, 1618. The Defenestration of Prague is a clear signal of a European war that will wreck the last hope of Christian pacification, and project into the minds of Catholic militants images of a bloody redemptive crusade and the renewed fervor of Tridentine orthodoxy. But only the successive, frightening apparition not of one but of three comets in the dark skies of Europe will adequately presage in all its horror the ferocious cruelty and Biblical duration of the Thirty Years' War.

March 1619. The Society of Jesus publishes, anonymously, a lecture on the comets that appeared during the winter. It was delivered by the thirty-six-year-old Father Orazio Grassi of Savona, professor of mathematics at the Collegio Romano. The publication, entitled *De tribus cometis anni MDCXVIIII disputatio astronomica habita in Collegio Romano*, wins the official validation of the Society's great university because it documents the scientific quality of observational astronomy cultivated by the order. Grassi bases himself on Tycho Brahe's modern ideas, given specificity by a new harvest of cometary observations made by the Society's observers scattered throughout Europe.

Thus, Father Grassi can offer a brilliant confirmation of the absence of an appreciable parallax for the comet: on the very same day, December 10, at Rome and Cologne, there has in fact been observed the same stellar concealment on the part of the wandering star. The exiguous size of the parallax, the constant motion of a planetary type, the lack of telescopic enlargement, lead Father Grassi to place the comet in a position between the moon and the sun: a celestial body in motion along a great circle, brilliant with reflected solar light, unlike what Aristotle had maintained. Once again, Jesuit astronomy gives the Collegio Romano an example of its open-minded freedom of research, as when it officially recognized the discoveries of the *Starry Messenger* and the sunspots.

June 1619. The only comet that Galileo had observed in his life figured in memories of his adolescence, when he was thirteen and saw from Florence the comet of 1577.[21] But, urged by several groups to declare himself[22] and

[21] See *The Assayer*, p. 190 (*Works*, VI, p. 314). The comet of 1577 had led to Tycho Brahe's observations presented in *De mundi aetherei recentioribus phaenomenis*, Uraniburgi 1588, pp. 89-159; and in *Astronomiae instauratae Progymnasmata*, Pragae 1602, p. 511-13.

[22] Giovanbattista Rinuccini, household prelate of Gregory XV, wrote from Rome that "the Jesuits have publicly created a problem, which is in print . . . and until now some of the Jesuits spread the rumor that this book knocks Copernicus to the ground" (*Works*, XII, p. 443). On the controversy over the comets between Galileo and Father Grassi, described in all the biographies of Galileo, see the preface to S. Drake, *Controversy of the Comets of 1618*, Philadelphia 1960; Shea, *Galileo's Intellectual Revolution*, pp. 75-108.

concerned with the threatening authoritativeness of the theses officially adopted by the Collegio Romano, he decided to reply through an intermediary. He inspired, and in large part wrote, the *Discourse on the Comets* given by his pupil Mario Guiducci at the Florentine Academy and published immediately. In it the thesis of the purely optical nature of this phenomenon is polemically presented, after rejecting the hypotheses of the ancients as well as those of Tycho Brahe and Father Grassi's *Disputatio*. Mario Guiducci had been a student at the Collegio Romano.[23]

July 12, 1619. "Admirable, miraculous, a new thing. . . . What never stops pleasing here is this attack on the Collegio Romano. . . . The Jesuits consider themselves deeply offended. . . ."[24] This was written to Galileo by Don Giovanni Ciampoli, a Florentine pupil, philosopher, and thirty-year-old literary man, now turned Roman under the protection of the brilliant Cardinal Maffeo Barberini. Ciampoli has only recently joined the Academy of the Lynceans.

August 1619. At the Collegio Romano, Father Orazio Grassi completes some experiments in physics for the reply he is preparing with respect to the Galileo-Guiducci *Discourse*. Present at these experiments, on the invitation of Father Grassi, are Giovanni Ciampoli and Virginio Cesarini, another Roman intellectual from the Academy of the Lynceans. For the historian, this is one of the first living images of Father Orazio Grassi, and it is not very flattering. Ciampoli says he has no doubts about the divergence between what Father Grassi says and what he writes.[25] Coming from such an able political observer and diplomat as Ciampoli, this judgment deserves consideration and induces distrust.

December 1619. The reply to the *Discourse on the Comets* is in the Roman bookstores. It was published in Perugia under the pseudonym Lotario Sarsi, behind which, as Ciampoli reveals, hides Father Grassi; it is entitled *Libra astronomica ac philosophica*. Now there is open warfare.

The pseudonym, according to the usual admonishment of the Society of Jesus in cases of serious scientific, theological, and political controversy, does not officially compromise the Collegio Romano. But the rumor is put about that the *Libra* has the full support and solidarity of the Society's curia. Indeed, the book is embellished by the permission of Father Grassi's superiors.

[23] See Galileo's letter to Cardinal Maffeo Barberini, June 29, 1619, accompanying the dispatch of the *Discourse on the Comets*, *Works*, XII, p. 461 ff. This *Discourse*, signed by Guiducci, as well as *De tribus cometis* (Romae 1619), by Grassi, are published in *Works*, VI.

[24] *Works*, XII, p. 465.

[25] See G. Ciampoli's letter to Galileo on August 24, 1619: "Certainly he [Father Grassi] speaks personally, as regards compliments, with great respect . . ." (*Works*, XVIII *Supplemento al Carteggio*, pp. 423-25). All the succeeding testimonies *de visu* confirm this first impression by Father Grassi. The meeting with Ciampoli and Cesarini is also confirmed by the *Libra* (*Works*, VI, p. 157) and by the *Ratio* (ibid., p. 471).

On the controversial and scientific plane, Galileo will find Sarsi a tough opponent. Against the paradoxical Galilean counterhypothesis, Sarsi appeals to the principle of authority. The authority of tradition: from the Pythagoreans to Tycho Brahe it has always been recognized that a comet is something real. The authority of the senses: it is sufficient to have seen a comet "only once," with the naked eye or a telescope, to understand that it is not a matter of the play of light. On the scientific plane, the counterhypothesis on the nature of comets is confuted by a whole range of refutations of Galileo's interpretive model, which he based on the difference between the physical supports of the rainbow phenomenon and those of other phenomena of solar light reflection. Father Grassi is an expert in the field of optics; he has already published, in 1618, a book entitled *Disputatio optica de iride*.[26]

The focus of the polemic shifts from the astronomical controversy to problems connected with the physics of heat, light, and fluids. In fact, the *Libra* offers a series of new and sensational experiments in physics, among which are those on the motion of air in a revolving basin, publicly repeated at the Collegio Romano before Roman supporters of Galileo.

But the *Libra* transcends the terms and style of a normal scientific dispute. It reveals the controversial and apologetic matrix that saturates all forms of Jesuit polemic. The mysterious Sarsi betrays an invincible propensity to introduce into the scientific dispute hypocritical conclusions and insinuations about his opponent's religious opinions.[27]

Sarsi recalls with veiled menace Copernicus's condemnation in 1616 and, even more serious, does not pass up the opportunity to say that Galileo's theory of comets is reminiscent of some ideas in *De cometis* (1618), by the Copernican and heretical Kepler, though in reality the theory is taken from Cardano and Telesio. Probably the implication is true, even if Galileo did not admit it, for Cardano and Telesio were authors who had the reputation of being materialistic atheists. Sarsi's hypocritical allusion to the "sterile and unfortunate philosophy" that inspired Galileo is a heavy piece of malice, a really low blow which irremediably poisons the controversy.[28]

February-March 1620. There are repeated Roman appeals to Galileo that he accept the challenge and bring down the "pride of the Jesuits,"[29] as the illustrious Doctor Johannes Faber, member of the Academy of Lynceans, writes. Also, Professor Giulio Cesare Lagalla of the Sapienza, a heterodox

[26] See *Disputatio optica de iride proposita in Collegio Romano a Galeatio Mariscotto*, Romae 1618.

[27] See L. Sarsi, *Libra astronomica ac philosohica* . . . , Perusiae 1619 (*Works*, VI, pp. 109-179); *The Assayer*, p. 45 ff.

[28] The first accusation of "telesismo" against Galileo was made in the Florentine piece of writing *Contro il moto della Terra*, attributed to L. delle Colombe (1611), in *Works*, III, pt. I, pp. 253-90, especially p. 253.

[29] J. Faber's letter to Galileo, February 15, 1620, *Works*, XII, p. 23.

Peripatetic, bigot, and renowned libertine, hated by the Jesuits and a good friend of Galileo in Rome, denounces the arrogant censorship that the Jesuits want to exercise over ideas.[30] From many quarters, Galileo is asked to react and set forth his "inventions in philosophical matters."[31]

May 15, 1620. The Congregation of the Index bans Kepler's *Epitome*, and suggests the necessary changes in *De revolutionibus* so that the work may become permissible for Catholics.[32]

May 10-18, 1620. During a sojourn in the ducal palace of Acquasparta, Prince Cesi's country house near Urbino, the active Roman nucleus of the Academy of Lynceans—Cesi, Ciampoli, and Cesarini—decides on "Operation Sarsi," in the epico-satiric style of the fiery literary polemics of the day.[33] They plan to launch, using the debate on comets as a pretext, an attack in the grand style, against the intellectual strongholds of traditional culture then dominant in Rome. At stake are not only Galileo's authority but, above all, the intellectual prestige and legitimation of the Academy of Lynceans.

July 17, 1620. Monsignor Ciampoli informs Galileo of the strategy agreed upon at Acquasparta, a strategy that Galileo will follow.[34] The rebuttal must be dedicated to the young Duke Virginio Cesarini. This is easily explained. A great Roman intellectual and aristocrat, the personal friend of the ambassador of Spain and of Cardinals Bellarmino and Ludovisi, Cesarini offers the best camouflage for the audacious undertaking. At the same time, Cesarini is a member of the Academy of Lynceans; by dedicating the reply to him, the compactness and strength of the Academy would be officially stressed. Ciampoli also suggests to Galileo that he exploit the pseudonym used by the author of the *Libra*, pretending to be unaware that Father Grassi is behind it, in order to have greater freedom of criticism.

September 17, 1621. Cardinal Bellarmino dies, surrounded by the odor of sanctity. He will carry with him, in the magnificent tomb decorated by Bernini—which is being prepared with an eye to a rapid canonization—all the secrets of Bruno's trial. Even if the reign of the next pope (Ludovisi) will more than ever be pro-Jesuit and pro-Spanish, to many Italian heretics and innovators the death of the greatest representative of the Counter-Reformation seems to open up new political and intellectual hopes. However, Bellarmino leaves to immortal memory his *Controversiae*, which will see dozens of editions throughout the seventeenth century.

[30] G. C. Lagalla's letter to Galileo, March 6, 1620, ibid., p. 26.

[31] A. Santini's letter to Galileo, April 3, 1620, ibid., p. 29.

[32] See *Works*, XIX, p. 400 ff. This *Monito per l'emendazione* and censure of Copernicus had been preceded, on May 10, 1619, with the condemnation in the Index of Kepler's *Epitome astronomiae copernicanae*; see *Elenchus librorum . . . prohibitorum*, Romae 1640, p. 211.

[33] G. Ciampoli's letter to Galileo, May 18, 1620, *Works*, XIII, p. 38 ff.

[34] Ibid., p. 43.

January-May 1622. Repeated requests come from Monsignor Giovanni Ciampoli, now the Secretary of Briefs, Duke Cesarini, and Filippo Magalotti, asking Galileo to fulfill his commitment to answer the Jesuits. The Academy of Lynceans would like to publish the reply within the year.[35]

Late October 1622. The long-hoped-for manuscript of *The Assayer*, the much awaited war machine against "the obstinate worshippers of antiquity,"[36] reaches Cesarini.[37] The second phase of "Operation Sarsi" begins: a collaborative revision of the work, as was in any event set down by the Academy's internal rules.

November-December 1622. Cesarini has copies made of the manuscript and makes numerous corrections of the original. We shall probably never know which ones, since no original copy of *The Assayer* has survived.[38] The corrected original manuscript is then sent to Federico Cesi.[39]

January 12, 1623. We are in the final phase. Prince Cesi, Monsignor Ciampoli, and Cassiano dal Pozzo, a well-known Roman "virtuoso," are also revising Galileo's text.[40] Cesarini writes to them: "We want to publish the work and we want to do it in Rome, notwithstanding the power of the adversaries, against whom we will arm ourselves with the shield of truth."[41] Immediately thereafter, however, Cesarini also guarantees that he can count on the "favor of the masters": far more reassuring protection than that mentioned before. Cesarini announces the prohibition of Campanella's *Apology for Galileo* and predicts that the clash will therefore be bitter. But nothing dissuades Cesarini, this former student of the Jesuit fathers, from the idea that *The Assayer* must come out in "the face of the Church, before the eyes of the congregations."[42]

January-February 1623. "Operation Sarsi" is by now an open secret. Cesarini has not been able to refrain from reading some parts of the manuscript to various acquaintances. In Rome, even the walls have ears; the Jesuits are informed, and have guessed the rest ("they have seen through everything").[43] Above all, they have seen that Galileo's book threatens to

[35] G. Ciampoli's letters to Galileo, August 2, 1620, ibid., p. 46 ff.; January 15, 1622, ibid., p. 69; February 26, 1622, ibid., p. 83 ff. See, also, V. Cesarini's letter, May 7, 1622, ibid., p. 89; and F. Magalotti's letter, May 7, 1622, ibid., p. 89 ff.

[36] Ibid. [37] Ibid., p. 99.

[38] See A. Favaro's preface to vol. VI of the *Works*; and G. Govi, "Alcune lettere inedite di Galileo Galilei," in *Bullettino Bibliografico e di Storia delle Scienze Matematiche e Fisiche* 14 (1881), p. 367, where the author affirms having held in his hands a manuscript of *The Assayer*, "an amanuensis, corrected in several places by Galileo."

[39] See V. Cesarini's letter to F. Cesi, December 22, 1622, *Works*, XIII, p. 102; and F. Cesi's letter to J. Faber, November 19, 1622, ibid., p. 100.

[40] See G. Ciampoli's letter to Galileo, January 7, 1623, ibid., p. 104; and V. Cesarini's letter to Galileo, January 12 1623, ibid., p. 105.

[41] Ibid., p. 106. [42] Ibid. [43] Ibid.

produce a scandalous denigration of the Collegio Romano. They respond preemptively: during the inaugural academic lessons, the professors at the Collegio Romano stigmatize the innovators with particular vigor. Cesarini is taken aback, but not intimidated: "Despite this 'excommunication' fired forth with such eloquence," he writes to Galileo, "I hope that your very noble speculations will have free commerce and applause in Rome." Moreover, *The Assayer*'s "philosophical freedom" suggested the idea of translating the book into Latin to ensure it the European circulation it deserved. This project is not realized, and *The Assayer*'s circulation outside Italy will be very small so far as one can see from Galileo's correspondence.

April-May 1623. The Assayer is printed.[44] We have no date for the printing which, as usual, must not have been more than a few hundred copies. Johannes Faber and Francesco Stelluti have also gone over the manuscript.[45] In all, there were six authoritative "censors," from the Academy of Lynceans and of different backgrounds, who could not have failed to observe any philosophical or theological aspect that would be too compromising for the common endeavor. Long before publication, Monsignor Ciampoli and Virginio Cesarini have arranged matters so that the ecclesiastical authorization is entrusted to a young professor of theology at the Minerva, the Dominican Father Niccolò Riccardi, a Genoan and a follower of Galileo destined for a rapid career in the Holy Office.[46] Father Riccardi's order, having had a serious conflict with the Society of Jesus in a very recent and burning theological debate, could ask for nothing better than the opportunity to license a book openly denigrating the Collegio Romano's authority.

The license bears the date of February 3. In reality, Father Riccardi had not granted *The Assayer* an authorization in the usual style demanded by the inquisitorial bureaucracy, but had bestowed on it dithyrambic praise—a kind of publishing blurb before its time, scandalously favorable and alluring. Father Riccardi, rather than confining himself to a statement that the work was not contrary to religion, as was his strict duty, proclaimed the orthodoxy of the book with an unrequired acknowledgment. Just as improperly, he praised the "many fine considerations belonging to our philosophy" and "the many secrets of nature" announced to the age, "thanks to the author's subtle and solid speculations," in which age the theological press-censor declared himself happy to have been born. The authorization for the imprimatur even goes so far as to smile at the effective irony of the book's title, mimicking the style with which Galileo put the *Libra* at the mercy of his

[44] F. Stelluti's letter to Galileo, April 8, 1623, ibid., p. 113. The editing of the edition entrusted to the poet Tommaso Stigliani includes, besides some mistakes, some changes in the text, as becomes clear from Guiducci's letter to Galileo, December 18, 1623, ibid., p. 161.

[45] V. Cesarini's letter to F. Cesi, January 28, 1623, ibid., p. 108.

[46] V. Cesarini's letter to Galileo, February 3, 1623, ibid., p. 109.

wit.[47] The signature of a brilliant Dominican theologian below so laudatory an imprimatur sealed *The Assayer*'s success even before publication.

Late July–August 6, 1623. A dramatic turn of events: Rome's papal throne is empty. At the Vatican, there are great maneuvers—first by Ciampoli and then by the Francophile Cardinal Prince Maurizio of Savoy—during the conclave. Duke Cesarini does his part at Piazza Navona with the Spanish ambassador and the pro-Spanish cardinals. The desired outcome is the election of Cardinal Maffeo Barberini, who, on August 6, after exhausting, all-night negotiations among the great cardinal electors, became Urban VIII.

At Jesuit headquarters on the Piazza del Gesú, the top figures in the Society, traditionally aligned with Spain, carefully evaluate the unknown aspects of the election of the former nuncio to Paris and of the imprudent ambition to remove papal politics from Spanish hegemony. The Jesuits are not worried by the petulant and noisy manifestations of jubilation on the part of literary innovators and forward-looking Roman aristocrats, galvanized by the election of a pope who is a refined intellectual friendly to Galileo. Their concern, rather, is over a line of cultural liberalization and an improvized policy that are counterproductive to the program of renewal and struggle set forth by the Council of Trent for the Counter-Reformation Church. The Society of Jesus, which is the most efficient instrument for carrying out that program, is not the victim of the narrow provincial and Roman view which conditions so many of its enemies in the Curia. The principal fronts of the Counter-Reformation struggle are neither the corridors of the Curia nor the salons of the Academy, but rather the plains and cities of Hungary and Bohemia, where the fathers of the Society, following the imperial line regiments, are triumphing. In Rome, they reconquer the churches profaned by Protestant rituals; they raise their banner, decorated by the Eucharistic symbol, over the monasteries of the decadent and corrupt religious orders, confiscating and turning them into colleges and centers of religious re-education, without concern about the Roman protests of the friars. The Jesuits' success in the chief theater of the religious war, is impressive: in the territories just wrested from the Protestants, whole populations are reconverted en masse to Catholicism, by every means, at all costs—even with solid coin, as Cardinal Bellarmino had cleverly suggested.

Strengthened by these victories, and by the political and religious realization of its worldwide dimensions, the Society of Jesus knows that fidelity to the empire is the best guarantee for the Counter-Reformation. The Society distrusts the new pontificate's reckless diplomatic overtures to such an

[47] "I regard myself as happy to have been born, when no longer with steelyard balance and crudely in large lots, but with delicate coin scales one weighs the gold of truth," *The Assayer*, imprimatur (*Works*, VI, p. 200). Sarsi in his reply will take sarcastic advantage of the Florentine expression "coin scales" (*saggiatore*) used by the Ligurian Father Riccardi.

unscrupulous adventurer as Richelieu, the new rising star in European politics. For the Jesuits, this is a delicate moment: they risk being isolated from the new regime and having to relinquish their traditional Vatican influence on political and cultural matters to the new intellectual entourage.

August 1623. The Church has not had such a young and open-minded pope for a long time. Barberini is a Florentine, a poet, and a brilliant, lively man. His horseback rides in the Vatican Gardens and his retinue of Florentine artists excite the aristocratic snobs and the Roman intellectual innovators. The literary innovators are in ecstasy. Their bard, Giambattista Marino, leaves Paris (where he has published with enormous scandal his poem, *L'Adone*) to come and pay homage with his presence to Rome's new regime. The academies will give him a triumphant welcome.

On August 12, the academies will march in the front rank of the sumptuous investiture procession from the Campidoglio to the Vatican. Agostino Mascardi, the liberal rhetorician of the Roman intelligentsia, will celebrate the new regime,[48] in which research and disagreement with traditional canons in literature and philosophy can develop freely and cause Urban VIII's new regime to shine as brightly as the stones that adorn his tiara. Meanwhile, the editors of *The Assayer* grasp the unexpected advantage of the new situation, putting on its frontispiece, together with the Lyncean Academy's coat of arms, that of Pope Barberini (Fig. 1).

The alliance between the Academy and the new power is not merely symbolic. Three Lynceans, three "lovers of the new philosophy," enjoy the closest personal relationship with top figures in the Church: Monsignor Ciampoli, canon of St. Peter's and gray eminence of the Secretariat of Briefs; Monsignor Cesarini, now Master of the Pontifical Chamber; and a layman, Cavalier Cassiano dal Pozzo, the great art patron, in the key position of secretary to the pope's nephew.[49]

September 8, 1623. Francesco Barberini, the twenty-six-year-old nephew of Pope Urban VIII, is appointed by Prince Cesi—with absolutely perfect timing—to the Academy of Lynceans. A few days later, Francesco's uncle, the pope, bestows on him the cardinal's purple.[50] As of October 7, the nephew, Cardinal Barberini, a splendid patron of the arts, will be the general superintendent over all secular and ecclesiastical questions in Rome.

Late October 1623. *The Assayer* is ready. The edition is full of errors; there are even a few changes made by the proofreader, the brilliant poet Stigliani. But aside from the author, nobody complains: the book is eminently enjoyable, most readable, the most brilliant that Galileo has yet written. In the

[48] See A. Mascardi, *Le pompe del Campidoglio per la S. di N.S. Urbano VIII*, Rome 1624, published again in *Discorsi moraldi sulla tavola di Cebete Tebano*, 1642.

[49] See F. Stelluti's letter to Galileo, August 12, 1623, *Works*, XIII, p. 121.

[50] Ibid., p. 132.

midst of all these events, Cesarini has not properly supervised the printing. To make up for it, he is the author of the book's very fine dedication to Pope Barberini, a dedication that sings the praises of the new papal cultural policy and is signed collectively by the members of the Academy of Lynceans.

October 27, 1623. In the Vatican, the official presentation of *The Assayer* is made to Urban VIII, before a serried and select representation of the Sacred College of Cardinals. We recognize the cardinal-nephew, numerous prelates from the Curia, and among them the friends of the Academy of Lynceans. Its founder, Prince Cesi, presents in homage to Pope Barberini a richly bound copy of Galileo's book, other copies being offered to Cardinal Francesco Barberini, the other cardinals, and friends. The book is in great demand everywhere.[51]

The pope's approval of *The Assayer* is thus publicly confirmed before the Church and the congregation, as Cesarini had hoped. Indeed, the official recognition, of great cultural significance, had been caught on the wing at a moment of intense euphoria—"in this universal jubilation of good letters," as Cesarini had written in his dedication to the pope.

The pope had not yet read *The Assayer* by his old friend Galileo. In the days following the official presentation, he had Ciampoli read him some passages, while at table: the fable of sound and other spicy sections. He was amused and full of admiration.[52] Ciampoli, like a good politician, hastened to suggest to Galileo that he strike while the iron was hot and take advantage of the moment of freedom that had come about so as not to deprive the world of his scientific speculation. The consent obtained thanks to a series of fortunate coincidences must be used to go forward, and the Academy of Lynceans wanted to benefit. Moreover, the dignity of the Collegio Romano had been sadly diminished.

We have said, and we will say again, that Father Grassi was right to feel insulted by the violent satire hurled at his pseudonym by *The Assayer*; for that satire took the form of intolerably ironic or gravely injurious personal attacks, which cast discredit and offense on the institution to which the author of the *Libra* belonged. Especially painful was the fact that those witticisms and cyanide-saturated passages were in everyone's mouth in Rome; even the pope, as we know, had not been able to refrain from laughing pleasurably over them.

Let us select one of these passages, by way of example. *The Assayer* compared his adversary's haughty attitude to "that of a lacerated and bruised snake having no vitality left except in the tip of its tail, but which nevertheless continues to wriggle, pretending that it is still healthy and vigorous."[53]

[51] See F. Stelluti's letter to Galileo, October 28, 1623, ibid., p. 142.

[52] V. Cesarini's letter to Galileo, October 28, 1623, ibid., p. 141.

[53] *Works*, VI, p. 268. The passage goes back to Galileo's handwritten marginal note on the *Libra*,

Galileo seemed sure that it was dead, or at least that is what he gave people to believe.

In any event, *The Assayer* is a successful book even before reaching the bookstores.

November 1623. At the Sun Bookstore, a meeting place in Rome for virtuosi, innovators, and libertines, *The Assayer* is displayed on the counters for new books.[54] Together with Galileo's book, there is *L'Adone* and another book hot off the press, also heterodox, which sells well in the new Roman atmosphere: the *Hoggidí* by the Olivetan Father Secondo Lancillotti. Like *The Assayer*, this essay on politics, history, and morality discredits the cult of tradition. Like *The Assayer*, it too is dedicated to Pope Barberini.

April 11, 1624. Virginio Cesarini, one of the architects of the success of Galileo's new book, dies. The official funeral takes place at the Campidoglio.

April 23, 1624. In quick succession, one sees the arrival of illustrious artists and intellectuals, such as the classical poet Gabriele Chiabrera, Cavalier Marino, and Nicolas Poussin, as well as the most illustrious scientist in Europe, Galileo. He is preceded by Mario Guiducci, who also figured, though in a minor role, in the victorious polemic on the comets to which this triumph is owed. The next day, Pope Barberini grants Galileo a private audience. This is the first of a series of very amiable meetings between the two Florentines. Even more friendly are the meetings with the cardinal-nephew. Galileo tastes again the Roman success of 1611, even though this time the Jesuits of the Collegio Romano dissociate themselves from the chorus of applause with an embarrassed official silence.

May 1624. Galileo takes advantage of the festivities and invitations, making contact with various cardinals in order to sound out whether there might be a real possibility of speaking again about Copernicus, as certain rumors in the Curia suggest. There is no official denial of Copernicus's condemnation, but there are many kindly assurances.[55]

June 8, 1624. With much encouragement and felicitations, the promise of a pension for himself and his son, gifts and blessings, Galileo departs for Florence with his mind on the new *Treatise on the Flux and Reflux of the Sea*. He brings Ferdinando II de' Medici a brief from the pope, full of high praises (written by his friend Ciampoli). It constitutes an official recognition: Galileo is called Pope Barberini's "beloved son."[56]

Galileo departs. But, wary about the assurances received and the noisy fes-

reproduced in the printed version of *The Assayer* with slight variations, perhaps owing to the editors: "that snake which having been cut into many pieces and dead, yet still wriggles the very tip of its tail, with the hope of making passers-by believe it is both alive and victorious."

[54] See *Works*, XII, p. 145. On the Sun Bookstore, see R. Pintard, *Libertinage érudit dans la première moitié du XVIIe siècle*, Paris 1943, I, p. 216.

[55] *Works*, XII, pp. 175 and 181. [56] Ibid., p. 183 ff.

tivities, he leaves Mario Guiducci, his pupil and devoted friend, in Rome. In Galileo's company, Guiducci has been propelled into the Roman circles that matter. Forty years old, a lawyer and literary man, an affluent gentleman, Guiducci is a man of the world: he is a good talker and knows how to cultivate social contacts.

Galileo entrusts him with two delicate tasks. First, in the autumn, he will receive a sensitive manuscript; after it has been checked word by word by their Lyncean friends, he will discreetly circulate it in Rome and take careful note of the reaction. This will be the reply to Monsignor Ingoli—a Copernican apology in a very measured tone, to test the Roman waters.

The second task entrusted to Guiducci by Galileo was to inform him in a timely and regular fashion about everything that might happen in Rome concerning *The Assayer*. Above all, this was a matter of spying on the Jesuit reaction, especially that of Father Orazio Grassi, to find out on what basis he was secretly preparing the reply which, as we shall see, he had immediately hinted at. Spying today is a common profession, paid for by the proper institutions. In seventeenth-century Rome, where everything was secret, spying was an invaluable art: an "art of the gaze" for discovering without being discovered the dissimulated intentions of the human spirit.

Signs of the Times

The rejection of dogmatic submission to the principle of authority in the field of philosophy; the vindication of a new language; the rights of research and free intellectual discussion against the prevarication of institutional culture—these were the contents that made *The Assayer* the manifesto of the new philosophy in Rome. The book was a literary sensation because, even more than the Jesuits, even more than Scholastic thought, it seemed to oppose a whole intellectual tradition. The telescope was the instrument through which one looked at the entire universe, and *The Assayer* was the manual that taught one to read the universe like a book.

When *The Assayer* was published, comets had by now gone out of fashion; very few were still interested in discovering who was right in that difficult and confused astronomical question. There was a new and much more impassioned controversy between Galileo and Sarsi, much more interesting and accessible from the intellectual point of view than the one over cometary motion. The new exciting subject of debate was the physics of phenomena perceptible to the senses. In the literary academies, it was continually repeated that the book of nature must be read by means of the senses. Never were the senses so much at center stage as in Urban VIII's Rome. The theme of the senses was congenial to the literati: Monteverdi's madrigals, Marino's verses, Poussin's and Lorrain's colors opened all the senses to the world. The sensorial perception of heat, smells, colors, and touch was a problem that led immediately to a debate on physics.

Sarsi, in the *Libra*, had directed the discussion to the physics of heat in order to demonstrate, *contra* Galileo, that if the comets were not exhalations set on fire by the friction of their movement against the celestial spheres, then the Aristotelian explanation of heat was exact. To discredit his adversary, Galileo, as was his custom, aimed his attack at the flank of physics and philosophy. Nature—the Aristotelian professors of philosophy said—has spoken through the mouth of Aristotle. What was the grammar of this language? For an Aristotelian scholar, physics, the science of nature, was the study of the changes and motions of material things. The terrestrial world is composed of matter. The data of the senses are the first one can know as well as the most reliable for penetrating the reality of the world before us.

Reality presents to the senses infinite differences of color, heat, smell, fluidity, and hardness, but all things have in common among them at least some qualities or fundamental forms; for example, things are cold or hot, dry or moist. For Aristotelian science, there are four fundamental elements: earth, air, fire, and water, which are composed of the same material and can be reciprocally converted, depending on the exchange of those qualities.

Nature, for an Aristotelian, was written in terms of sensible qualities. All these qualities—heat, hardness, color, and odor—were inherent in a substance, were real qualities or substantial forms. Only a miracle would have been able to make a quality subsist separate from its own substance.

Also, when the elements combine to form more complex substances, the qualities that accompany them combine in turn to form the qualities of the resultant compositions. Matter was conceived of as a mode of being, and the grammar of the scientific language that described it was a complex combination of the names of real qualities.

The rules of this grammar were those of Aristotle's logic: a language of mere names tied to conceptual variables. To pronounce a scientific demonstration in Aristotelian language consisted in searching for a proposition having a subject associated with a predicate. Since the world was written in qualitative characters, each of these signs could be traced back to a qualitative concept. Thus, a body in rapid motion became heated because, by moving, it received from the warm and humid air the quality of being hot. This grammar of substantial forms and real qualities was already very complex as it stood. In the seventeenth century, however, there was added a series of new hidden qualities, such as magnetic attraction, viscosity, and chemical affinity, which the studies of alchemy and physics had introduced into scientific language.

Matter, read through this grammar of names (or logic of names), was never something in itself with respect to its properties. It was constantly a mode of being: to be hot or cold, to be odorous or to be colored. In their turn, moreover, heat, odors, or colors resulted from real modes of being.

It is not difficult to see that this language of real qualities was a game of conceptual acrobatics which, however, always remained tied to sensible experience. The advantage lay in the possibility of avoiding recourse to imaginary, invisible structures to explain phenomena and the properties of bodies. For the rest, Democritus's idea that there exist invisible elements of matter was absurd for an Aristotelian because bodies were, like time, space, and motion, forms of the continuity, potentially susceptible to infinite division. It was equally absurd to claim to study natural changes with a quantitative method: though mathematics could serve to describe abstractly some data of experience, as in geometry or musical harmony, still it could not grasp the causes of observable phenomena.

Like Campanella and Cesi, Galileo also adopted, contrary to Aristotle's grammar, the slogan which said that the book of nature was open before our eyes. Everyone could see around him the pages of this fascinating book. But what was the cipher that allowed one to read it? *The Assayer*'s originality consisted in the proposal of deciphering that book's signs, which when brought together formed its text. According to *The Assayer*, in order to decipher these signs one must see in them "triangles, circles, and other geometric shapes, without which means it is impossible to understand."[57] The goal is to decipher the book of nature, to discover its secret code, to decode the illusory appearance in order to know the mysterious laws that govern it. Here, then, is another great intellectual suggestion—evocative, extremely captivating and immediate to its readers.

We are in the seventeenth century, the century of the gold of the Cabala, of exegesis, of ultrasophisticated systems of ciphers concealing beneath irreproachable forms the most delicate diplomatic messages. Everyone interprets, deciphers, makes up anagrams, and combines. It is the *grand siècle* of the combinatorial and the linguistic: the world is populated by a collection of invisible signs. The truth is never an appearance, but a refined game of signs that dissimulate or that permit one to decipher.

Thus, the philosopher of nature becomes the possessor of the same art as the moralist and the politician. He too is a secretary, the secretary of nature, "Nature's secretary" [English in original—TRANS.] as a satire by John Donne will say;[58] he, too, can describe nature only if he is able to decipher its apparent signs.

The cipher proposed by *The Assayer* was not difficult, as were the formulas

[57] See *Works*, VI, p. 561 ff.

[58] J. Donne, "Away thou . . . humorist," in *Poems*, edited by H. Grierson, London 1933, p. 129.

of the alchemists: it was enough to know Euclidean geometry. Every intellectual virtuoso, every innovative literary man, saw the philosophy of nature open up before him and felt it to be his own.

However, the external signs of nature, except in rare instances, do not reveal themselves to the sight as geometric shapes. It was necessary to continue one's reading of *The Assayer* until paragraph 41, where it begins to speak about the physics of heat, in order to understand how one must use the cipher proposed in the famous page on the book of nature.

The problem of heat had been advanced by Galileo right from the initial phase of the polemic on comets. Aristotle in fact had supposed that a comet was an exhalation lit by its motion, like a falling star. Father Grassi, though he had introduced new hypotheses on the nature of comets, still maintained, as we have said, the traditional exegesis of the Aristotelian doctrine of "motion" as the cause of heat, as attrition, endowing it with a host of literary and legendary examples and quotations, all equally convincing for him.

Galileo proposes a different exegesis of the Aristotelian proposition according to which "motion is the cause of heat." In truth, he probably knew Telesio's ideas on the subject; but, since this author was on the Index, he was right not to quote him. In any case, for Galileo, heat is produced when the friction is so strong between two bodies as to detach particles of matter. The production of heat was thus associated with the emission of very thin parts of matter.

The hypothesis was purely theoretical; Galileo did not have at his disposal any observation on the loss of mass in a body heated by friction. To illustrate his idea, he had used, in Mario Guiducci's *Discourse*, improvised observed examples, which the *Libra* had no difficulty in dismissing. The author of the *Libra* might have been satisfied with refuting them on the experimental level, but he did not. In fact, Sarsi mobilized against the frail Galilean examples an appeal for the authority that should be accorded the classical authors.

This was the powerful argument of official culture; and its lightning bolts were not reserved only for scientific debates, but were hurled regularly against innovators in poetry, music, and the drama. The reigning method of debate—appeals to the principle of classical authority—was, obviously, the style of argumentation in those great theological controversies which form a background to all of seventeenth-century culture.

The principle of the classical authors' authority was so indisputable in the eyes of the author of the *Libra* as to make him abandon even minimal critical caution as he indiscriminately quoted sources in support of his idea in physics. But what for him was a strong argument—more authoritative than his own empirical arguments—was in Galileo's eyes his adversary's weak flank, the side more vulnerable to the weapons of intellectual polemic and literary irony.

Galileo at this point must surely have thought of Machiavelli, who had taught that enemies must either be flattered or crushed. He decided to crush his adversary, firing from *The Assayer* his most dependable and destructive weapon against the scientific honor of that most illustrious cultural institution which was the Society of Jesus: the weapon of ridicule was aimed straight at their devotion to the authority of tradition.

Now, for the Jesuits, this principle was something more sacred than a quotation that might be criticized. It was a value of a religious nature and a stronghold in the struggle against heresy. The very model of authority was that of the tradition of the pedagogical Church.

When Galileo criticized the recourse to the authority of a great many authors—observing that in philosophy it was not much help to line up a series of writers, like one draft horse after another, but that one should run free and alone like "Barbary horses" at Carnival time—such a fulminating witticism could not help but recall (this was 1624) the same sarcasm that Lutheran theologians directed against the appeal to authority in connection with the most controversial and delicate Catholic dogmas. Eliminating the principle of devout deference to the authors of the past was tantamount to opening up a clear road for the consideration of hypotheses, old or new, in an entirely different light.

There were some chapters of the book of nature to which unaided experience drew attention in qualitatively illusory and misleading ways. Their decipherment, according to Galileo, must be of a rational nature. Hence, the language of nature must be studied on the basis of universal notions of a new grammar.

The New Grammar of Physics

The signs of nature, so as to be interpreted with the certainty of demonstration, must be recognized by the certainty of a consideration that can be put in mathematical terms.

The passage in *The Assayer* containing the elements for reading the book of nature is very famous:

> I say that whenever I conceive of any material or corporeal substance, I am compelled of necessity to think that it is limited and shaped in this or that fashion, that it is large or small in regard to other things, that it is in this or that place, at this or that time, that it moves or is immobile, that it touches or does not touch another body, that it is one, a few, or many; nor can I by any stretch of the imagination separate it from these conditions. But that it is white or red, bitter or sweet, sounding or mute, of pleasant or unpleasant odor, I do not feel compelled in my

> mind to perceive it as necessarily accompanied by such conditions. On the contrary, if we were not assisted by our senses, perhaps reasoning and imagination would never apprehend these qualities. Therefore, I think that tastes, odors, colors, and so on as regards the object in which they seem to reside are nothing but pure names and reside only in the feeling body, so that if the animal is removed, all these qualities are taken away and annihilated.[59]

In truth, when Galileo wrote this famous page, he was not thinking of the mysterious and fascinating book of nature, but of one of the most read and commented upon books of Aristotle, *On the Soul* (at paragraphs 428b and 429a, relating to the problem of imagination), upon which this page of *The Assayer*[60] commented very polemically:

> "Perception of the special objects of sense is never in error," one reads in *On the Soul*, "or admits the least possible amount of falsehood. That of the concomitance of the objects concomitant with the sensible qualities comes next: in this case certainly we may be deceived; for while the perception that there is white before us cannot be false, the perception that what is white is this or that may be false. Third comes the perception of the universal attributes which accompany the concomitant objects to which the special sensibles attach (I mean e.g. of movement and magnitude); it is in respect of these that the greatest amount of sense-illusion is possible." [Translation of *De anima* by J. A. Smith in *The Basic Works of Aristotle*—TRANS.]

The Assayer overturned this hierarchy. Indeed, Galileo a) attributes geometric and mechanical properties only to the imagination or intuition of matter; and notes b) that the qualitative properties of matter could not be conceived of "for discourse and imagination" without the senses. He then maintains c) that if the sensible qualities exist only in sensation and in the material substance, then taste, odor, and color are mere names with respect to the substance in which they seem to reside.

On the Soul privileged as real the direct sensorial perceptions (color, etc.) and devalued the perception, only indirect, of motion, size, and shape. The intellectual of the seventeenth century immediately caught the glaring change of perspective contained in that page of *The Assayer*.

Galileo had not written *The Assayer* just to reply to his adversary. He also

[59] *The Assayer*, p. 261. For a recent discussion on this passage, see R. E. Butts, "Some Tactics in Galileo's Propaganda for the Mathematization of Scientific Experience," in Butts and Pitts, *New Perspectives on Galileo*, pp. 59-85.

[60] On the Aristotelian distinction between individual sensibles and common sensibles (in *De sensu* 4, 442a), see Crombie, "The Primary Properties and Secondary Qualities in Galileo Galilei's Natural Philosophy."

wanted to meet a strongly felt request, as we know, on the part of his friends and editors in Rome. So comets disappear quite soon from the horizon, and Galileo's book becomes what he had been asked to write; that is, a polemical treatise on style in physics. "Style"—in poetry, history, the biographical and narrative genres, and drama—was being discussed everywhere in 1624. It is fascinating to see that even science evinced an interest in this fundamental preoccupation with language. There was a new semantic program, an epistemologist reading *The Assayer* today might say. The intellectual of the seventeenth century said "new philosophy," and what was new was the style.

Galileo proposed a new language in physics. This was not at all a question of neologisms, but rather one of new definitions and rules. In the first place, he suggested a new way of talking about physical objects in general. Physics is the study of matter. Matter is defined by universally recognizable material properties. Physical objects, as the direct sensations of things, sensations that in no way are susceptible to measurement, are therefore purely unique phenomena. Physical objects are mathematical and mechanical properties described by words like shape, number, distance, motion, impact, and so on.

When these properties are not directly observable, the physicist must search for them beneath the macroscopic appearances that conceal them. Speculative imagination? No, "certain demonstrations."

In point of fact, *The Assayer* proposed to supplant Aristotelian physics by translating its predicative propositions, hinged on the experience of qualities, into a new language: from "the fire is hot" to "the fire transmits the sensation of heat." Such translation was no small matter, since it went from a language modeled on everyday common sense to a more elaborate and analytical, richer, and more rigorous language.

There were in fact two levels of words here. First of all, there were "names," that is, words like heat, red, and sweet, which have value for the individual sensation, but not for scientific knowledge. And then there are the material properties, words like shape, motion, and so on, which are universally and mathematically knowable.

Many are the ways to tell the story of the book of nature—poetry, painting, and music—in which the senses are privileged. But if one wishes to read that book in the style of physics, one must use only words in the latter category. However, one still must explain, through propositions endowed with these words and this material content, just what transmitted, say, the sensation of heat.

Atomistic Intuitions

To demonstrate that Aristotelian qualities are only sensible stimulations erroneously related to the objective world, Galileo used metaphors. One of these was the tickling produced by a feather under the nostrils or on the bottoms of the feet. Tickling is a sensation. The feather does not in itself have this property, which it produces only in proximity to particularly sensitive organs. Would it not be just as absurd to assign physical reality to the color of an object as to attribute an intrinsic and real stimulating quality to a feather?

Like all the metaphors that propelled the readers of the day into ecstasies, this too was in truth a rhetorical argument which was not very concerned with the demands of reason and credibility. Between a tactile sensation (such as tickling) and a visible perception there were no evident analogies. From a rhetorical point of view, however, the image was suggestive. The nose, the hands, and other seats of sensibility having been eliminated, the possibility that sensations might be caused by the motion of very thin particles of matter, endowed with shape and striking the senses—"very thin and flying minims," "smallest quanta"—becomes convincing. The five senses (touch, taste, smell, hearing, and sight), then, are activated by the penetration of the pores and the meatus passages of the respective organs by particles of material elements (earth, water, fire, air, and light) in more or less rapid motion.

Once again Galileo advanced in the conceptual garb of the Hermetic tradition. One such traditional idea—that of Renaissance Hermeticism, concerning the correspondence between the human microcosm and the macrocosm—had very recently been presented again by an illustrious university doctor at Pisa, the Portuguese Esteban Rodrigo de Castro, in his book *De meteoris microcosmi* (published in 1621 at Florence).[61] Also traditional was the idea of keeping the four Aristotelian elements. The novelty lay in the mechanical pattern of this correspondence, which proposed (or, better, anticipated) a physiology of perception based on the movement of matter.

Also, with respect to Galileo's statements of ten years earlier, the corpuscular theory of thermal action was more elaborate. Heat is produced by the dissolution of a body into very tiny parts. In the case of its sensible perception, this is not a matter of the modification of a property or condition, but of the penetration of the flesh with greater or lesser intensity. A body is dissolved by "fiery minims," distinct from the microscopic elements of various heated bodies. *The Assayer* does not make clear in what way these fiery particles dissolve bodies, or how they reside in bodies when they are still. In any event, this is a matter of microscopic particles of raw material, universally diffused by means of a disintegrating function made possible by their dimensions and their great velocity.

[61] See Stephani Roderici Castrensi, *De meteoris microcosmi libri quatuor*, Florentiae 1621. A student at the University of Coimbra, disciple of Fracastoro, Alberto Magno, and Agricola, De Castro was chief physician at the Studio of Pisa.

We have already seen, in the preceding chapter, that at the frontier of the most extreme resolution of matter, at the point where heat becomes incandescence, light is created. Galileo reserved the word "atom" and the concepts of infinitely small dimension and infinitely great velocity only for luminous, infinitesimal particles of discontinuous material, capable of penetrating sight.

These pages of *The Assayer* in which a premechanistic theory of sensation is presented, as well as the preceding pages, have been at length and variously interpreted by the historians and philosophers of science. Inevitably, one name flows from the commentators' pens: Locke. But John Locke, as everyone knows, expressed his more or less similar ideas at least a half-century after *The Assayer*.

It is difficult for us today to understand fully the strength and novelty of these pages: we would have to forget Locke and turn to the realism of common sense. Not only that, but we would have to add a solid tradition of Scholastic culture, of philosophical and theological knowledge, and somehow submit to the great authoritativeness and authority which that tradition represented. If it is difficult, if not impossible, for us to grasp the novelty of these statements, it is just as difficult (but perhaps less precluded) to realize the reactions that *The Assayer* must have provoked in its contemporary readers.

This was the rejection of a philosophy—one that was inextricably connected with the Catholic religion and the reigning mentality. It was also the re-evaluation of marginal, condemned, rejected ideas. As we have seen, besides the refutation of *On the Soul*—that is to say, the sacred text of Scholastic university philosophy—*The Assayer* immediately summoned up before contemporary eyes two names: Democritus and Ockham.

The "Characters" of the Universe

Let us consider, first of all, Democritus. For when reading Galileo's very famous passage about the book of the universe, one cannot help but think automatically of Democritus.

Of course, the metaphor of the universe as a bible of nature, to be read according to the faith of reason and experience, was very traditional and widespread—to all purposes, it was a slogan—among Roman and Catholic "virtuosi." It was the graphic aspect that distinguished Galileo's book of nature from the others. In fact, *The Assayer* also carefully specified the kind of characters. But to speak of characters, or letters, in the book of the universe was for the men of that time, men of humanistic and theological education, to repeat a Democritan symbolism already present in Cicero's theology. Cicero, in *De natura Deorum*, had assimilated atoms to gold and bronze characters of different shapes but of the same type, which by the pressure of weight could become inserted in the body of nature's word-objects.[62] The classical subject of graphic characters harked back to Aristotle, who had cited it to illustrate polemically the Democritan atomism in Epicurus's version.[63]

The statement that Galileo offered concerning the type of "graphics" in the book of the universe was strictly scientific. It was visibly a declaration of scientific atomism. As a matter of fact, the same idea—that of natural elements as geometric characters—had only recently been taken up again, this time by Giordano Bruno.

In *De minimo*, Giordano Bruno also had imagined that in order to read the universe it sufficed to combine a rather small number of characters, or letters, as with natural language. "And, just as in the case of letters, it is not necessary that there be many kinds and shapes of minims in order to form innumerable species,"[64] Bruno had observed in connection with these characters. We shall soon deal with what he thought about the geometric forms of those characters.

Today, we perceive quite easily the many differences between Bruno's philosophy and Galileo's, and there is no point in enumerating them. For contemporary readers of *The Assayer*, however, those graphic characters of nature brought to mind the geometry proposed both by Bruno and by Francesco Patrizi for a violently anti-Aristotelian physics of Pythagorean discontinuity. For a physics like that of *The Assayer*, in which problems of the physical-geometric infinite were only hinted at, that geometry was enough. Bruno's ideas on the infinite were something else, completely inadequate from the mathematical point of view.

Behind *The Assayer* and beyond Bruno, of course, as one continued to read Galileo's book, one caught a glimpse of those Democritan ideas quoted in Galen's *De elementis*: color and taste were opinions, "atoms and the void the truth." The same truth appeared immediately under Galileo's pen, as well.

De elementis was a widely read and studied work. When confronted by a theory such as the one on heat, smell, and taste which Galileo proposed, it was enough to have read a manual from a philosophical *cursus* to have the name of Democritus on the tip of one's tongue.

Another work often cited in the universities, and widely read outside of them and for a long time, was Lucretius's *On the Nature of Things*. Now, when Galileo warned the reader not to think of the reality of color, for ex-

[62] See Cicero, *De natura Deorum* II, 37.

[63] See Aristotle, *On Generation and Corruption* 315b, 14. See R. Cadiou, "Atomes et éléments graphiques," in *Bulletin de l'Association Guillaume Budé*, October 1958, pp. 54-64.

[64] G. Bruno, *De minimo, De minimi existentia*, Frankfurt 1590, in *Opera latine conscripta*, 3 vols., Naples and Florence, 1886-1891, vol. I, pt. III, p. 140.

ample, who would not immediately remember Lucretius's self-same criticism:

> of whatever color the bodies might be impregnated, do not believe that the elements of their substance are tinged with this color. . . . [T]he atoms are all colorless, for if they are endowed with different forms which permit them to produce all the tints and to vary them *ad infinitum*, by the play of combinations, the respective positions, and the movement that they impress on each other reciprocally, then you will easily explain why what was black can immediately become a marmoreal whiteness, like the sea that changes from black to white at the approach of a storm.[65]

The Assayer, however, had said that in natural philosophy it did not help to make use of poetic images, such as this one by Lucretius. Moreover, Galileo's atomist theory was preceded by a page of solid logic, in order to refute the objective value of sensible qualities. Galileo did not want to "probe the essence" with a universal metaphysical ontology like that of the Renaissance visionary philosophers. Galileo was a mathematician, not just a philosopher. Hence, he wanted to distinguish what one could know objectively and quantitatively from what could not be said scientifically. But that methodological vigilance which makes us cite Locke must, for Galileo's contemporaries, have recalled the warning on the criteria of knowable evidence which Epicurus had imparted: "If you reject any single sensation and fail to distinguish what is opined from what is evident on the basis of sensation and affections, and every act of attention by the mind, you will also confound all other sensations with your foolish opinion, and thus you will reject every criterion."[66]

With the criteria Galileo had stated, *The Assayer* could "probe the essence," whatever was physically knowable, of the atom. In fact, he interpreted the Aristotelian doctrine of "motion as the cause of heat" in terms of the movement of fiery minims and the emission of particles of substance. This exegesis, too, immediately brought to mind a book with a title similar to Lucretius's—Telesio's *De rerum natura iuxta propria principia*. Indeed, Telesio proposed the same exegesis of the doctrine of "motion as the cause of heat": the emission of particles of fire and other elements, both in the sky and on earth.[67]

A reader who might have had at his disposal a rich and up-to-date library, and who had the possibility of knowing, say from the catalogue of the

[65] T. Lucrezio Caro, *De rerum naturae*, bk. II, p. 703 ff. and 842 ff.

[66] Epicurus, *Opere*, edited by G. Arrighetti, Turin 1960 (3rd ed. 1973), 2, p. 54 ff.

[67] B. Telesio, *De rerum natura iuxta propria principia liber primus et secundus*, Romae 1565 (2nd ed., Naples 1570), pp. 80 and 147.

Frankfurt Book Fair, what was being printed beyond the Alps, could have put *The Assayer* on the same shelf where he now kept, together with the perennial Scaligero[68] and Zabarella, Campanella's *De rerum sensu et magia*, the alluring *Philosophia naturalis adversus Aristotelem*, which the French doctor Sebastien Basson had recently published (1621) in Geneva. If he had missed the first edition of *Philosophia Epicurea, Democritiana*, the amusing manual of atomist and anti-Aristotelian aphorisms by Nicholas Hill, he could however find the second edition, published at Geneva in 1620. Going back to the same year (1620), there was also the more serious atomistic *Exercitationes philosophicae* by David Gorlaeus, published in Leiden.

But it was not necessary to obtain all these works full of appropriate references in order to comprehend the corpuscular allusions in *The Assayer*. It was enough to have purchased or borrowed a recent book, published at Florence in 1621, by the pope's chief physician and philosopher at the University of Pisa, the Lusitanian Esteban Rodrigo de Castro. Entitled, as we have said, *De meteoris microcosmi libri quatuor*, and certainly known even to a frugal reader like Galileo, this is an important book unjustly neglected by historians.

In De Castro's book were all the necessary references to place *The Assayer*: the theory of atoms, graphic elements, or (according to Cicero's theology) letters. Present were all the useful quotations from Lucretius in order to understand how there could not be qualities without atoms. It explained the difference between atoms and Aristotelian minims, the idea that "heat is substance, not quality," the correspondence between macrocosm and microcosm—all this, of course, in order to demonstrate that Aristotle and Galen had "impugned," but had by no means "expunged," Democritus's atomist theory.

For a less up-to-date reader, who had simply gone through the normal philosophical studies at the university, without continuing to follow the debates of the new atomist fashion, it sufficed to recall, if he had attended a Jesuit college, the lessons in Father Suárez's or Father Pereira's manual. It was enough: when one read those famous and most suggestive pages of *The Assayer*, the word "names" used by Galileo for sensible qualities opened a large peephole through which to view a mnemonic table of the most often repeated quotations.

Indeed, it made one think instinctively of the same criticism of Aristotle's metaphysical realism developed by Ockham's "nominales." And yes Galileo insisted on associating our intuition of substance with the idea of quantity, traditionally an Ockhamist association. One learned about it and com-

[68] See G. C. Scaligero, *Exotericarum exercitationum liber quintus decimus de subtilitate ad Hieronimum Cardanum*, Lutetiae 1557, p. 31 ff. The theory of motion as the cause of heat was illustrated by the heating of the arrow in flight owing to the friction of the air, an example used again by Sarsi in the *Libra*.

mented on it *ad infinitum* on the university benches, where Galileo also had learned to know at least the name—so far as we know—of Ockham. But it will help to remember that Ockham had many fortunate editions in the sixteenth century, and that in those years—as we shall see—his name was at the center of the most modern theological discussions.

The Assayer's new grammar of physics, distinguishing between pure names and objectively knowable concepts in an intersubjective manner, recalled Ockham's logical grammar and his distinction between names that refer to things (*intentiones primae*) and names that refer to other names, such as species, gender, and accident (*intentiones secundae*). Ockham had affirmed that none of these names of the second type corresponded to anything real, and that metaphysical entities or substances (whiteness, for example) had no empirical reality, but were the fruit of a "personal" supposition. The attribution of reality to such pure names was unfounded: all the Aristotelian propositions hinging on names that stood in the place of nothing were false propositions. "One cannot know with obvious proof that there is whiteness," William of Ockham's *Summa totius logicae* instructed.

The rejection of Aristotelian tradition on the part of *The Assayer* seemed to be inspired by Ockham's fourteenth-century polemic against Scholasticism, in the name of the freedom of philosophy against the abstract speculations of the Scholastics. *The Assayer* seemed to want to celebrate the tricentennial of Ockham's incrimination in 1324 before the doctors of theology at Avignon, reaffirming the foundation of free inquiry in experience, unhindered by Scholastic metaphysics and its theology.

In addition, the medieval Ockhamist philosophers, as is known, were clearly in favor of theories in physics—owing to their identification of substance with quantity—that were close to the atomism of Democritus and of Plato's *Timaeus*. Democritus and the *Timaeus* had illustrated the forms of atoms.[69] Galileo, in contrast, after his grand declaration on the geometric characters of the book of the universe, remained quite vague on this point. How could one insert microscopic matter into those circles, triangles, and so on?

If Galileo had been a follower of mathematical Platonism, he would not have been so elusive and would have had to discuss critically the *Timaeus*'s speculations on the solid geometry of atoms. It is this gap in the apparent "Platonism" of the book of the universe that perplexes us today.

But the contemporary reader was perhaps more disposed to connect this declaration not to the *Timaeus*, but to another, more recent proposition—one that associated the circles, triangles, and other shapes of Euclid's plane geometry with microscopic structures of formless matter that had no rigor-

[69] See *Timaeus Platonis sive de Universitate interpretatibus M. Tullio Cicerone et Chalcidio*, Lutetiae Parisiorum 1563, p. 1250; and the Stephanus edition (1578), III, p. 31b ff.

ously defined form and were only by intuition roundly shaped. This was what had been proposed in 1591 by Giordano Bruno's *De minimo*. It was one of those theories which today's textbooks of the history of philosophy readily call "confused lucubrations of the Pythagorean type." For contemporaries, however, such theories may have seemed less confused—indeed, perhaps too simple.

In *De minimo*, moreover, we have seen that Giordano Bruno had thought of nature as a written language. Mixed bodies were, like words, composed of elementary bodies, as if these were letters or characters. These characters in their turn were composed of points. The points were the atoms of material elements.

Hence, in *De minimo* the letters or characters are full of points. But these letters have geometric forms: "he who considers things as a physicist will see that the minims are the triangle and the cone."[70]

Bruno's aphorisms were never too explicit. But, fortunately, *De minimo* explained itself better with respect to shapes: triangles, squares, and circles were formed by atoms within, which filled up these shapes, save for infinitesimal portions between the tangent and the atom's circular rim. A physicist could add a "very subtle ethereal substance" to fill those infinitesimal spaces within the geometric letters, between one atom and the next. Bruno had spoken of ether in his cosmological dialogues; here he spoke of a certain substance that "agglutinates" the atoms.

For Bruno, atoms were also animated. Not so for Galileo: they moved because pushed by heat, collisions, and their own weight. Nonetheless, it was difficult to read *The Assayer* without thinking of those Brunian geometric lucubrations on the physics of the discontinuous.

From this point of view, the image of the book of the universe written in geometric characters evoked the image of another book, one whose pages were like nets with very small geometric meshes in the form of circles, triangles, and squares. These nets were thrown boldly by *The Assayer* into the "infinite ocean" of experience—"away from the shore,"[71] far from the usual routes—in order to capture, within the meshes of mathematical description, atoms invisible to the eye.

It was a journey full of adventure. The new Christopher Columbus, the discoverer of new worlds, sang Johannes Faber in the propitiatory ode published at the beginning of *The Assayer*—or, as in the other introductory poem, written by the academician Francesco Stelluti, a guide to the infinite wealth of that invisible world "through those immense abysses."

[70] Bruno, *De minimo*, vol. I, pt. III, pp. 332 and 179.

[71] *The Assayer*, p. 267 (*Works*, VI, p. 352).

Bad Characters

"What you now teach us / you did not obtain from ancient charts, from modern devices / from the friendly stars," Francesco Stelluti's verses added. For if some names and some works came readily to the lips of *The Assayer*'s readers, to say which name and which work may have inspired Galileo's atomist intuitions is a desperate and futile undertaking. Atomism, as Gaston Bachelard says, is a doctrine that cannot be transmitted. From a mechanistic viewpoint, the typical originality of the proposals in *The Assayer* is proof of this. But in Galileo's case we are obliged to cite one source, because this was the source *par excellence* of his work and his life as a scientist.

Galileo was a Copernican on strict parole. From 1616 on, it was impossible for him to be a Copernican in his astronomy because the theologians of the Holy Office had declared the heliocentric doctrine and Copernicus's ideas on the earth's motion to be absurd and heretical. *The Assayer* respected that prohibition. However, Galileo was not forbidden to be a Copernican in physics, which meant to eliminate the difference between the celestial and terrestrial worlds, between the different modes of the propagation of light and the production of heat.

The hypothesis of atoms—material atoms, not mathematical points—which Galileo used to continue his battle on the front of physics, was already before his eyes in Copernicus's *De revolutionibus*, book 1, where Copernicus, perhaps recalling his medical studies, spoke of the corporeity of atoms. In the sixth chapter of the first book, entitled "Of the Immensity of the Sky in Relation to the Size of the Earth," Copernicus confuted the geometric argument against the earth's motion vis-à-vis the universe, in terms of their respective dimensions. A definitive geometric argument to prove the earth's motion in relation to the fixed stars would have been the measurability of their annual parallax. This would not be accomplished until three hundred years later, at the time of Bessel's complex measurements in 1837.

To protect himself from this criticism and to explain the apparent absence of parallaxes in the fixed stars, Copernicus insisted on the immensity of the sky, its "indefinite magnitude" with respect to the earth. In the context of that immensity, the varying of the distance of the earth from the sun became imperceptible to the observer's senses. To illustrate the lack of visual perception of extremely small deviations with respect to the huge dimensions of the universe, Copernicus offered an analogy taken from the other end of the scale of natural dimensions. He shifted from the universe to the microscopic world:

> As conversely occurs in the smallest and inseparable corpuscles [*minimis corpusculis ac insectibilibus*], which are called atoms, not being percepti-

ble [even if] duplicated or multipled several times, they do not immediately form a visible body, but can be multiplied until at last they are sufficient to come together in a size that can be seen. Thus, also on this place the earth, although it is not the center of the world, its distance [from the center] is not comparable in particular with the sphere of the fixed stars.[72]

This was an elementary, occasional atomist intuition that Copernicus did not develop; but it was eloquent enough to make the point that atoms belonged to the Copernican system and that the step was short from the infinity of the universe to the infinitely small atom. Giordano Bruno had already taken that step. Now, twenty years after Bruno's sentence, *The Assayer* was once again attempting it.

Democritus, Epicurus, Lucretius, Ockham, Telesio, Bruno, Campanella, and Copernicus: *The Assayer*'s philosophy of nature summoned up a gallery of not very recommendable authors. Those among them who were Catholics were even less recommendable and quotable in a work published in Rome, for all had been condemned by the Church.

In a period when every work of philosophy strove, with more or less success, to have "Catholic" authors as its official defenders, *The Assayer* instead displayed before the exegetical and deciphering eyes of its contemporaries (in dissimulated but quite visible form) the signs of pagan and Catholic authors who, to a man, reeked of heresy or atheism. Galileo worked only at the philosophy of nature, but he was not even minimally suspected of moral and irreligious Epicureanism. Nonetheless, he made his own all the "errata veterum philosophorum" (errors of ancient philosophy), as Father Pereira's classic Jesuit manual of philosophy called atomism.[73]

"Philosophers of this type" (e.g. Democritus, Epicurus, and Lucretius), the great Tridentine Dominican theologian Father Melchior Cano had authoritatively established, could not be reconciled with defense of the Christian religion.[74] Even if Galileo, as a devout Catholic, did not at all intend to speak of the soul? Perhaps atomism was equally dangerous whether or not one went into questions of atheism and immorality.

And perhaps Vincenzo Viviani's belated regret actually wished to indicate to us, through dissimulated allusions, this direction of the Galilean tragedy and these causes of his master's subsequent methodological and philosophical caution. But Galileo, just before the publication of his brilliant book *The Assayer*, in early October 1623, did not seem at all worried. On the contrary,

[72] N. Copernicus, *De revolutionibus orbium caelestium*, edited by A. Koyré, F. Alcan, Paris 1934, p. 84.

[73] See B. Pereira, *De communibus omnium rerum naturalium principiis et affectionibus*, Romae 1576, p. 162.

[74] F. M. Cano, *De locis theologicis libri duodecim*, Salamanticae 1563 (2nd ed., Lovanii 1564), p. 531.

he was full of optimism. From his house at Bellosguardo, perhaps thinking already of his book on the flux and reflux of the sea, he wrote to Prince Federico Cesi: "I am turning over in my mind things of some moment for the republic of letters which, if they do not come to fruition in this marvelous conjuncture, there will be no point—at least with respect to what is expected on my part—in hoping to meet its like ever again."[75]

[75] Galileo's letter to F. Cesi, October 9, 1623, *Works*, XII, p. 134 ff., especially p. 135.

THREE. THE "MARVELOUS CONJUNCTURE"

The Telescope to this age unknown
for you, Galileo, was the composed work,
the device which to the senses of others, though far away
brings the object closer and made much larger
. . . .
And with the same lens . . .
you shall see each atom distinct and close up.*

Carnival Rites

February 6, 1626, was Holy Thursday, the last day of Carnival. A great crowd strolled through Rome's streets and piazzas. From one basilica to the next, large processions dressed in multicolored costumes crossed each other, chanting psalms behind the precious baldaquins and glittering banners of the confraternities, both those of Rome and those which had come to Rome from all across Italy and Europe. Different color-tones marked those processions: the bishops' violet, the cardinals' crimson, the diplomats' black and the black dresses of the charity ladies of the Roman aristocracy, the brown and gray sackcloth that covered the many pilgrims, the Swiss Guards' blue and ochre, and the white embroidery of the canons' surplices and the noblemen's silk ruffs. That year, the sumptuous festivities of the Jubilee replaced the great Roman Carnival in the streets and piazzas. The enormous influx of pilgrims, the predictable brawls among the various confraternities, and the danger of contagion had the result that "with the approach of the Carnival season the Holy Pontiff Urban prohibited all masks and all other worldly pastimes."[1] However, the spectacular papal procession on the morrow, all the way to the Church of Santa Maria in Trastevere, would amply repay the mass of visitors for the lack of the traditional "Palio" of "Barbary horses" and, above all, the exhilarating spectacle of "two-legged beasts": naked Jews,

* Del Telescopio a questa etate ignoto / per te fia, Galileo, l'opra composta, / l'opra ch'al senso altrui, benché remoto / fatto molto maggior l'oggetto accosta. . . . / . . . E col medesimo occhial . . . / vedrai da presso ogni atomo distinto."—Giambattista Marino, *L'Adone*, 1623, "Canto X: The Wonders," 43-44

[1] On these days of Carnival, see T. M. Alfani, *Istoria degli anni santi dal loro principio al presente 1750*, Florence 1750, p. 185. Also see V. Prinzivalli, *Gli anni santi. Appunti storici con molte note inedite tratte dagli archivi di Roma*, Rome 1899, p. 101 ff. For information on the Holy Year of 1625 and on Descartes' presence at Rome in Cardinal Barberini's pro-Galilean circles, see A. Baillet, *La vie de Monsieur* Des Cartes, 2 vols., Paris 1691, vol. II, p. 122.

forcibly and derisively crammed with food, who on every Carnival raced down the Corso.[2]

In the cosmopolitan throng of visitors who watched the great mystical spectacle of the Roman Jubilee, we recognize also, perhaps on that same day, but certainly during those weeks, a French pilgrim. This thirty-year-old gentleman—cultivated, curious about everything, an adept mathematician and a devotée of natural philosophy, his head full of ideas and questions, and an admirer of Galileo since his college days—will soon be talked about: René Descartes.

The musical floats being prohibited, the inevitable blasphemous pantomime at the expense of the ghetto Jews prohibited, homilies were recited instead of the chaos of Spanish comedians and strolling players in Piazza Navona and Campo dei Fiori. So, when night fell, only the brawls among pilgrims drunk on free wine dispensed from the confraternities' wine cellars animated Rome's piazzas.

Forbidden the piazza, Carnival instead took its usual course, only slightly mitigated by more sober and noble rites, behind the illuminated façades of the palaces. At the Collegio Romano and the Roman seminary, the Jesuits, great lovers of the theater, presented plays from their rich repertory. Among the most requested was a sacred musical drama, recently composed by Father Orazio Grassi and staged for the first time in 1622 with marvelous scenic machinery of his conception: *The Apotheosis of Saints Ignazio and Saverio*.[3] But the plays most often performed were usually the "recitation acts" and the prose tragedies of Father Ortensio Scammacca: religious tragedies having an apologetic purpose and characterized by a great abundance of Aristotelian rules applied to holy and Biblical subjects. Included were highly emotional lives of the saints, sensual tortures of Jesuit martyrs in Japan and America, and allegories—all aimed polemically at pagan and licentious subjects fashionable among unbelieving, hypocritical, or simply nonconformist high society.

But the traditionally most suggestive theater apparatus that the Collegio Romano offered for public admiration, every year at Carnival time, was a silent stage set of extraordinary light effects. Even the pope will go, on Sunday, "with a modest retinue to adore the Venerable object displayed with sumptuous magnificence by the Jesuits."[4] If we are not mistaken, even today

[2] See A. Ademollo, *Il carnevale a Roma nei secoli XVII e XVIII. Appunti storici con note e documenti*, Rome 1883; F. Clementi, *Il carnevale romano nelle cronache contemporanee*, Città di Castello 1938; and J. B. Thiers, *Traité des jeux et des divertissemens qui doivent être defendus aux Chrétiens selon les règles de l'Église et les sentiments des Pères*, Paris 1686.

[3] "There was repeated a very noble tragic action, which played three times in the Roman court, always to renewed applause." G. V. Verzellino, "P. O. Grassi," in *Memorie degli uomini illustri di Savona*, 2 vols., Savona 1886-1891, vol. II, pp. 347-51.

[4] See Alfani, *Istoria degli anni santi*, p. 466 ff.

one can see such a theater represented on the ceiling of the sacristy of the Church of Gesú (Fig. 15). But to see the real thing was something else again. Inimitable owing to "the extraordinary stage apparatus and the number of lights,"[5] that stage set "was illumined by a splendor of light and by the reflections of hidden lights, these works being very beautiful in appearance." It was a deceptive appearance, and this was the marvel of the scenic machinery: a play of lights and glitter, a series of beams reflected by invisible mirrors and a complex system of hidden lamps; an ethereal light, without any luminous source visible to the spectators. It reverberated, with blinding flashes and reflections, on the single object that occupied the scene: a great sunburst monstrance. No other spectacle of the Roman Carnival captivated people's eyes as much as that exciting spectacle of the worship of the Sacrament.

In the palaces of the academies, at the Pamphili, Colonna, and Corsini houses, one went instead to "act and sing" the inaudible and intellectual "harmonic novelties" of madrigalesque plays: "the modern Delight against the harmonic rigor of the past." Just that year Adriano Banchieri had published his new comedy, *Virtuoso ritrovo accademico del dissonante*. Since the new regime tolerated political satire, anti-Spanish plays or comic operas were given.

Among all others, the celebration that caused the biggest sensation that evening was the one given at Montegiordano Palace, not far from Piazza Navona. That Thursday too, as for some months, the Academy of the Desirous Ones gathered in Orsini's fine palace. The protector of the Academy, a personage of great rank—Prince Cardinal Maurizio of Savoy, son and ambassador of Carlo Emanuele—did the honors of the house. He was one of the most prodigal and splendid patrons of art in all Italy.[6]

At his side was the secretary of the Academy: Monsignor Agostino Mascardi, the acclaimed representative of Roman intellectual society. "La meilleure plume qui fut," he would later be called by the libertine Gabriel Naudé.[7] Mascardi was a restless and anticonformist rhetorician, a lover of literary disputes and of the appanages of the cardinalitial courts, which vied for him as for a thoroughbred horse. He was the pet of the Barberinis and was profoundly detested by the Jesuits. In fact, only a few years before they had expelled him from the Society of Jesus for his literary excesses. Known

[5] On the Eucharistic stage scene organized by the Old Congregation of the General Communion at the Collegio Romano, for the Forty Hours of the days of Carnival, see D. Bartoli, *Della vita del padre Nicolò Zucchi*, Rome 1683, p. 38. The technical aspect of the lights was so elaborate and complicated that, at least in one case, it caused a serious fire.

[6] See I. della Giovanna, "Agostino Mascardi . . . Maurizio of Savoy," in *Raccolta di studi critici dedicati ad A. Ancona*, Florence 1901, pp. 117-26. On the Cardinal of Savoy's expenditures and the decoration of St. Eustachio commissioned by him with A. Tassi, see G. B. Passeri, *Vite di pittori, scultori e architetti . . . morti dal 1641 al 1673*, Rome 1773, p. 110 ff.

[7] On the pro-French Roman situation, consult the book by R. Pintard, *Libertinisme érudit dans la première moitié du XVIIe siècle*, vol. I, p. 209 ff.

for his adventurous escape from Rome, Mascardi could officiate publicly in the Roman salons thanks only to the tolerance of the new regime.

In the palace courtyard, around a fountain, were stationed the magnificent turbaned figures of a Turkish guard, and this announced sumptuously that the celebration was honored by the presence of their master, the most powerful man in Rome, the Cardinal-nephew Francesco Barberini. With such guests, the social success of the evening was obvious, as was its political significance. In fact, the palace of the Cardinal of Savoy was a meeting place for the pro-French group, of which the prince had been the recognized leader ever since, as protector of the French Crown, he had pointedly left Rome in protest over Paul V's pro-Spanish and pro-Jesuit policy.

The Cardinal of Savoy had returned to Rome for the conclave during the summer two years before. He had then been the great elector of Urban VIII, achieving on behalf of Maffeo Barberini the union of two opposed parties (Borghese and Ludovisi) who were agreeable to Paris, where the new pontificate was still considered "a miraculous turn." Not even the Valtellina war, in fact, had substantially changed the pro-French attitude of the papacy, which was preparing an ambitious diplomatic initiative aimed at the two blocs in Paris and Madrid.

The Cardinal of Savoy received in his palace the success and congratulatory homage that was due to a figure of such influential power. Another high-ranking personality that festive evening was Cardinal Magalotti, recently named secretary of state by Pope Barberini; the true gray eminence, however, was Monsignor Giovanni Ciampoli, the pope's secret adviser and the guiding spirit in the new Curia.

That evening, half the Curia was present. Many prelates had attended out of conviction, or convenience, pretending to be unaware of the slanderous rumors that the Spanish embassy had spread about Rome concerning the Cardinal of Savoy's extravagant private life. All of the academies were being attacked and censured: for instance, the vicious gossip about Cardinal Deti, and the Academy of the Orderly Ones protected by him, spread accusations of immorality. Even so serious an academy as the Lynceans was subjected to criticism.[8] This, at bottom, was understandable: the proliferation of literary academies had created meeting places for lay intellectuals and men of power, introducing a spirit of cultivated discussion and opinion. In short, it had freed intellectual life from the Counter-Reformation's monopoly of cultural institutions.

As a rule, though, this preoccupation was unjustified and wrongly slandered the innocent social frivolity of meetings like those of the Humorists

[8] On the Roman academies, see G. Tiraboschi, *Storia della letteratura italiana*, 2nd ed., Modena 1783, vol. VIII, pt. I, p. 43 ff.; and M. Maylander, *Storia delle Accademie d'Italia*, 5 vols., Bologna 1926-1930, vol. V., p. 270 ff.

and the Orderly Ones, whose influence was not felt beyond the setting of their literary skirmishes. Ephemeral and refined poetic contests were staged between the followers of Petrarch and those of Marino, the lovers of literary conceits and the classicists, for the pleasure of an elite composed of high society, the rich, the cultivated, and the fashionable, who responded with an uncertain longing for both novelty and conformism.

However, suspicions concerning the supporters of intransigent institutions were somewhat justified in the case of the academy of the Cardinal of Savoy. On the surface, it did not differ from the others. It had the same easygoing literary vocation as the other intellectual and high society cenacles. Its name gave off an impertinent aroma of pure "delight." In fact, thirty years before, the Desirous Ones were the first theatrical company authorized to act in public after a long and severe prohibition of all "spectacles."[9] Yet, if one looked closely, it was not an academy of poetry like that of the Humorists, the Fantastics, or the Orderly Ones. It was a creation of the new regime, whose new political image and cultural refinement it ostentatiously exhibited.

Of course, style was discussed by the Desirous Ones; however, rather than poetry recitations, there were discussions about moral philosophy and history. It was an academy of moral and political sciences such as Rome had never seen. The Desirous Ones were the intellectuals of the pope's party. They were, in the great majority, people who had transferred to the Court of Rome from Bologna, where the Cardinal Legate Maffeo Barberini had come to know their qualities. From Bologna, a rich cultural center animated at the beginning of the century by the Carraccis in art and the Manzinis in letters, came Mascardi, Count Virgilio Malvezzi, the Florentine Giovanni Ciampoli (who in Rome had become Pope Barberini's cultural attaché), Tommaso Stigliani, and Luigi Manzini (adviser to the Cardinal of Savoy).

The meetings in Montegiordano Palace also saw the Romanized Florentine intellectuals Giulio Rospigliosi and Giuliano Strozzi. Prince Savelli and his secretary, the poet Pier Francesco Paoli, the jurist and poet Marcello Giovanetti, the critic Matteo Peregrini, Cavalier Cassiano dal Pozzo, who had commissioned Poussin, the great Roman "virtuoso" and cultural secretary of the cardinal-nephew, and Cavalier Giorgio Coneo (the Scot George Conn), his diplomatic secretary, were among the principal participants in this initiative, begun only in the autumn of the previous year.[10]

However, the main figures during that initial phase were two young, brilliant intellectuals whom Monsignor Mascardi had called to his side. The first

[9] See A. G. Bragaglia, *Storia del teatro popolare romano*, Rome 1958, p. 64.

[10] See A. Mascardi, *Saggi accademici dati in Roma nell'Accademia del Serenissimo Principe cardinale di Savoia*, Venice 1630.

was Virgilio Malvezzi, thirty, already renowned for his *Discourses on Cornelius Tacitus*. A follower of Tacitus in politics and a Senecan in moral philosophy, Count Malvezzi was one of those Counter-Reformation lay intellectuals, realist and disenchanted, who, like Traiano Boccalini, opposed the Livian historiography of the Jesuit universities and found in Tacitus (and, after him, in Machiavelli) a theory of power more in keeping with the present state of affairs.

In Rome, incomparable theater of tyranny and politics, Count Malvezzi was completing his education and maturing his choices. Later on, when the political balance shifts, Malvezzi will also choose Madrid. But now he was finishing his medical studies and his studies of judicial-political astrology—an area of the science of politics full of libertine temptations and anti-Catholic prophecies, in which old and new condemned authors (the atheist Cardano and the heretic Campanella) had been prominent.

The other *enfant prodige* of the Academy of the Desirous Ones was a very young and eloquent Roman philosopher of high social rank: Marchese Pietro Sforza Pallavicino. After Cesarini, he was the rising star in the Roman intellectual world, the best pupil in Father Vincenzo Aranea's philosophy courses at the Collegio Romano.

Aristocrats or courtiers, the intellectuals of the academy of the Cardinal of Savoy were "virtuosi" with a traditional humanistic-philosophic background who tried to form an image of greatness patterned on literary virtues. They were intellectuals whose opinions were not yet fully formed, and all were galvanized by the idea of being the witnesses of a "New World," the title of a successful poem by Tommaso Stigliani. They were not "strong minds," innovators of the stamp of a Bruno or a Campanella. Nor were they extremist celebrators of the new poetic: in literature as in politics they were moderates, Catholics aware of their privileged relations with a pontificate full of promises for renewal. They felt a need for a break with the past and sensed the crisis of knowledge in their time.

They were men of an old world in search of a different grasp of reality. The instruments at their disposal were the historically eternal ones: moral and literary. But travel accounts, such as the famous letters of the Roman aristocrat Pietro della Valle, gave the taste for discovery and natural curiosities an active role in their intellectual lives.[11]

Thus, inevitably, their rhetorical celebration of the virtuoso discovery of a new world led them to associate the old name of Christopher Columbus with the new name of Galileo. During those years, the association of Columbus and Galileo became a literary commonplace. Behind the celebrative rhet-

[11] See P. della Valle, *Viaggi . . . descritti in 54 lettere famigliari*, 3 vols., Rome 1650.

oric, however, a cultivated opinion was being created, one that discerned a relationship between two historic events, feeding the myth of the discovery of nature.

So it was that Galileo found his most enthusiastic followers among literary innovators and, above all, those moralists and historians of Savoy's academy who created Roman intellectual opinion. The traditional literary and philosophical garb must be abandoned in order to embrace new criteria. The theme of the book of nature against the principle of authority had quickly become their intellectual slogan. This could be seen on that Carnival night during which, at the Academy of the Desirous Ones, a topic was recited, and the topic was *The Assayer*, which lent itself marvelously to such an occasion. In one of its first pages, Galileo had described the atmosphere of Carnival, when it is permissible "to speak freely about everything."[12] Thus, taking advantage of the impunity of that Holy Thursday evening, Monsignor Mascardi, in place of a musical play by Banchieri, had staged Galileo's intellectual polemic in the form of a provocative lecture singing the praises of *The Assayer*'s ideas.

That night almost everyone was there. Only Mario Guiducci was absent, and so he will be able to give Galileo only a hazy report of this important literary and social event.[13] But the lecture's text has been published, even if perhaps not in its entirety.[14]

Let us not criticize Guiducci for his absence that night. Usually, meetings in Montegiordano Palace were devoted to learned speeches by political writers and erudite philosophers. (The last time, for example, the Marchese Pallavicino had spoken, expatiating with affectation on politics, philosophy, theology, and the nobility of the will.)[15] Moreover, the chosen orator that evening, was not even a literary man of the first order. His name was Giuliano Fabrici, and he was known in Roman poetic circles as a poet "of occasion"; that is, one of those virtuosi of rhyme who were capable of reeling off on the spur of the moment a lyric on any object or event presented to them—a kind of experimental poet.[16]

Fabrici's title for the reading was "The Literary Man's Ambition." In present terms, the subject was intellectual arrogance based on the principle of Aristotle's authority and prophetic smugness.[17]

Galileo had directed his scorn at those uncritical followers of ancient authors who froze the philosophy of nature in the forms of a dogmatic "commentary." Fabrici took up again, generalizing it, the criticism against those

[12] *The Assayer*, p. 15 (*Works*, VI, p. 219 ff.).

[13] See M. Guiducci's letter to Galileo, February 8, 1625, *Works*, XIII, p. 253.

[14] G. Fabrici, "Dell'ambizione del letterato," in Mascardi, *Saggi accademici*, pp. 71-89.

[15] Father Sforza Pallavicino, "Se sia piú nobile l'intelletto o la volontè," ibid., p. 50 ff.

[16] See *Works*, XIII, p. 253. [17] Fabrici, "Dell'ambizione del letterato," p. 74.

"who with greater assurance believe more in what they have read than in what they have always seen." It was the metaphor of the book of the universe open before our eyes. But Fabrici developed the metaphor in a much more polemical sense, denouncing the form of Scholastic treatises which began with a list of quoted classics. For Fabrici, this sort of bibliography was a Swiss Guard deployed in defense of the official philosophy:

> The discovery of a falsity in a book revered by them does not seem less of a sacrilege than burning down a temple. Experiences, which are the letters of nature and the words of God, are so hated by them that, closing their eyes so as not to see, they immediately open their lips in order to degrade them.[18]

As always on these official occasions, the orator evoked for his listeners, without naming names but in a very obvious manner, the poisonous polemical reactions of the Collegio Romano Jesuits against *The Assayer*. Galileo, in fact, was praised immediately after, as usual being compared to Christopher Columbus. And together with Galileo was praised a famous Calabrian alchemist, who had to be Campanella though he was not named.

At this point, Giuliano Fabrici launched a frontal assault on Aristotle:

> Philosophy should study the great text written by God, where the book is the world and experience the characters, and should not be subjected to the law of a litigious piece of writing which after two thousand years of interpretation is still not understood even by those philosophers who have sworn to believe what it dictates. And in sum, man should confine himself to pondering opinions with the weight of reason and not authority, while today he speculates stunned by the writings of one who taints him more now that he is dead than when he was alive.[19]

The "weight of reason" was thus thrown into the scales of time. Two thousand years of culture *ex libri* were finished. Out of the crisis of traditional knowledge began the culture of a new world. His discourse rife with reforming afflatus and Christian optimism, the lecturer was certainly not a pessimistic and cynical unbeliever, like so many Roman libertines hiding behind irreproachable religious habits.

This was the tone that suited the Academy of the Desirous Ones, fashioned in the cultural style of the reforming Barberini pontificate. It was an Augustinian culture animated by an optimistic vision of man, open to modern knowledge and in search of a renewed rather than a Counter-Reformation Catholicism. In keeping with the sensibility of the "virtuosi" who were listening to him, Fabrici had given a mystical version of the Galilean theme

[18] Ibid. [19] Ibid., p. 77.

of the book of nature. Now, precisely this "Christian" interpretation of the Galilean polemic was the most radical, contentious, and open-minded aspect of the intellectual Roman echo of the ideas in Galileo's book. Not only did the innovating intellectuals defend *The Assayer* in order to benefit their search for new styles in poetry and history, against the canons of the official culture, but they also broadened the consequences of the new philosophy to the point of foreshadowing a more authentic Catholic culture. Thus, their idea of a reconciliation between faith and philosophy clashed intolerably with the sentiments of those whose task it was to defend a much different reconciliation, stipulated on the basis of authority.

Indeed, around *The Assayer* a "literary" controversy was forming which would split the world of Roman Catholic culture. Both sides cited books inseparably linked to their positions. On one side was the book of Biblical revelation and the book of nature, both written by God. On the other were the declarations of the Council of Trent and the books of Aristotle. On one side was the guarantee of a contemplative faith; on the other, the guarantee of the tradition of the teaching Church. But such considerations would have been inappropriate to a literary meeting at Carnival time. Thus, Fabrici chose a tone of pungent mockery aimed at Galileo's adversaries.

Galileo had compared the cultivators of authority to ducks incapable of following the flight of eagles. Fabrici compared them to gourds filled with an insipid liquid, but these gourds, if emptied and properly prepared, could be used to hold a refined wine from another vineyard.

We could go on and on with metaphors of this sort. But it will suffice to recall one, used by Fabrici, introducing Galileo as the new Prometheus. The image was certainly fulsome, but it served to call up for the audience the symbol of Prometheus used in *De fato* by Pomponazzi, the innovative, anti-Scholastic philosopher who had challenged Scholastic culture with his mysticism and his dream, still so topical, of separating reason from Thomist theology.

The affected and pompous rhetoric of the poet Fabrici skipped pleasantly, without once becoming too heavy, from metaphor to salacious allusion, from moral tale to personal anecdote. The rules of academic pleasure cleverly guided the author's eloquence toward the sure goal of his audience's consensus: the enlightened princes of the Church, cultivated prelates, refined "virtuosi," moralists, and poets with moderate sentiments.

It was, let us not forget, a Carnival evening in Rome at a time of festive devotion, of Jubilee. The license of that evening was wholly "poetic," in opposition to the reigning intellectual rules. On the other hand, there could not be even the shadow or mask of an anticonformism concealing a misconceived and dissimulated irreligious infraction. The request for intellectual freedom, the vindication of newness, was not to be confused with impiety. In his conclusion, the poet Fabrici made a point of reminding his audience that the other face of intellectual arrogance was the attempt "to hang onto the ropes of heaven"; whereas the charismatic presumption—"an intolerable and ridiculous ambition"—prevaricated in the name of theology, the former prevaricated in the name of intellectual authority. The Desirous Ones were not visionaries, "Narcissuses of their own genius." They knew "that one could not do anything without the assistance of divine power."[20] God's fecundating power, as Pierre de Bérulle's new French spiritual and reforming theology had taught them to believe, was behind every sign of nature.[21] But this did not mean that one should appeal to God indiscriminately. Only in those cases which "because they are marvelous seem to exceed the ordinary forces of nature"—that is, the great miracles of faith—must one call upon that power.[22] But within the limits of nature the "admirable thoughts" of human reason were sufficient; they, too, were acts of faith and humility before the created world that is to be known.

When Fabrici concluded his lecture, there was a very long round of applause. That public ovation on the part of extremely important figures in the Curia and cultivated Roman opinion was the most striking manifestation of the consensus that *The Assayer* could count on in Rome.

It was an authoritative consensus. The Academy of the Desirous Ones was certainly not an official Catholic institution comparable, even distantly, to the Collegio Romano; nevertheless, it had a clean bill of health and the necessary connections with those in power to give it the hope of causing an innovative cultural line to prevail at the summits of the new pontificate, which promised to be long and modern.

In fact, the Desirous Ones had a rather clear political awareness of this privileged moment for "talents" and "novelties." They officially declared their candidacy as cultural exponents of the new pontificate, eliminating the intellectual tutelage and control traditionally exercised by the Jesuits.

Aware of the importance of a political turn, these Catholics of the new culture nourished credible aspirations that were reciprocated by the new power. In the inaugural speech of the Cardinal of Savoy's academy Agostino Mascardi, the official spokesman of the pro-French intellectuals and sup-

[20] Ibid., p. 80.

[21] See P. de Bérulle, *Discours de l'état et des grandeurs de Jésus*, Paris 1623. Bérulle's theological masterpiece, published at the same time as *The Assayer*, had become the manifesto of mystical French theology, the heir of contemplative Carmelite spirituality and Oratorian culture; the Christological revolution of this new theology was symbolized, among other things, by the heliocentric transformation of modern astronomy.

[22] The Eucharist, image of the unity and divine fecundity of the Word incarnate will be reflected in the French school in the theological works of Scotist and Neo-Platonic inspiration in Italy.

porters of Urban VIII, had already illustrated the intellectual exigencies, "the desires of so many literati, who now are rising again."[23] He explained that these desires were above all turned to the defense of the search for truth in science, and he underlined the importance that Plato had attributed to mathematics for rulers of a state.[24]

Needed was "a wise conjunction of the academy and the court," Count Malvezzi had pointed out, a new understanding between "the virtuosi and the city"[25]—the virtuosi, not the old repressive and inquisitorial cultural institutions. This state of mind and affairs echoed through the refined Italian prose of these Galilean clerics, these Romans or Romanized literati.

The sciences which in aulic language they proclaimed they officiated over were not dominated by speculative theology, "the queen of the sciences," but by the Bible. *For in Holy Scripture all sciences are contained*, such that each of them reveals a part of Scripture and is illuminated by it. Thus proclaimed one of these clerics of the academy, Alfonso Pandolfi—also a poet influenced by Tacitus and a rhetorician of the Biblical concordance of knowledge. "The world is a poem," he declared with Plotinus. "Man is God's poem," he repeated with St. Paul. But with St. Augustine he reiterated that "if what the philosophers say is true and consonant with our faith, then it may be converted to our use, although it might be contested by unjust possessors."[26]

It is easy to see the intense reflection of these reforming sentiments on high Barberinian society. These reflections crossed back and forth between the Academy of the Desirous Ones and the artistic Academy of San Luca, and then spread over the canvases in the studios of Via Paolina, where Poussin prepared works commissioned by the Roman virtuosi (e.g. *The Death of Germanicus*, a great quotation from Tacitus destined for Cardinal Francesco Barberini). The Biblical works centered on the figure of Moses and, later, on the great cycle of *The Seven Sacraments* (Fig. 12) commissioned by Cavalier dal Pozzo for the collection in his palace on Via dei Chiavari. These scenes were faithfully inspired by the Biblical text, just as the treatise on the seven sacraments by Cavalier Coneo, a lay diplomatic consultant of the cardinal-nephew, and the Biblical theology and Mosaic cosmology of Monsignor Agostino Oreggi, the pope's personal theologian,[27] were mystically modeled on Biblical fundamentalism.

[23] Mascardi, *Le pompe del Campidoglio*, p. 96.

[24] See A. Mascardi, "Dell'aritmetica" and "Della geomatria," in *Discorsi morali*, p. 213.

[25] See V. Malvezzi, 'Ragioni per le quali i Letterati credono non potere avvantaggiarsi nella Corte," in Mascardi, *Saggi accademici*, pp. 13-33.

[26] A. Pandolfi, "Che nella divina scrittura si contengono tutte le scienze," ibid., pp. 131-43, especially p. 134 ff. and 136.

[27] See G. Conn, *Assertionum Catholicarum libri III*, Romae 1626; A. Oreggi, *De opere sex dierum*, Romae 1623. It should be noted that in this work of Mosaic Neo-Platonic physics, Oreggi denounces the impious nature of Aristotelianism, like Campanella (pp. 170-72). Stipulating according to the Bible the existence of light without subject, Oreggi's book affords him full freedom to philosophize on the

These were academic and Biblical images in the style of an austere, cultivated, intellectual Christianity: sober archaeological setting, parsimony of gesture, intense inner devotion. This taste met the aspirations of an educated Catholic elite of refined culture. There were no allegories or vortexes of fluttering images; no deceptive effects, iridescent sensual colors, or smiling ecstasies as in the incipient devout representations of the Jesuits.

The sensuality and eroticism that played so great a part in devout literature and art were here reserved for the gracefulness of pagan mythological representations, the bacchanals of the frescoes in the Farnese Palace, or the octaves of Marino's poem *L'Adone*—studied classicist effects of an innocent, unreal literary world. But in Poussin's majestic sacred images, in Virginio Cesarini's or Maffeo Barberini's devout and moralistic poetry, there was another mystical "grace," closer perhaps to austere Protestant innerness, but certainly very far from the intoxicating sentimentalism of Jesuit Catholic expression.

This was the political, cultural, and religious background that had made *The Assayer* the literary sensation of the new pontificate, an intellectual intervention and manifesto that completely transcended the initial framework of one of the many scientific debates Galileo engaged in. It would be impossible to understand the profound, very complex resonance of *The Assayer* without understanding that climate and these men. They were literati; they did not know mathematics, and understood almost nothing of the difficult astronomical problems of the Galilean battle, but they mobilized against that work's opponents.

The apparent frivolity of these literary manifestations which sang the praises of the spirit behind *The Assayer* has led historians to minimize their importance and to suppose that their support was orchestrated purely for the purposes of literary debate. But this is only a superficial appearance. In reality, Fabrici's lecture had been an act of courage, which protected Galileo from the campaign of attacks unleashed by the Collegio Romano against the new philosophy, as we shall see in the next chapter.

Now, if one asks what were the effects of the consensus, we could reply, as did Manzoni, that they were "nothing and at the same time miraculous."

"ethereal matter" of light (p. 9). Nicolas Poussin, closely tied to F. Barberini and Dal Pozzo, and an admirer of Campanella's philosophy, expresses in his sacred painting the ideals of the Roman virtuosi when he abandons every allegorical reference to connect antiquity to Biblical Christianity. Dal Pozzo's series on the *Sacramenti* offers a heterodox religious iconography polemically faithful to the Scriptural version and to the historical reconstruction that will inspire Phillipe de Champaigne's painting at Port Royal. See J. Busquet, "Les relations de Poussin avec le milieu romain," Sh. Soners Rinheart, "Poussinet la famille Dal Pozzo," and J. Vanuxem, "Les tableaux sacrés de Richeome et l'iconographie eucharistique chez Poussin," in *Nicholas Poussin, Colloque International du C.N.R.S.* (Paris 1960), pp. 1-18, 19-30, 57-70; F. Haskell and S. Rinheart, "The Dal Pozzo Collection," *Burlington Magazine*, July 1960, pp. 318-26; N. Poussin, *Sacraments and Bacchanales*, Edinburgh 1981.

Someone had said publicly what many in Rome were thinking but did not dare say openly. A psychologial and moral climate had been shattered, a heavy atmosphere of pliant complicity and silence, an ambiguous feeling of danger. Intellectual conformism had been broken with an experience of free speech that did not have a precedent in Rome. Galileo was no longer alone. He could count on an official support in Rome, no longer only whispered in the ears of friendly prelates in the corridors of the Curia.

Indeed, the Galilean polemic against the Collegio Romano was no longer a personal matter; it was a political event. Virginio Cesarini's dream of shifting the traditional Roman equilibrium was being realized. But with what possibility of evolving further?

This is the point; the moment was favorable. The civil and moral exigencies produced enthusiasm, but did those virtuosi truly have the qualifications to transform their restless need for renewal into an intellectually innovative turnabout?

No, they did not: they were "virtuosi," and "virtuosi" who issued from the Aristotelian philosophy of the Jesuit colleges. Their humanistic culture was dressed up in rational and scientific argument as though in costume for a Carnival masked ball. They sang Galileo's praises, but their true mental clothing was that of traditional philosophy. They lacked a new philosophy, the knowledge and books that would give them wings and transform them from the rhetoricians they were into men of modern culture.

The Libraries of the New Philosophers

In Rome, at the moment of "marvelous conjuncture" for Galileo, the Academy of the Desirous Ones made opinion. The Academy of Lynceans made the books.

The men in Montegiordano Palace were not unlike those in Cesi Palace, on the nearby Via Maschera d'Oro, where the Lynceans met. Indeed, in many cases, they were the same men. They were aristocrats and prelates, humanists in their education and vocation, having in common moderate and religious sentiments.

But at Cesi Palace there were neither social amenities nor politics. There was, without fuss, an intense activity devoted to ideas and publishing programs for the library that Prince Cesi had collected. That library was akin to a temple. One would have to go back to the monasteries of the Middle Ages to find an intellectual universe in which the written word—books and manuscripts—played as crucial a role as it did during the first decades of the seventeenth century, the epoch of the great library as monopoly of traditional culture. As instruments of intellectual monopoly, the great libraries created at the beginning of the century expressed the strength and prestige of traditional humanist and theological culture, which was forging new instruments of erudition and exegesis: the most modern weapons for sustaining, on all intellectual fronts, the effort of Catholic reform and religious struggle.

Great libraries, such as the Vatican or the very recent Ambrosiana Library in Milan, with its annexed museum, were organized as centers of research. But in Rome there is a profusion of libraries animating studies of every type: the library of the Oratory of Vallicella (open to the public since 1581), the Angelica of the Augustinians, the new Aniciana Library, the libraries of the Chigi, Pamphili, and Altieri families, and finally the library being set up by the Barberinis without consideration of cost. And then there is the library of the Collegio Romano, the largest Jesuit library in Italy.

But one must not look upon the libraries only as fulcrums of theological and humanist culture. If one looks at them as instruments of information and of the elaboration of new ideas, then the dominant motif of the events of the written word becomes the control exercised by the special ecclesiastical institutions of the Counter-Reformation. Thus, along with the richness and publicity of the libraries of official culture—one thinks of the Ambrosian Library and the freedom of access it granted to instruments of traditional knowledge—there is a corresponding persecution and extreme discretion in regard to the written word of the new authors considered dangerous. It is a picture with great contrasts of light: on one side the written word is flaunted, on the other it is hidden and persecuted. At the gates of the cities, messengers and merchants are searched for new books; bookstores are watched and policed; bequests to libraries are not granted without scrupulous inquiries; the catalogues of international fairs are under control of the omnipotent Congregation of the Index, which collaborates with the Holy Office in the work of surveillance and intimidation of authors, publishers, bookstore owners, and private libraries.

If it is noticed that in some Italian cities the system of prevention and persecution has gaps, and that dangerous books elude the controls and are circulated, then Cardinal Bellarmino, faithful servant of the Church, is called to the Congregation of the Index, so that by dint of his untiring, efficient zeal the severe regulations concerning books are respected, and works that might otherwise have escaped Roman control are singled out and denounced. In fact, books have been declared by the Council of Trent to be the vehicles of heretical infection in Catholic countries. Aside from the Waldensians, there is no longer even one center of Protestant infection in early-seventeenth-century Italy. Now the heretics across the Alps count only on supporters who are invisible and above suspicion, infected with disbelief, philosophical atheism, moral pessimism, materialist naturalism, and libertine literature. These are the evils of Catholic Italy that the Index has been called upon to oppose with the impulse given it by Cardinal Bellarmino and the support of the inquisitorial apparatus.

Of course, in the secret libraries of collectors, in diplomatic pouches (and in those of compliant ecclesiastics), through foreign friends and booksellers who know the right routes, the prohibited and suspect books circulate and are exchanged. But they are rare and expensive; this is a clandestine market, illicit and often dangerous. One must go back to the Middle Ages to find another period in which the fate of ideas was so tied to the material precariousness of the written page.

It was in this landscape of suspicion and subterfuge that Prince Federico Cesi had understood the connection between the idea of a new philosophy, based on reading the book of nature, and the official creation of a scientific library where, to the texts of the classic tradition, would be added the more modern authors and also some entirely new books. And he said so officially. From the beginning, Federico Cesi had declared, with a provocation which solicited scandal, that his academy was devoted solely to the search for natural truths independent of theological and political controversy.

To read the book of nature, as all the innovators said, the books of the ancients were in fact no longer sufficient. During the previous century, there had been an explosion of modern knowledge: botany, zoology, anatomy, and alchemy had experienced the most striking developments, even more striking than astronomy and mathematics.

If the Lynceans were to catalogue, interpret, reproduce, and illustrate the book of the universe, they needed herbariums, pharmacopoeias, works of magic, medicine, and geography, and a philosophy of nature. To the works of the classical mathematical disciplines, such as mechanics, astronomy, and geometry, must be added the books and manuscripts of disciplines still in their infancy. All this corresponded to the great collecting proclivity of the century, but also to the will to extract from that library an encyclopedia of knowledge capable of supplanting that of the traditional culture.

The Academy of Lynceans thus competed with Jesuit culture, imitating its very techniques of success: the discipline and solidarity of the Society's members, the absolute dedication to the aims of research—even to the presumed Lyncean celibacy—the pronounced international vocation, the will for proselytism and decentralization. Just as a Jesuit priest could not publish a book unless it had been reviewed by at least three theologians of the Society, so a Lyncean had to submit to the reading of several members of the Academy any book that he hoped to publish under its aegis and collective approval.[28] The fact that the Academy of Lynceans was by statute precluded

[28] See F. Cesi, *Linceografo*. This precious manuscript, regulating the bibliographical library and publishing activities of the Academy of Lynceans, has been described repeatedly: by Odescalchi, Favaro,

to religious of the orders (not to the secular clergy), one of the rare norms rigorously respected, resulted evidently in an automatic opposition of wills between these two cultural organizations having such similar methods and difficult aims.

This competitive similarity and difference was particularly obvious when it came to the role assigned to the written word. The Jesuits covered the European map with very up-to-date libraries. They had modeled their *summa* of knowledge on the Vatican's organization: the *ratio studiorum*. On a much smaller scale, their Lyncean competitors urged the creation of scientific libraries in several Italian cities. The thematic catalogue of the library in Cesi Palace was the index of the work that would have to reform the traditional *summae*. The Lyncean encyclopedia—encyclopedia, a fashionable word among innovators all over Europe—was Prince Cesi's great intellectual and publishing project.

There was something of Bacon, Comenius, Bodin, and Mersenne in the missionary will of this "curious" Roman patron of the arts who, instead of hanging out in the literary salons, retired to the country to study insects and fossils, accumulated manuscripts of natural philosophy, and dreamed of encyclopedic projects. He had much less philosophy than Bacon, infinitely less erudition than Comenius and Bodin, and a nonexistent mathematical foundation compared to Mersenne. Yet this "minor" character had certain abilities for organization and practical accomplishment that did not make him look bad in respect of his contemporaries. All of them, from Bacon to Campanella, had longed for academies, libraries, the publishing of books. Federico Cesi had the merit of doing all this; and, after all, he was Galileo's publisher in Rome.

At his palace in Rome, together with a small museum of fossils, whale bones, stuffed fish, and astrolabes, he had gathered a library, which was the vital nucleus of his academy. He had begun to accumulate the new books twenty years before, when the first attempt at an academy had been dispersed by calumnies of heresy and black magic. The Flemish doctor Jean Eck, one of the founding associates, had been forced to take refuge beyond the Alps after being branded a heretic. And from Prague he had sent to Cesi Palace, if they were not there already, Ramus's books of philosophy, the precious herbariums and works of the Bohemian botanist Adam Zaluzianski von Zaluzian, the works of Brunfels, Bock, and Fuchs, Paracelsus's books of alchemy and natural secrets, as well as those of certain German followers of Paracelsus, Andrea Libavius's books of medical chemistry, and Quercetanus's vegetable and mineral pharmacopoeia. Later on, the physician Johan-

and later by G. Gabrieli, "La prima biblioteca lincea o libreria di Federico Cesi," in *Rendiconti della Regia Accademia Nazionale dei Lincei. Classe di scienze morali*, ser. VI, 14 (1938), pp. 606-625, especially p. 621 ff.

nes Faber would leave to the library on Via Maschera d'Oro his rich collection of works on anatomy, botany, and alchemy.[29]

But since 1616, since the time when Galileo had joined Prince Cesi's Roman group, the library had tried to cover the "new and heterdox" area of natural philosophy. In fact, in 1613 there was a truly amazing acquisition: Antonio Persio's bequest of books and, above all, his unpublished scientific manuscripts. Persio had been a pupil of the great Bernardino Telesio, the celebrated author of the metaphysics of nature which the Church had condemned. Abbot Antonio Persio, who had published in 1590 some of his teacher's works on natural philosophy, was also suspected of heresy by Rome. He had been able to live in Rome and even to support Campanella's cause with impunity only because he had enjoyed the protection of such powerful aristocratic and cardinalitial dynasties as those of Orsini, Caetani, and above all Cesi. In fact, he lived in the palace of Cardinal Cesi, Federico Cesi's uncle, right across from the Holy Office's palace.[30] The library of Abbot Persio, who had taught at Padua, was not only rich in published works of natural philosophy and theology, but bequeathed three manuscript volumes of a work entitled *De natura ignis*, a treatise of philosophy entitled *De ratione recte filosofandi*, an unpublished book on Lullo's logic, a manuscript arguing against Galen's medicine, and other unpublished works on theology, philosophy, and politics.[31]

Obviously, this catalogue of unpublished works was very promising in terms of setting off an anti-Aristotelian offensive. Abbot Persio was notoriously an anti-Aristotelian for many years. It was also known, however, that he was not even slightly suspected of harboring Copernican ideas, for Antonio Persio had declared himself a supporter of Galileo in everything "but his opinion of Copernicus."

[29] The dispersion and loss of Cesi's library on the Via Maschera d'Oro (later in the Albani Library) renders quite uncertain any reconstruction of the archives and library originally collected by Federico Cesi. Nonetheless, there exists a descriptive catalogue of the published works, described and reproduced in some small part by Gabrieli, "La prima biblioteca lincea," pp. 613-19, together with a description of the materials in Cesi's natural history museum. On the ideals at the origin of the constitution of the Cesi library, see id., "Federico Borromeo e gli Accademici Lincei," in *Atti della Pontificia Accademia delle Scienze dei Nuovi Lincei* 7 (1934); and id., "Scritti di Giovanni Faber Linceo," in *Rendiconti della Regia Accademia Nazionale dei Lincei* (1934), p. 283 ff.

[30] Antonio Persio was among the witnesses to the astronomical meeting organized at Monsignor Bonaventura Malvasia's villa on the Janiculum by Federico Cesi, on April 14, 1611, in the course of which Galileo described the experience of the luminescence of the Bologna rock. Persio had published a *Liber novarum positionum adversus Aristotelem*, Venetiis 1575; the *Trattato del'ingenio dell'huomo*, Venice 1576; and, in 1590, Bernardino Telesio's pamphlets of natural philosophy on comets, rainbows, and heat. See, on Persio's life, G. Gabrieli, "Notizie della vita e degli scritti di Antonio Persio Linceo," in *Rendiconti della Regia Accademia Nazionale dei Lincei. Classe di scienze morali* (1933), p. 477 ff. On Persio's philosophy, E. Garin, "Nota telesiana: Antonio Persio," in *Giornale critico della filosofia italiana* 3 (1949); and id. *La cultura filosofica del Rinascimento italiano*, Florence 1961, p. 432 ff., in which are published extracts from *Apologia pro Bernardino Telesio adversus Franciscum Patritium*.

[31] F. Cesi's letter to Galileo, March 22, 1612, *Works*, XI, p. 285.

But even though Persio was not a Copernican, Prince Cesi, an admirer of Telesio, had for some time now hoped to ensure the publication in Rome of Persio's unpublished works. Indeed, he quite rightly felt that these books on the philosophy of nature "will be very bothersome to the Aristotelians,"[32] as he had revealed to Galileo when telling him in advance about the idea of publishing them.

In 1611, after Galileo's triumphant journey to Rome, Antonio Persio had died, naming Prince Cesi as the executor of his will. The latter, defying the criticisms and censures of official culture and intransigent circles, decided to do what until that moment he had prudently avoided: he inscribed the name of Antonio Persio as a member of the Academy of Lynceans. Above all, he tried to ensure the security of Persio's library in the Via Maschera d'Oro palace in order to publish the precious manuscripts of the Telesian philosopher.

The Academy of Lynceans at that moment in 1612 had very audacious publishing plans: Galileo's polemical letters on sunspots and Persio's physics. The first project was quickly realized, but the second was blocked. Federico Cesi had not much difficulty in moving Persio's library (in January 1612) from the palace of his uncle, Cardinal Bartolomeo Cesi, to the palace on Via Maschera d'Oro where the Academy held its meetings.

It was a great coup for the Lyncean library. But some months later, in May of 1612, when Prince Cesi tried to sound out the reactions of the Holy Office as to the project of publishing Persio under the Academy's aegis, he found the road barred. Cesi had begun by submitting to the Holy Office the list of Persio's unpublished works. However, authorization for that catalogue was not granted, and only after a full year of postponements and insistences was it finally possible to obtain it and print the catalogue. The Holy Office had opposed much resistance to it, believing quite rightly that those works were "greatly contrary to Aristotle."[33]

Cesi considered himself warned and did not try to insist on his publication project. Persio's scientific and philosophical works remained unpublished in the Cesi Palace's library. But since the catalogue had been published, everyone knew what a prestigious center of research in the "new philosophy" the Via Maschera d'Oro library now was.

In any case, it had been a setback. Federico Cesi had hoped to publish Persio in order to stir up the Roman waters, as a maneuver of alleviation in regard to the official culture, and so to reduce the stifling pressure of Aristotelian conformism. Galileo's ideas would have benefited from this al-

[32] See F. Cesi's letter to Galileo, May 19, 1612, *Works*, XI, p. 298.

[33] F. Cesi's letter to Galileo, May 26, 1612, *Works*, XI, p. 303. See A. Persio, *Index capitum librorum . . . de ratione recte philosophandi et de natura ignis et caloris*, Romae n.d. [1613]. A copy of the first six books of *De natura ignis* remains in the Lyncean Archives (ms. Linceo V and VI). See Gabrieli, "Notizie della vita e degli scritti di Antonio Persio Linceo."

leviation. But the Holy Office and official institutions, confronted by that challenge to official Aristotelian philosophy, had said no even to the powerful Prince Cesi.

From that moment on, as we shall see later, Cesi's library found itself in the gunsights of Rome's Counter-Reformation institutions as a dangerous arsenal of suspicious works. It was 1612. Copernicus will be condemned only four years later.

Federico Cesi's Encyclopedia

But Prince Federico Cesi stood fast. He made the project his own and developed it to bring to light, in a much more organic and systematic form, a philosophy of nature. At bottom, the defeat had not been so serious: Persio's unpublished works must surely be stuffed with misty Scholastic notions. One could revise and improve that publishing project and, with the collaboration of new forces in the Academy, prepare an entirely new work.

A title with a Renaissance flavor was chosen: *Theatrum naturale*, like the *Théâtre de la nature* published by Bodin in 1597. Just as the Persio publication, which would have had at its beginning the philosophical treatise *De recte philosophandi*, this new encyclopedic project included at the beginning a philosophical treatise in the form of a preliminary discourse entitled *Speculum rationis*, on the art of the gaze and ratiocination in scientific experience. Furthermore, this preliminary discourse contained a precise, synoptic table of contents. It was not yet a tree of knowledge; it was instead a critical catalogue arranged by subject, of the Cesi Palace's library. It was a "mirror" revelatory of the features of the new knowledge sponsored by the Academy of Lynceans. We know about this project and the preliminary discourse of the Cesi encyclopedia because a Cesi manuscript containing the table of contents and that discourse ended up, one knows not how, in the National Library at Naples.[34]

In the Cesi encyclopedia would figure different aspects of the book of nature, corresponding to the different sections—*Frontespicia*; physics (*Physicomathesis*); cosmology (*Coelispicium*); meteorology (*De aere et thaumatombris*); biology (*De mediis naturis in universo* and *De plantis imperfectis*)—all connected to a bibliographical and bibliotechnical treatise (*Bibliologia*) as well as a program of academic editions (*Linceografo*). One can date this program at around 1615, after the futile attempt to publish Persio's works and after Galileo's enthusiastically received visit to Rome.[35] The *Physico-mathesis*

[34] This is the third part of ms. volume XII E 24 of the Lyncean papers in the Vittorio Emanuele National Library, Naples, published and discussed in G. Gabrieli, "L'orizzonte intellettuale e morale di Federico Cesi illustrato da uno Zibaldone inedito," in *Rendiconti della Regia Accademia Nazionale dei Lincei. Classe di scienze morali*, ser. VI, 14 (1938), pp. 663-725.

[35] In any case, at the end of 1622 Cesi wrote to Faber about his *Specchio della ragione* and *Theatrum*

contained, under its felicitous neologism, the polemical ideas and fascinating experimental programs that had characterized Galileo's fulgurant official entrance into natural philosophy and the Academy of Lynceans: atomist ideas on heat, corpuscular suggestions on light and magnetism.

"Atoms" is the listing for a key article on microphysics (which would have been developed in the Cesi encyclopedia by means of a special section entitled *De corporibus invisibilibus*) that embraced air, magnetic forces, muscular strength, and odors. Other datable traces of Galilean influence on the grand project were the corpuscular study of heat and of the hot light and cold light of the luminescent rock in the famous Roman experiment of 1611. On all these "physical paradoxes which appear completely contrary to accredited dogmas, we will establish logical reasonings and experiments"[36]—certainly a very ambitious program.

At any rate, the Cesi encyclopedia was intended to be a work that broke with institutional knowledge, and atomist ideas were the only ones capable of bringing into question the entire scaffolding of natural philosophy and Peripatetic medicine. This was a fascinating and very difficult program for a group of scholars whose backgrounds were almost entirely literary and philosophical: to study, through experiments in alchemy, metallurgy, and medicine, a corpuscular theory which until that moment had been linked with the mysticism of the *Corpus hermeticum*, the full panoply of qualitative and occult ideas in the animistic arsenal. This was, at bottom, the program of the Cesi encyclopedia: to channel into the new *Theatrum naturale* the entire Hermetic, magical, alchemist library without being contaminated by the traditional metaphysical speculations (sympathies and analogies) or by the purely qualitative principles of Renaissance naturalism and the Hermetic tradition. The corpuscular perspective in physics guaranteed Galileo's Roman followers against these metaphysical perils. But these followers were few and did not have much familiarity with experimental observation, not to mention their deficient mathematics.

Prince Cesi's program was not, however, so presumptuous as to embrace the classical disciplines. Mathematics, astronomy, and mechanics had been wisely omitted because they were beyond the competence of Federico Cesi and his collaborators. For these matters, there was Galileo. Galileo, the sole mathematician and sole astronomer worthy of the name in the Academy of Lynceans, was in fact given full responsibility for "remaking" the macro-

naturale in such terms as to let it be understood that this was a matter of a program already worked out for some time and now in the execution phase, thanks to the announced completion of *Plantaria distribuzione methodica*; see Gabrieli, "L'orizzonte intellettuale e morale di Federico Cesi," p. 683. In 1618, Cesi will communicate an extract of *Coelispicium* to Cardinal Bellarmino, as will be discussed below. The dating of these convergent realizations in the encyclopedia program allows us to place the project, as we have done, around 1615-1620.

[36] See Gabrieli, "L'orizzonte intellettuale e morale di Federico Cesi," n. 32, p. 672.

world: the astronomic and mechanical systems. The Roman Lynceans took on the task of deciphering the sublunar book of nature, which was congenial to their backgrounds as "curious about nature." That micro-world which Galileo had only sketched in *The Assayer* and which was congenial to their education as philosophers. But even so, the program was very ambitious and was a flight forward with respect to the state of knowledge and the intellectual personnel on which Prince Cesi could count.

Left to his own resources, Prince Cesi was actually able to realize only a small part of the *Theatrum naturale*, of which he published only the two sections on botany and zoology, *Tavole fitosofiche* and *Apiario*,[37] each bearing in its title the indication that it was a section of the projected encyclopedia work. He had composed other parts, on geology, but they did not get beyond the manuscript stage.[38] To write the other sections, the very important ones on physics and chemistry, required new forces, new philosophers.

Virginio Cesarini's Lucretian Poem

Despite its enthusiasm and its ambitions, the Academy of Lynceans remained seriously deficient in intellectual personnel capable of developing the most promising research programs in Galileo's physics: the corpuscular theory of natural elements, heat, and light. And yet, that program was essential and urgent—above all after 1616, when the condemnation of Copernicus blocked the development of the Galilean and Lyncean program in astronomy.

The great majority of the Lyncean Academy's members, until that moment, comprised literary men and jurists. Until 1616, Federico Cesi had counted on a limited number of physicians and naturalists (Johannes Faber, Fabio Colonna, and Giovanni Schreck) and then, but only for a short time, on mathematicians such as the Roman Luca Valerio, professor at the Sapienza, and the Neapolitan Francesco Stelliola.[39] In order to forge ahead, it was necessary to strengthen the Academy's scientific and philosophical component with young and brilliant philosophers. Usually, however, those to be found in Rome had come out of the Collegio Romano and remained in the elite orbit of Jesuit culture. Between the Collegio Romano and Cesi Palace there was bad blood to say the least, even if personal relations may have been marked by correctness and courtesy, at least up to a certain date.

[37] Se F. Cesi, *Apiarium*, Romae 1625; and *Phythosophicarum Tabularum . . . pars prima* (in press in 1630, while Cesi was still alive), definitively published in F. Hernandez, *Rerum medicarum Novae Hispaniae Thesaurus*, Romae 1649.

[38] *De Laserpitio*, F. Cesi's ms. in the Library of the Botanical Institute and Garden of the University of Padua, described in Gabrieli, "L'orizzonte intellettuale e morale di Federico Cesi," p. 714.

[39] See G. Gabrieli, "Il carteggio scientifico ed accademico fra i primi Lincei (1603-1630)," in *Memorie della Regia Accademia Nazionale dei Lincei. Classe di scienze morali*, ser. VI, vol. I (Rome 1925).

At the Collegio Romano, there reigned respect for the Aristotelian tradition in philosophy, and the principles of Catholic faith were constantly present, to be put before every other preoccupation in one's research. At Cesi Palace, in contrast, reigned an atmosphere of individual rationalism; there a research free from theological controversies and preoccupations was brazenly professed, almost as if to hide the well-known dangerousness of the studies and friends with which Prince Cesi had always surrounded himself.

At the start there had been the "black magician" Eck. Then the Academy had stopped bragging publicly of counting among its members Della Porta, the patriarch of Italian natural magic. Della Porta was not officially condemned by the Church, but he was by the Jesuits. In their special Index for internal use, which was the work against natural magic purposely prepared by Father Martin del Rio, Della Porta was singled out as a heretic who should be banned, like Pietro d'Abano, Agrippa, Lullo, Paracelsus, and Bodin.[40] Then came Persio and, finally, Galileo.

In fact, between the Collegio Romano and Cesi Palace there was an undeclared war of competition, each trying to snatch up the most promising intellectuals. Immensely favored, from all points of view, the Jesuits initially had an easy game of it, and Federico Cesi suffered serious defections. In 1611, the Helvetian naturalist Schreck (called Terrentius), who had been entrusted with the important task of editing the *Tesoro messicano*, an edition of great prestige for the Academy, went over to the enemy camp, becoming a Jesuit missionary and even taking with him the Academy's telescope. All Cesi could do was expel him. In 1616, at the delicate moment of Galileo's admonition, the Academy also expelled Luca Valerio, who had defected and tried to involve the entire Academy in a condemnation of Copernican theory. To make up for this, new forces arrived first from out of town and then from within Rome. And these new acquisitions will be key figures in the "marvelous conjuncture" of the Academy of Lynceans at the moment of *The Assayer*.

The first new adherent did not have to be converted to Galilean ideas. He was a twenty-five-year-old Florentine priest, fresh from his philosophical and juridical studies in Padua, Bologna, and Pisa. In 1615, Giovanni Ciampoli presented himself at Cesi Palace at the behest of Galileo, whose admirer and pupil he had been for the past decade.[41] Ciampoli had the protection of Cardinal Maffeo Barberini, pontifical legate at Bologna. He also had a Pindaric, oratorical, and magniloquent poetic vein, which seemed designed ex-

[40] See Martin del Rio, *Disquisitionum magicarum libri sex* (1606), published Moguntiae 1624, p. 9; and B. Pereira, *Adversus fallacies et superstitiosas artes libri III*, Lugduni 1603.

[41] See D. Ciampoli, "Monsignor Giovanni Ciampoli, un amico di Galilei," in *Nuovi studi letterari e bibliografici*, Rocca San Casciano 1900, pp. 5-170; A. Favaro, "Amici e corrispondenti di Galileo. Giovanni Ciampoli," in *Atti del Regio Istituto Veneto di Scienze, Lettere ed Arti* 62 (1902), pt. II, pp. 91-145.

pressly for the lofty language of Curial diplomacy. With these talents, the cardinals' protection, and his Pisan diploma in civil and ecclesiastical law, this young Galilean philosopher seemed destined for a brilliant career in the Vatican. It was realized even more rapidly than could be foreseen.

Don Giovanni Ciampoli had very close relations in Rome with a young philosopher of the high Roman aristocracy, with whom he had in common the same literary propensity for poetry of a classicist style: Duke Virginio Cesarini. In 1615 Cesarini was just twenty, but nonetheless well known—and deservedly so for his prodigious erudition, intellectual vivacity, and profound religious feeling.

Duke Cesarini, Federico Cesi's cousin, was a product of the Jesuits and an intellectual with orthodox ideas in philosophy. As a matter of fact, in his early youth he had studied philosophy at the court of Ranuccio Farnese in Parma, in a university environment infiltrated by the followers of Averroism, a small colony of not really orthodox Paduan Aristotelianism.[42] An accident while horseback riding had forced the young duke to interrupt his regular studies at the university, where he had already distinguished himself by his intellectual gifts. But in the seventeenth century, when an aristocratic intellectual fell ill, there was always a Jesuit spiritual director at his bedside. Also during Virginio Cesarini's long convalescence, Jesuit priests from Parma helped him complete his philosophical *cursus* in a more orthodox theological light.

The gifts of the young philosopher must have been heralded in Rome, because when Duke Cesarini returned there in 1610, Cardinal Bellarmino took him under his prestigious intellectual protection with the flattering intention of starting him off on the path of philosophical apologetics. Cesarini was a precociously serious scholar averse to extremist literary provocations and their frivolous and fashionably outspoken manifestations. The tuberculosis that since adolescence had undermined his chronically sick body also tinged his vocation and his religiosity with a Senecan, dolorous, and meditative morality. This did not pass unobserved in Roman intellectual and aristocratic circles.

Bellarmino, who was also impressed and who knew how to recognize the value of that singular marriage of intellectual and moral qualities, encouraged him by publicly flattering him as a new Pico della Mirandola. This praise circulated through Rome and had a premonitory value.[43] Cardinal Bellarmino had made a choice in that demanding and authoritative associa-

[42] See A. Favoriti, "Virginii Caesarini Vita," in *V. Caesarini Carmina*, Rome 658, pp. 1-30.

[43] See ibid.; A. Gottifredi, *In funere Virginii Caesarini, Oratio*, Romae 1624, p. 19 ff.; Erytraeus (Father I. N. Rossi, S.J.), *Pynacotheca imaginum . . . illustrium virorum*, Coloniae 1695, I, p. 59 ff. On Cesarini's literary personality, see E. Raimondi, "Paesaggi e rovine nella poesia di un virtuoso," in *Anatomie secentesche*, Pisa 1966, pp. 50-72.

tion, since Pico della Mirandola, the natural Catholic magus—acceptable to the Jesuits—had been one of the greatest intellectual opponents of Averroism and of naturalistic, anti-Christian materialism.

At that moment the Holy Office, guided by Cardinal Bellarmino, was seriously preoccupied with the atheism of the official Paduan philosopher Cremonini, whom the indulgent protection of the aristocracy of the Venetian Republic still gave refuge from the proceedings brought against him by the Inquisition. Then there was the apostate Vanini, the heretic Campanella, and the whole dangerous spread of a disbelieving libertinism which worried Cardinal Bellarmino much more than the diffuse literary and moral libertinage of the cultivated classes. But the repressive work—and Cardinal Bellarmino knew this very well—was impotent when confronting this state of affairs if it lacked an intellectual counteroffensive against the challenge of philosophical irreligiosity.

For these reasons, Cardinal Bellarmino directed the culture and talents of Duke Cesarini toward the great task of writing a philosophical and theological work on the burning problem of the soul's immortality. This was a program of up-to-date research in philosophy, of erudition and polemic against atheism and naturalism, which had even infiltrated the salons of the Roman palaces. To Cardinal Bellarmino's perspicacious eye, that theme appeared perfectly suited to the intense, melancholy sensitivity of this young intellectual believer.

With the expected assiduity, Duke Cesarini devoted himself to gathering sources for this work; yet, the hopes that had been authoritatively put in him were disappointed. In 1616, as we know, Galileo had come to Rome. Federico Cesi took him to a series of lectures and propaganda debates in support of the new philosophy, held in various Roman palaces. One of these was the beautiful Renaissance palace of Duke Cesarini's family; there Federico Cesi introduced the young prodigy and cousin to Galileo. Now Galileo was a seductive talker and a daring experimenter. A young philosopher who did not have an academic reputation to protect could not easily escape the irresistible enchantment of the rational arguments from Galileo's "restless brain," his "natural talent" for making, "with minimal effort, matters that were very difficult and recondite seem easy and obvious," as *The Assayer* will recall in connection with the exciting experiments in physics performed in the Cesarini house.

Those experiments, above all, opened Duke Cesarini's eyes to science and to the importance of direct experimental study of the great theoretical problems. Galileo, on that day, in order to illustrate the earth's various motions according to Copernicus, had rotated a ball in a water-filled basin, which was also twirled about. With this simple experimental apparatus, he had charmed the "literati" in the Cesarini house, demonstrating not only the probability of terrestrial rotation, but also the "simple appearance" of a third motion falsely attributed by Copernicus to the terrestrial axis.[44] Great speculations were within the reach of experiments accessible even to someone who was not a mathematician.

The encounter with Galileo resulted in implanting the experimental vocation in Virginio Cesarini's mind, as had been secretly hoped by Federico Cesi, who was quite determined to carry to the end the conversion of his Aristotelian cousin. Cesarini's conversion to the new philosophy could not be a superficial, enthusiastic decision. Federico Cesi has left us a record of the successive stages of the abandonment of Aristotelian categories, through the long discussions among Cesarini, Ciampoli, and Cesi himself that took place in the favorite haunts of the Roman virtuosi's new philosophy: the archaeological collection in Cardinal Bartolomeo Cesi's palace at Bargo Santo Spirito; Federico Cesi's botanical garden at Via Maschera d'Oro, among the herbariums and fossils of the Lynceans' library. It was a pilgrimage through the open pages of "the most beautiful book of all, that of the world and nature written by God," Cesi writes, and at the end of that contemplative pilgrimage the hoped-for result was achieved: persuading that talented Catholic philosopher to abandon the philosophy of rational entities and to set off toward a new religiosity, along the true road "for philosophizing and arriving at the knowledge of the things of nature."[45] He is not seduced by the Copernican novelties superficially praised by literary extremism. Virginio Cesarini remained what he was, a reserved and moderate intellectual, a classicist and moralistic poet, a true believer. Cesarini was converted by the religious and intellectual afflatus of a truthful reading of the book of nature written by God.

When in 1618 Duke Virginio Cesarini was officially entered, together with Monsignor Ciampoli, in the Academy of Lynceans, thus making it officially known in Rome that he had abandoned the paths of orthodox Aristotelianism, the news created a great stir.[46] On entering the Academy, Cesarini signed the official act of his conversion, a sensational open letter to Galileo. It had the tone of an abjuration and requested a provocative pedagogic initiative of broad intellectual scope on Galileo's part. The philosopher Cesarini asked Galileo to teach a "more certain logic, whose syllogisms, based on natural experiments and on mathematical demonstrations, will at once open the intellect to consciousness of the truth and shut the mouths of some vain and pertinacious philosophers whose science was opinion and, what is worse, other people's and not their own."[47]

[44] *The Assayer*, p. 214 ff. (*Works*, VI, p. 157).

[45] See the official Lyncean eulogy prepared on the basis of F. Cesi's memoirs by J. Reck (Riquius), *De vita viri praestantissimi Virginii Caesarini Lyncaei*, Patavii 1629.

[46] See Gottifredi, *In funere Virginii Caesarini*, p. 10.

[47] V. Cesarini's letter to Galileo, October 1, 1618, *Works*, XII, pp. 413-15, especially p. 413.

The Jesuits, who saw that prestigious intellectual creature of theirs pass over to the enemy camp, will never forget this betrayal. Later, they will present Cesarini's conversion to what they called "the new and sensible [as in sense experiences'] philosophy" as a desertion caused by youthful infatuation and an abduction carried out by Galileo's Roman friends.[48] In Rome, with the acquisition of the new aristocratic Pico della Mirandola, the Academy's prestige rose enormously, as did the legitimation of the Academy with respect to the centers of power. For Cesarini was not just any literary man in the pay of a prince or a cardinal: he was a personality in high Roman society and in the political world. He had influential relations with the party then in power; he had the ear of the Spanish embassy and was a friend of the Ludovisi family. In 1621 Gregory XV, the new Pope Ludovisi, had brought Cesarini into the Vatican as privy chamberlain. Moreover, in the same period Don Ciampoli, who had already been taken on as Cardinal Ludovico Ludovisi's secretary, became a monsignor in the Vatican at the Secretariat of Secret Briefs to Princes—through the intercession of Cardinal Maffeo Barberini. These Lyncean positions in the Curia were tantamount to a public legitimation of their academy.

Between 1618 and 1624, the Academy of Lynceans easily recruited new followers among the Roman intellectuals: the naturalist Carlo Muti, Cesi's friend; the philosopher, doctor, and amateur astronomer Claudio Achillini; and the great collector and art patron Cassiano dal Pozzo, who entered the Academy in 1622. But the two most direct collaborators of Federico Cesi's encyclopedia project were in fact Virginio Cesarini and Giovanni Ciampoli. Both had a philosophical education, and both were autodidacts in mathematics. Just as Ciampoli in 1609, after having heard Galileo speak, had rushed to study Euclid on his own, so did Cesarini, after Galileo's visit in 1616, decide to "apply himself wholly this winter to mathematics." So Ciampoli informed Galileo on the effects of proselytism because of the "marvels of his intellect."[49]

Abandoning the books on the soul's immortality, Cesarini discovered the books on astronomy and mathematics in Federico Cesi's library; but these directed the studies of the neophyte to the more accessible terrain of natural philosophy. In this field, *The Assayer*, dedicated to Cesarini and edited by him, will give Galileo's answer to the philosophical and methodological request that Cesarini himself had in 1618 put forward in the name of Rome's Galilean community. But at that moment, Galileo's "curious" Roman pupils already knew which programs of research were to be developed.

Federico Cesi's encyclopedia had listed them as the principal problems of *Physico-mathesis*: "with arguments and experiments as shall establish the

[48] Gottifredi, *In funere Virginii Caesarini*.

[49] G. Ciampoli's letter to Galileo, December 31, 1616, *Works*, XII, p. 300; and V. Cesarini's letter to Galileo on the same day, ibid., p. 299.

physical paradoxes (*paradoxa physica*) that appear completely contrary to consecrated dogmas," as already mentioned. There were several of these "paradoxes" opposed to "dogmas": from the famous experience of the luminescent rock of Bologna to thermoluminous phenomena, the study of the heating of fluids, magnetism, and the transformation of substances. During his Roman visits, Galileo had suggested, as in his books, a return to the atomist or corpuscular study of these phenomena.

The need to delve deeper, to clarify in its multiple aspects the theory of the minimal parts of matter, involved for the Lynceans the shift from qualitative speculation to the quantitative considerations suggested by Galileo. The obligatory road to travel was one on which Galileo did not want to, or did not know how to, commit himself: chemistry experiments aimed at understanding whether and how the ultimate corpuscles of substances maintained their characteristics. Were they atoms or minims?

The great problem of the essential constitution of matter was that of the permanence of forms in compounds. It was, in other words, the problem of the relative permanence of the qualities of heterogeneous minims—each with characteristics typical of the original substance—or of atoms of homogeneous material. It seemed that the study of chemical transformations would allow for a better understanding of this delicate question.

In practice one had to go to the crucible and alembic, to calcine, cook, and distill in order to grasp the "lofty secrets" of the chemical transformations of bodies and compounds. If the observable minimal particles preserved their elementary qualities, it would be possible to investigate at least the quantitative properties of the transformations: the movement and positions of the corpuscles. A kinetic theory of matter could advance experimentally, through practical chemistry, to a quantitative stage. Corpuscularism would thus no longer be a speculation, but rather an experimental base capable of supplanting the Aristotelian theory of the elements in physics and medicine alike.

Virginio Cesarini devoted himself to this research. In his library, side by side with books of poetry, theology, and Scholastic philosophy, now stood new works, ancient and modern: Fracastoro and Paracelsus, Cardano, Telesio, Gilbert, Agrippa, and Bruno; books on natural magic, chemical philosophy, and medicine by Faloppio, Della Porta, Lenius, and Fioravanti.

Like all the chronically ill, Cesarini went from one doctor to another, made a cult of medical discussion and ended up knowing more about medicines than his own doctors. He knew the simples, the miraculous opiates, remedies for the sufferings of a physical pain that was a reality of the time and that we today cannot comprehend. He knew how to prepare secret mineral prescriptions, discussed iatrochemical and philosophical principles. He was linked with the famous Roman doctor Curzio Clementi, by "whom he was initiated into chemical secrets, after which he began to devote himself with greater attention to experiments in that art which more than any other thing would help them overturn Aristotle's fundaments."[50] Cesarini is a patient of the Lyncean physician Achillini, knows Faber well, and has connections with Giulio Cesare Lagalla, a doctor and philosopher—officially Aristotelian, but really a disquieted Neo-Platonic natural philosopher—who has for some time studied the phenomenon of Bologna's luminescent rock.[51] In Bologna he becomes the friend of the pharmacist Pierre de La Poterie, who introduces him to the even more recondite secrets of the alchemical art,[52] and perhaps he also knows the Roman doctor Giulio Mancini, physician at the Santo Spirito Hospital, who is also an astrologer and Epicurean. He certainly writes to Naples, where Campanella is in prison, to ask him for a bibliography.[53]

All the information we have on the philosopher Cesarini shows him intent on "spying out nature's lofty secrets."[54] His friend Monsignor Mascardi, whom Cesarini knew since his Parma days and who in Rome frequents the Academy of the Humorists, provides us with more precise information: Cesarini "even performed chemical distillations and with exquisite diligence experimented in order to see with his own eyes those transmutations both in simples and minerals which are difficult for the speculative intellect to comprehend."[55]

During his researches, Cesarini, as was discreetly rumored, had discovered in the philosophy of nature something "unknown to Aristotle." What had Cesarini discovered? We do not know. Perhaps he did not know very well himself. We do know, however, that he had begun writing a poetic commentary on Lucretius's *On the Nature of Things*.

We can understand what Stoic and melancholy resonance in Lucretius impelled Cesarini, as before him Torquato Tasso, to recognize himself in the Latin poet who had committed suicide. But Lucretius's book had been condemned by the Church at the Fifth Lateran Council and then condemned again, so much was it circulated and read, by the Florentine synod of 1518.

Virginio Cesarini was a student of the "curious" sciences, tempted by Platonic ideas; but he was certainly not an Epicurean libertine, even if his curiosities, his readings, and researches had weakened his Aristotelian roots. He

[50] Favoriti, "Virginii Caesarini Vita," p. 7.

[51] On the relations between Cesarini and Lagalla, see Reck, *De vita viri . . . Virginii Caesarini*.

[52] Favoriti, "Virginii Caesarini Vita," p. 9.

[53] See *Lettere di Tommaso Campanella*, Spampanato, Bari 1927, pp. 216 and 275; and G. Gabrieli, *Bibliografia lincea*, II; "Virginio Cesarini e Giovanni Ciampoli con documenti inediti," in *Rendiconti della Regia Accademia Nazionale dei Lincei. Classe di scienze morali*, ser. VI, 8 (1932), pp. 422-62, especially p. 430.

[54] See the lyrical euology of Cesarini by F. Testi, *Poesi liriche*, Venice n.d., p. 91.

[55] See A. Mascardi, "Per l'esequie di Virginio Cesarini," pp. 3-23 in *Orazioni e discorsi*, Milan 1626, p. 9; and *Prose volgari*, Venice 1653, p. 355.

had been reconfirmed in the Curia by Pope Barberini, and people spoke of him as a cardinal *in pectore*. Perhaps these complex and contradictory human motives led him to leave among the manuscripts of his library the poem that he had written *Lucretium imitatus*. Or perhaps he had imitated Marsilio Ficino, who after having written a commentary on Lucretius, burned it, so it is said. Virginio Cesarini had in any event made a secret pact with himself.

In 1620, at Acquasparta, victim of a very serious relapse, Cesarini had made his will. He must have remembered then how Cardinal Bellarmino had linked him with Pico della Mirandola. Just as Giovanni Pico della Mirandola had willed his Hermetic library to the care of his brother Anton Maria with the obligation of not giving it to any religious order, but had also asked to be buried in the Dominican habit, so Cesarini left his books and manuscripts to his fraternal friend Giovanni Ciampoli, with the restriction of handing them over to the library of the Academy of Lynceans when it was definitely constituted. But Cesarini left his body to the Society of Jesus, having obtained from the father general permission to be buried in the Jesuit habit.[56]

The contradictions of Virginio Cesarini's human experience appear clearly with his death on April 11, 1624. On April 13, he was given an official funeral in the Church of Santa Maria in Aracoeli, on the Campidoglio, at which Jesuit priests officiated. The funeral eulogy was recited by Father Alessandro Gottifredi, professor of rhetoric at the Collegio Romano, who sung the praises of the Christian virtue of the poet who did not let himself be seduced by the pagan fashion of the literary innovators.[57]

Prince Federico Cesi did not intervene or participate in any way in the ceremony.[58] However, a eulogy of a much different kind was recited a month later, on May 5 at the mourning-draped Academy of the Humorists, by Agostino Mascardi, who remembered his friend above all as a man who was "curious" about nature and as a philosopher who had become "almost a pure skeptic" in moral philosophy.[59]

Some days later, on May 11, Johannes Faber wrote in alarm to Federico Cesi asking him to move Virginio Cesarini's library to a safe place: "Your Excellency should see to it in all ways that we recover Signor Virginio Cesarini's books . . . and it will also be necessary to see to it that we have the license for the forbidden books, and that the Friars do not steal them from us."[60] The "Friars" were the functionaries of the Holy Office who were pre-

[56] Cesarini's testament has been published in its entirety in Gabrieli, "Virginio Cesarini e Giovanni Ciampoli," pp. 445-47.

[57] See Gottifredi, *In funere Virginii Caesarini*.

[58] G. Gabrieli, "Una gara di precedenza accademica nel Seicento fra Umoristi e Lincei," in *Rendiconti della Regia Accademia Nazionale dei Lincei. Classe di scienze morali*, ser. VI, 11 (1935), pp. 235-57.

[59] See Mascardi, "Per l'esequie di Virginio Cesarini," n. 52.

[60] J. Faber's letter to F. Cesi, in Gabrieli, "Una gara di precedenza" p. 242. On Cesarini's death and

paring to carry out a survey of dangerous papers at Cesarini Palace (if they had not already done so) or to investigate the catalogue. But three days before this alarm at Acquasparta, the famous symbolic testament which left "books and writings" to Monsignor Ciampoli was opened. From this moment on, nothing will ever be heard again about Cesarini's manuscripts, among which, perhaps, was the unpublished Lucretian poem. In 1624, Giovanni Ciampoli was a gray eminence in the Curia, with solid friendships at the Holy Office's palace: it must not have been difficult for him to avoid prohibitions and sequestrations and to take Cesarini's books and manuscripts into his library.

Giovanni Ciampoli's Library

As an intellectual, Ciampoli was completely different from the fragile and melancholy Cesarini. The latter had always been the victim of aristocratic desperation, of an inability to resolve the contradictions between old and new, knowledge and faith. Ciampoli was an extroverted and pragmatic personality, a celebratory and pompous poet.

Florentine, impulsive, and brilliant, Ciampoli was a great operator, the protagonist of a rapid career in the Secretariat of Briefs from which he corresponded with cardinals, princes, and intellectuals. Sumptuous banquets—slandered as "orgies"—were given by Monsignor Ciampoli at the Vatican when one of these personalities came to Rome on a visit.[61]

His talents as a skillful diplomat were recognized. In the shadow of Cardinal Magalotti and of the splendid but politically inept Cardinal-nephew Franceso Barberini, he guided the Vatican's foreign policy toward the hoped-for pro-French rapprochement. His invisible intervention must have played an important role in preventing the bellicose Spanish maneuvers from gaining the upper hand at the time of Valtellina. His bombastic prose secured the diplomatic mediation between Richelieu and the successors of Charles V.

His anti-Jesuitism favored the creation of the Congregation of the Propagation of the Faith. The secrecy of the cardinals' balloting, which facilitated the election of Urban VIII, was another intelligent innovation owed to Giovanni Ciampoli, one that still exists today.

But Ciampoli was not only a powerful and ambitious prelate, nor only a

---

his library, see Faber's letter to Cesi, April 13, 1624, *Works*, XIII, p. 171. The eyewitness account reported by Faber in this letter leads one to believe that the cause of death might have been tuberculosis: "the lungs stuck to the ribs . . . the liver almost cirrhotic." Cesi was then at Acquasparta with Galileo.

[61] See Anonymous, "Vita di monsignor Giovanni Ciampoli," in G. Targioni-Tozzetti, *Aggrandimenti della scienze fisiche in Toscana*, Florence 1780, II, pp. 102-116; G. Negri, *Istoria degli scrittori fiorentini*, Rome 1722, pp. 272-77; and p. 210 of the Targioni-Tozzetti edition.

stately court poet. He was not only the official mediator between Cesi Palace and the Vatican, but also a friend of the "new philosophy," as the pope will say much later. He, too, shared in the Roman Galileans' program for the reform of knowledge.

Monsignor Ciampoli had a rich personality. He is a prelate skilled in history and theology, a believing intellectual; a neo-Tridentine reformer more than a post-Tridentine Catholic. The "signs of the times" convince him that Pope Barberini's pontificate will give an unprecedented turn to the Church, making it a universal spiritual institution for modern men. Galileo, in his eyes, is the modern Christian philosopher who will supplant the old and pagan Aristotle at the peaks of Catholic culture. During the years of the "marvelous conjuncture," Ciampoli's optimism could not be darkened by shadows. He is convinced, on the basis of historical and rational arguments, that Aristotle's world is finished. His enthusiasm aroused, he became preoccupied, like Mersenne, by the need to fill the void of the old world in ruins with a new Catholic culture, so as to prevent the spread of the atheistic and libertine skepticism of so many hypocritical conformists.

The Assayer's triumph, which he so expertly knew how to orchestrate, confirms the hopes in the "marvelous conjuncture" of the modern pope. There is reason to believe that it was during the 1620s that Ciampoli, under the guidance of Galileo, Father Castelli, and Father Cavalieri, began covering so many sheets with his scientific studies. On the basis of their subject matter, which we know, we can say that these works may well have been destined for Prince Cesi's encyclopedic enterprise.[62]

The titles that have come down to us, in fact, seem to be those of chapters in a Latin treatise on physics and mechanics ("De motu et motoribus a velocitate," "De motu et de loco et de luce," "De quiete et motu," and "De magnete"). Ciampoli had also written in Italian an essay entitled "Physics of Nature" ("Of the Soul," "Of Light," "Of the Arts in Terms of Nature"). He had filled notebooks with essays on the subject of mechanics, on the resistance of ropes, on pumps and inclined planes.

Giovanni Ciampoli had a scientific relationship, well before 1629, with the Galilean Jesuit mathematician Father Bonaventura Cavalieri. We also perhaps owe to this scientific relationship a series of mathematical manuscript titles on both solid geometry and "quantity and the infinite." During the same period, Ciampoli was in direct communication in Rome with Father Castelli, a student of mechanics and physics, author of studies on heat and sunspots. A trace of that period of common scientific fervor can perhaps

[62] The catalogue of the Ciampoli mss. was published in an appendix to the testamentary dispositions in Ciampoli, "Monsignor Giovanni Ciampoli." There are given the titles and number of the notebooks relating to each one of them, which permits us to estimate the considerable amount of Ciampoli's scientific and philosophical work contained in these papers.

be seen in the title of Ciampoli's Italian dialogue, "Of the Sun and of Fire," which Castelli's ideas on the emission of light and thermoluminous absorption may have inspired.

However, his delicate duties in the Curia, in regard to which he could not perhaps compromise himself too much with the new philosophy, kept Giovanni Ciampoli from carrying to the end and publishing the scientific researches already undertaken or still being prepared. For the moment, those notes and manuscript projects—together with the scientific and personal correspondence with Galileo and the writings of Virginio Cesarini—remained in the voluminous mass of manuscripts in Giovanni Ciampoli's library.

Hurried Academic Steps and False Diplomatic Steps

A pragmatic and politic mind, Monsignor Ciampoli was now chiefly concerned with transforming the "marvelous conjuncture"—intellectual and political in nature, created by the new pontificate and the success of *The Assayer*—into an authentic cultural renewal in Rome. He became the "patron" of the Galileans in Rome and exploited his privileged relations with the new power, ably carrying out the work of enlarging the consensus and the institutional and university legitimation surrounding the new philosophy.

The new Sapienza College was the tangible result of the transforming and enduring effects of the "marvelous conjuncture" of 1623. In Rome, the intellectual and social prestige of the Jesuits' Collegio Romano had for some time overshadowed the ancient Roman university of the Sapienza—deserted by the sons of the aristocracy, who went there to obtain their doctorates in law or to carry out their studies in medicine, but who preferred the Collegio Romano's prestigious degree in intellectual subjects such as rhetoric, theology, and philosophy.

In 1624, Professor Giulio Cesare Lagalla died; he was the Sapienza Aristotelian who had attracted the Jesuits' scorn owing to his restless heterodox ideas in physics and cosmology, cultivated in Galileo's shadow.[63] Lagalla had shaken the conformism of the Sapienza, but with his philosophical confusion, his libertinage, and his bigoted attitudes, he had easily allowed the Jesuits to increase the discredit surrounding philosophical instruction at the Sapienza.

Ciampoli's influence was decisive in renewing the Sapienza's cultural image and in making it a competitive university with respect to the Collegio Romano, setting it on the path of a very modern pedagogy and endowing it with prestigious and brilliant professors. The project began to be realized

[63] See G. C. Lagalla, *De coelo animato dissertatio*, Romae 1622.

thanks to the renewed prestige of the two most representative courses: rhetoric and natural philosophy. In 1628, the illustrious and open-minded intellectual Monsignor Mascardi—acclaimed spokesman for the Roman virtuosi and innovators, secretary of the Desirous Ones—was appointed to the new chair of rhetoric. Mascardi was a literary man and a polemicist of unquestionable success, the spokesman of the academies, a fine scholar of historiography and moral philosophy, and a protégé of the Barberini family.

The presence of this lively and restless, political and intellectual personality ensured the Sapienza a modern and provocative reputation in its teaching of the moral sciences.[64] The hopes of seeing prevail a decided renewal of Roman culture with the new pontificate had already been confirmed by the summoning from Pisa of an authoritative representative of the new philosophy. In 1616, Father Benedetto Castelli had been appointed to the new chair of mathematics at the Sapienza. Castelli was Galileo's direct pupil and, in opposition to the Aristotelian culture of the Tuscan universities, had been his spokesman in the first controversies over natural philosophy.

Father Castelli gave the prestige of a Roman university chair to the entire Academy of Lynceans, since with a special suspension of the regulation barring priests from the Academy, Prince Cesi had "associated" the new professor of the Sapienza with the Lynceans' undertaking in a purposely created position. Now the Academy no longer had any problems in proselytizing. In March 1625, immediately after the manifestation in favor of Galileo at the Academy of the Desirous Ones, the Lyncean Academy had welcomed among its members the lawyer Guiducci, who stood in the front ranks during the controversy between Galileo and the Collegio Romano. In the same period was taken into consideration the candidacy of Cavalier Giorgio Coneo, an intimate friend of Ciampoli and Dal Pozzo and a protégé of Cardinal Francesco Barberini.

During those years, Monsignor Ciampoli carried out Galilean propaganda in the Curia that won over Vincenzo Capponi, Urban VIII's chamberlain, to the cause. But, above all, Ciampoli won for the Academy of Lynceans a member whose intellectual and social prestige was equal to that garnered ten years before with the joining of Duke Virginio Cesarini.

After Cesarini's death, the new promise of philosophy within the cultivated aristocracy was the Marchese Pietro Sforza Pallavicino, the precocious intellectual star of the Collegio Romano. In September of 1625, his doctoral thesis on philosophy—personally recited before the pope and his own protector, the Cardinal of Savoy, in the Great Hall of the Collegio Romano—had delighted the Roman intellectual community and the professors of the Collegio, who therefore put legitimate hopes in this pupil of theirs.[65]

[64] See A. Mascardi *Dell'arte istorica* (1636) and *Prolusiones ethicae* (1639).

[65] See *De Universa Philosophia a Marchione Sfortia Pallavicino . . . publice asserta*, Romae 1625, whose publication was inspired by Father Vincenzo Aranea of the Collegio Romano.

Marchese Pallavicino was already known, however, for his friendship with the virtuosi of the academy of the Cardinal of Savoy and for his sympathy with Giovanni Ciampoli's ideas. Pallavicino was regularly in the Curia, where at the Congregation of Rites he had begun an ecclesiastical career that everyone predicted would certainly be prestigious.

Thus, no one was very much surprised when, in early 1629, Marchese Pallavicino also became a member of the Academy of Lynceans. He brought to the new philosophy the support of his authoritative social and intellectual renown, on which his professors of theology and Scholastic philosophy had mistakenly thought they could count. Also in 1629, Father Cavalieri took over the chair of mathematics at Bologna: the most prestigious scientific university teaching position in papal territory had been cornered by a Galilean.

In the course of the 1620s, the "marvelous conjuncture" of *The Assayer* was transformed into a prolonged, favorable season. The new philosophy was in the Curia, the university chairs, the academies, and the families of Roman high society.

An exiguous number of initially marginal "curious" had adeptly converted the success of a book into a movement of ideas and men. To many observers of Roman affairs, Rome's cultural and moral climate seemed so favorably changed as to confirm Galileo in his daring intention of launching the campaign of Copernicanism in an official manner, with the approval and benevolent control (but, above all, the personal protection) of the pope. There was no longer a unified and hostile Aristotelian party. The Galileans could in fact turn to their favor the relationships and equilibriums of power, not to mention the profound dissensions among the different religious and cultural Catholic components. The spectacular support given to *The Assayer* by the Dominicans in the Holy Office, in polemic with the Jesuits, continued and was an important buttress.

Father Niccolò Ridolfi, Master of the Sacred Palace at the time of *The Assayer*, had in 1629 become father general of his order. His successor was *The Assayer*'s great admirer, Father Niccolò Riccardi—Galilean, authoritative professor of theology at the Minerva College, and Giovanni Ciampoli's friend. Other Dominicans present in Rome during the 1620s (including Father Giacinto Stefani, the Roman provincial, Father Michele Arrighi, and Father Raffaello Visconti) agreed in their support of the success of the Galileans, or at least were not hostile.

The front of religious culture was in disarray; it was no longer a solid group. Among his Roman friends, Galileo could count Father Orazio Morandi, father general of the Vallombrosian Order and amateur cultivator of astrology and the occult sciences. Intellectual tensions concealed in the jealousy of the traditional overweening power of the Society of Jesus were vented in support for Galileo, who had dared to challenge and deride the omnipotent and indisputable scientific prestige of the Collegio Romano.

The most sensational official support was that of the lively scientific and intellectual community of the Order of Regular Minor Clerks, an order that, together with the Theatines, benefited from the special protection of the new pontificate and was rapidly gaining ground at the Propagation of the Faith.

In 1626, Father Raffaello Aversa, illustrious professor of philosophy and theology at the College of Regular Minor Clerks in Rome, presented to Galileo the astronomical research of the scientific community of the Regular Minor Clerks of Castel Durante, in the state of Urbino, expressing his personal admiration and that of his brothers for Galilean astronomy as well as his desire to begin a scientific collaboration.[66] Father Aversa was an ultratraditional Scholastic. But it was precisely for that reason that he indirectly arrived at taking up the defense of *The Assayer* against Father Grassi.

In fact, in 1627 the second volume of his *summa* was issued: *Philosophia metaphysicam physicamque complectens*. A long chapter of this book dealt with the problem of the comets. An adamant Thomist, Father Aversa denounced the modern theories of Tycho Brahe and Father Grassi, which negated the Aristotelian theory of comets. A much more indulgent treatment, however, was reserved for the ideas of *The Assayer*, which, as we know, were an erroneous but excusable version of what Aristotle had thought.

We shall see later just how precious the sympathy of the Regular Minor Clerks was to Galileo in combatting his adversaries. The Regular Minor Clerks contributed in large measure to the intellectual, religious, and political personnel of that ecclesiastical institution which gave international scope to the restorative action of the new pontificate: the Congregation of the Propagation of the Faith, founded in 1622 by Gregory XV, but which Urban VIII could legitimately consider his own creation.

It was at the Propagation of the Faith that the new philosophy, again thanks to the mediation of Monsignor Ciampoli, won over a prestigious follower, the Capuchin Father Valeriano Magni, the "tall monk."[67] A Milanese who, however, had always lived in the empire, Father Magni was in 1624 the provincial of his order for Bohemia, Austria, and Moravia. A great competitor with the Jesuits in the restoration of Catholicism in Bohemia and a consummate diplomatic emissary, he was called in 1626 to the Propagation of the Faith and became the ambassador-at-large for papal foreign policy at Prague, at the Conference of Pinerolo, and in the difficult negotiations over the Mantua succession. Besides the same diplomatic problems, he shared with Monsignor Ciampoli an anti-Aristotelian intellectual vocation wedded

[66] See Father Raffaello Aversa's letters to Galileo, June 1, 1626, *Works*, XIII, p. 325 ff.; and July 6, 1626, ibid., p. 329 ff.

[67] On Father Valeriano Magni's support of the Galilean group, see *Works*, XV, p. 215, and XVI, p. 386. Later on, we shall find more substantial evidence of the relations between Ciampoli and Father Magni.

to an Augustinian theological conception. Father Magni furnished from his side the Franciscan tradition, the Christian anti-Aristotelianism of Saint Bonaventure, and an Ockhamist vision favorable to atomism in physics and to the mystical renewal of the new Catholic theology.

So it was not by chance that these two intellectual personalities, representative of the Catholic culture of the first half of the seventeenth century, Monsignor Ciampoli and Father Magni, had the same protectors: Pope Barberini and, later, King Ladislao IV of Poland. In fact, they shared indentical ideals and had in the Jesuits common critics and adversaries.

It was the weakness of these adversaries, more than any other factor, that was decisive in the 1620s in establishing the climate of "marvelous conjuncture." So long as the political isolation of the Jesuits continued, in the atmosphere of the pro-French turn of the Barberini pontificate, that climate would endure. And, indeed, until the end of the 1620s, this favorable relationship of forces was maintained.

Of course, that equilibrium, which made the scales dip to the disfavor of the Jesuits and in favor of the innovators, was a delicate one. The smallest misstep could compromise it irremediably. But, to the great joy of Galileo's Roman friends in that fortunate period, it was actually the Jesuits who in the spring of 1629 made the misstep, a scandalously compromising one. For, after two years in power, Cardinal Richelieu, whose internal and external policy was ambiguous or even openly out of line with the struggle against heresy, increasingly preoccupied a Society of Jesus concerned about his deleterious influence on the ill-advised pro-French turn undertaken by Vatican diplomacy.

The Jesuits were right to have been preoccupied by Gallic arrogance (to use a modern expression), which in 1610 had dared to make the Parliament of Paris condemn Cardinal Bellarmino (a condemnation not executed) for his thesis on the pope's temporal power. With even greater reason, in 1616, the Jesuits feared the obvious political ineptitude of Cardinal-Legate Francesco Barberini, who in almost three years of the pontificate had done nothing but collect books, paintings, and diplomatic failures.

The most significant failure, that of the sumptuous and inconclusive legation to Paris the year before, had revealed the complete political incapacity of the refined cardinal-nephew. The Jesuits had been content to round things out with the scandal of an anonymous anti-French attack, previously printed in Rome and left in Paris after the legation's precipitous diplomatic retreat.[68]

[68] See Anonymous, *Avertissement au Roi très chrétien*, Francheville 1625, incorrect place indication of the pamphlet printed in Rome and unanimously attributed by historians to Father Eudaemon-Johannes. Father Sommervogel, S.J., *Bibliothèque de la Compagnie de Jésus*, 12 vols., Brussels and Paris, 1890-1932, vol. IV, cols. 981-97. It suggests, however, that the author may have been the Bavarian Jesuit Father, Jacques Keller.

Despite setbacks, however, the pro-French tendency persisted, and policy in the Curia was made by the party of innovators led by the opportunistic and extremely pleasure-loving Cardinal of Savoy, who held in the palm of his hand the pope and his cardinal-legate. Furthermore, using his protective power, he ratified provocations against those institutions and books which defended the most orthodox Counter-Reformation.

To break their isolation, the Jesuits decided to repeat the anti-Gallic operation of the previous year—but this time more slyly, though officially and in Rome. This is how the famous Santarelli scandal came about, one of the most serious diplomatic crises of the 1620s. Father Antonio Santarelli was a Jesuit who had taught rhetoric and theology at the Collegio Romano. In 1625, he had prepared a voluminous manual of international inquisitorial law, which the Cardinal of Savoy had the surprise of seeing dedicated to himself.

This was the *Tractatus de haeresi, schismata apostasia*.[69] The dedication underscored the Turinese Jesuit education of the Cardinal of Savoy, extolled the heroism of his religious zeal (which had already made the Savoyan throne famous in the struggle against the pagans and heretics), and praised the traditional devotion of the Court of Turin to the Jesuits. The book harked back to Cardinal Bellarmino's theses on the pope's spiritual and temporal power, the subjection to it on the part of princes, and the pontiff's infallibility. It threatened excommunication for those who did not denounce heresies.

This in itself was pretty heavy stuff, but it seems that when the book arrived in Paris, two pages had been substituted for their counterparts in the copy deposited with the Holy Office, in such a way as to emphasize better the juridical principle according to which the pope can depose even a king for tolerance of heresy, disobedience, and laws contrary to proper mores.[70]

As we shall see on another occasion, in the seventeenth century the Jesuits were accustomed to handling books with the same dexterity with which magicians manipulated packs of cards in the alfresco theaters of Piazza Navona. But this time, the provocation had been manufactured at Piazza del Gesú: the book had had the usual internal triple approval, to which was added that of General Vitelleschi, who in this operation put himself at personal risk.

It was worth the chance, however. The hoped-for break in diplomatic relations with Paris would have produced extremist pressures from the Spanish embassy, which was demanding the pope's military intervention in Valtellina, using the troops gathered in Rome as a threat. Indeed, Cardinal

[69] A. Santarelli, *Tractatus de haeresi, schismata apostasia* . . . , Romae 1625.

[70] At least this was the argument presented at the Vatican by Paris. See Anonymous, *La grande et memorable deffaite de cinquante et deux mille Turcs, avec ce qui s'est passé à Rome entre le pape et le général des Jésuites touchant le livre de Santarelly*, Paris 1626 (three copies at the Paris National Library).

Richelieu seized the opportunity to gain control over France's powerful Jesuits. The French reaction was very harsh, but instead of becoming provocative and turning against the pope, it struck at the Jesuits. Condemned to the flames by the Parliament of Paris on March 15, 1626, Father Santarelli's book was even more seriously censured on May 12 by the Theology Faculty of the Sorbonne, and then by the faculties of the most important universities in France. The French general staff of the Society of Jesus was called before Parliament and signed a formal disapproval of the theses put forward by Santarelli, a disapproval dictated under duress on March 16, 1626. Pope Barberini disapproved of the book, but tried to defend his challenged authority. Piazza del Gesú went into reverse, putting out an expurgated edition of the book. But it was too late; the diplomatic crisis was in full swing. Instead of turning against the pro-French party, however, it went against the very people who had started it.

The following month, moreover, so as not to drag loyalties between Rome and France into the dispute, Pope Urban VIII decided to disavow officially this incitement to schism, summoning Muzio Vitelleschi, father general of the Jesuits, to the Vatican. If we are to believe the published contemporary account (May 16, 1626) of that singular scene in the Vatican, the pope, before the cardinals and prelates of the Curia, publicly summoned and reprimanded the father general, accusing him of having unleashed the crisis. It was the former nuncio to Paris who spoke: "Vois ne vous contentez pas de me nuire en France, vous me voulez encore déchirer en Italie" (You are not satisfied with harming me in France, you also want to dismember me in Italy).[71] Father Vitelleschi stood there, not knowing what to say.

The pope withdrew the troops from Valtellina and disarmed those in Rome. Later the crisis of Valtellina will be definitively settled, but for now the extremist protests went unheeded. The Santarelli affair had been for the Spanish embassy and the Jesuits an obviously crude misstep, completely counterproductive. During the following years, the Jesuits in their controversies will again substitute pages in their books, but the top people at Piazza del Gesú, as we shall see, will be more cautious about compromising themselves officially.

Throughout the 1620s the Society of Jesus could only bewail its former intellectual and political hegemony in the golden years of the pontificates of Paul V and Gregory XV, when Cardinal Bellarmino made the law. Now the pope no longer had a Jesuit as his personal theologian. "His Bellarmino"—as he liked to say—was Monsignor Agostino Oreggi, who had indeed been a pupil of Bellarmino's, but had not because of this become a fervent supporter of the Jesuits. And when, in 1627, he was presented with an application for Bellarmino's canonization, the same pope who had immediately

[71] Ibid., p. 11.

ratified the canonization of St. Ignazio (wished for by his predecessor) decided this time to take all the time necessary to reflect, establishing a new standard on the spot: to begin the beatification process, fifty years had to pass from the death of the candidate. It was an affront that the cardinal's hagiographers prefer to forget.

If one could not immediately make Bellarmino a saint, certainly one could ensure the Jesuits a successor to his post in the Sacred College of Cardinals. But not even this reasonable request was satisfied. To see a cardinal's cap set on a Jesuit head, we will have to wait for the very belated nomination of the great theologian of the Collegio Romano, Father Giovanni de Lugo. But that will not take place until 1643, in a radically changed political and cultural phase.

And yet, during this first decade of the pontificate, Urban VIII had not been miserly with the purple, and had not bestowed it only on his own family. Monsignor Cesarini had died a cardinal *in pectore*, and Monsignor Ciampoli was also a candidate. In 1627, Urban VIII caused a sensation by making a cardinal of Pierre de Bérulle, the "new theologian" from France, the mystical reformer of the faith, a great political and theological antagonist of the Jesuits.

The difficult moment of the Society of Jesus, deprived of its traditional power of influence, made the Galileans and innovators of the pope's party dream of a strength greater than what they possessed, in order to oppose the new Scholasticism and the intransigence of the men and institutions charged with the defense of Counter-Reformation Catholicism. If one looks closely, however, the "marvelous conjuncture," owing to a whole series of very favorable circumstances, had distanced the adversaries, but had not in the least diminished their importance in the historic mission of the Tridentine Church. The admirers of *The Assayer* had not won out over the power of their critics. They had only benefited, thanks to an exceptional and unrepeatable political conjuncture, from the power of impunity, as on a day of Carnival. But this social and intellectual impunity lasted only for the fleeting moment of a few days, and other Roman rituals showed that there was also limits to other sorts of impunity.

FOUR. DARK LIGHTS

In the infinite there are worlds and heavens
(as others contend) which the literary furor
from opposite sides leads to war.
. . . .
Some strain and wear out a vain intellect
in turbid and murky confusion
of infinite parts; and on these
the mad mind emptily reasons
and strives to separate. Others yet generate
from this bodily mass of shapes various aspects:
subtle fire becomes an acute pyramid,
stable earth turns into square shapes,
and he composes almost twenty faces
in the vague and light sublime breathing air
and eight shapes in water and wants weight and body
to have vain shapes, and without motion and weight,
be given to four elements in various guises.
. . . .
and here and there to the unseen objects
human ingenuity cannot find an open path;
and in those scenes he often still adumbrates,
and in others, in too much radiance, the light bedazzles.*

Judicial Rites

On the morning of December 21, 1624, a large crowd of Romans and pilgrims, who had come to Rome from all over Europe for the opening of the Holy Year, had converged on the square before the Church of Santa Maria sopra Minerva. The carriages of the cardinals and the authorities found difficulty pushing their way to the cordon of Swiss Guards which blocked access to the churchyard. The ceremony in the Dominicans' large and ancient church could begin only after some delay.

It was the feast day of St. Thomas the Apostle, the doubting disciple. It had seemed appropriate to the Supreme Tribunal of the Inquisition that the

* "Ned infiniti sono i mondi e i cieli / (com'altri afferma) che d'opposta parte / il furor letterato adduce in guerra. . . . / . . . Altri un vano intelletto affanna e stanca / ne la confusion torbida e mischia / de l'infinite parti; e quinci indarno / la mente pazza s'argomenta e 'ngegna / di separarle. Altri corporea mole / genera di figure i vari aspetti: / di piramide acuta il sottil fuoco, / di quadre forme poi la stabil terra, / di venti quasi facce il vago e leve / spirante aer sublime egli compone, / e d'otto l'acqua e vuol che peso e corpo / vane figure, e senze moto e pondo, / diano a quattro elementi in varie guisa. . . . / . . . e quinci e quindi a' / non veduti oggetti / non trova ingegno umano aperto il varco; / e ne' veduti ancor sovente adombra, / negli altri, al troppo lume, i lumi abbaglia."—Torquato Tasso, *Le sette giornate del mondo creato* [The Seven Days of Creation], 1607, II, 83-91, 142-54, 164-67

stately ceremony of the sentencing in this memorable trial, the most severe and spectacular sentence ever pronounced in Rome during the seventeenth century, should be held on that feast day.

After getting out of their carriages, the cardinals and the secular authorities again had to forge a path through the immense crowd that packed the church, in order to reach the end of the central nave. Here a courtroom had been set up for the occasion of this exceptional trial: a wooden enclosure, as tall as a man, whose entrances were protected by Swiss Guards. In consideration of the crowd, these special security measures were not excessive.

A stepped platform for the court rose on three sides of the sort of wooden cage which was the courtroom. On the left sat the judges of the Holy Office, among whom could be recognized the Cardinal-inquisitor Desiderio Scaglia; on the right sat the Sacred College of Cardinals, the Prefect of Rome, and the civil functionaries. High-level prelates and aristocrats followed the proceedings standing in front of the dais.

Set and script were those customary for such ceremonies. The only actor in an exceptional position was the accused, at the center of the court. He was not the ordinary defendant, as could be seen from the portrait depicting him that had been hung on the pulpit. He was a man of about sixty, and the portrait showed him dressed in black, holding up a tunic. His name, written beneath his image, was that of an aristocrat and important churchman, since it was accompanied by the title of archbishop. But he was even more noted as a theologian, scientist, and intellectual of European renown.

Yet, not even these illustrious titles would have caused such an uproar if it were not for a circumstance unprecedented in human memory. Indeed, it amazed everyone as if it were an unreal and stupefying spectacle, even in a city accustomed to this kind of judicial ritual.

Marco Antonio de Dominis, the accused, had been dead for three and a half months. The Holy Office was not condemning a man, but rather his corpse, which at the moment was awaiting sentence before the judges in a coffin blackened with pitch.

There was an emotion-charged silence in the Minerva church. A priest mounted the pulpit and read in Italian, and a "chanting voice"[1] so that he could be heard out on the square, the text of the sentence, which declared that the accused was a pertinacious heretical apostate. Long applause emphasized the public's approval of this ritual celebrated in the style of the medieval Inquisition. For many of the curious, that applause was above all a

[1] A. Bzowski (Bzovius), *Annalium Ecclesiasticorum*, tome XVIII, Coloniae Agrippinae 1627, p. 174 ff. On Father Bzowski (1567-1637), see J. Quétif and J. Echard, *Scriptores ordinis praedicatorum recensiti*, II, Lutetiae Parisiorum 1721, p. 488 ff. The sentence of the De Dominis trial had been announced in the *Avviso* of October 19, 1624 (Vatican Library, Barb. 2818), as reported by L. Pastor, *Die Geschichte der Päpste*, vol. XIII, 1920, and trans. into Italian in *Storia dei papi dalla fine del Medio Evo*, vol. XIII, Rome 1931, p. 625.

release, exorcizing the anxiety created by a macabre and solemn ceremony. But for the majority, aristocrats and ecclesiastics who had gathered to demonstrate their support for the Church's institutions and its most intransigent ideals, that applause expressed a mandate for renewed severity against the challenge posed by heretics and innovators.

According to the Inquisition's trial laws, he who has been recognized as a pertinacious heretic, unlike a confessed, penitent heretic, could not escape sentencing as an impenitent heretic. Thus, in such a case, the sentence inevitably involved being burnt at the stake.

Stripped of his ecclesiastical prerogatives, De Dominis was expelled from the Church. The condemned man's portrait and remains were carried out to the churchyard and put on a cart together with his theological and scientific books. The ceremony then wound its way between two wings of onlookers to the Campo dei Fiori, where the burning of those books, that portrait, and De Dominis's exhumed body was the horrifying conclusion and principal attraction. It is permissible to think that the solemn and dramatic funeral ceremony would have pleased the condemned man, who while alive had so loved and sought out spectacular ceremonies.

For its contemporaries, however, the conclusion of the De Dominis affair was a dark and unexpected lightning stroke on the horizon of Urban VIII's new pontificate. Its dark light illumined Rome's other face: the face of the institutions loyal to a resolute, intransigent Catholic restoration and to the Bellarminian ideal of a pontifical monarchy. In man's memory, a comminatory sentence *praesente cadavere* and executed *post mortem* could not be recalled.

There was, true enough, the case of Pietro d'Abano, the spiritualist and astrologer who had died in 1315; he was exhumed and burnt—whatever was left of him—forty years later, as the result of a new trial of his writings. But those were the distant days of medieval Christianity. In more recent times, there had been another famous case of a *post mortem* trial and sentence, that of the heretic John Wyclif, who had died in 1384 but was condemned in memory by the ninth session of the Council of Constance, in 1415, for having affirmed that in the Eucharist the bread and wine could not be annulled or transubstantiated. He was an impenitent heretic, guilty of an exceptionally grave subversive heresy—hence, the Church's severity. It was also true that the sentence had been carried out later, in England. But in the nature of things it was only a symbolic execution: Wyclif's ashes were dug up and scattered in a tributary of the Avon. It was true, but it had happened a long time before and in England; not in the country of the "new world" and not in Rome. The Congregation of the Holy Office, by deciding to proceed with that incredible judicial pantomime, had meant, with a historic sentence, to give a show of severity and a warning.

Indeed, it was a historian, a great historian and a witness to the ceremony, who left us all the details of the De Dominis affair. Besides, he was at that moment the official historian of the Church and the only person in Rome with access to all the archives. He was Father Abraham Bzowski, a Polish Dominican who lived in Rome and was entrusted with carrying on the writing of the *Annals of the Roman Church* begun by Cardinal Baronio, who had died in 1606. In 1624, Bzowski was writing volume XVIII; but when he witnessed that great event, he decided it was too important to ignore and interrupted his narration to insert a piece of court reporting on the case that had so excited and impressed his contemporaries.[2]

Modern historians have been less sensitive than Father Bzowski.[3] The historians of Galileo, in particular, have remembered of De Dominis only his ideas on optics and on the tides. It was Delio Cantimori, the great historian of sixteenth-century Italian heretics, who pointed out the importance of the De Dominis case for understanding the historical and intellectual ambience of the Counter-Reformation crisis in which Galileo happened to work.

Unfortunately, on this point, Cantimori has left us only a passing reference in an essay, interrupted by his death, that developed no further than the notes published posthumously by Eugenio Garin.[4] In the fire of De Dominis's body, Delio Cantimori saw consumed the last political illusions about the pacific reunification of Christianity.[5] The ideal of Christian unification, completely unacceptable in the climate of the Thirty Years' War, had been officially advocated by De Dominis, and also unofficially admired by such a great utopian as Kepler. But this ideal was only the general political consequence of the ideas that brought De Dominis before his accusers. The roots of that position were theological, and since heresy had to be pulled up by its roots, it was these ideas that were punished. Father Bzowski confirms this; but, above all, it is confirmed for us by the special character of that trial's historic conclusion.

From this point of view, the De Dominis affair reveals much more than one might suspect about Roman methods and about what went on behind the scenes: the other face of the Barberini pontificate. What is more, the De Dominis affair allows us to know exactly, better than any abstract judicial

[2] See, besides Father Bzowski's account, Eudaemon-Johannes, "Epistola de relapsu, morte poenaque M. Antonii de Dominis," in G. Stenghel, *Libri duo de duobus apostatis sive duae praenenses* . . . , Ingolstadii 1627, pp. 556-74.

[3] See in the absence of an adequate biography of De Dominis, A. Tamaro, *La Venetie julienne et la Dalmatie*, Rome 1918-1919, III, pp. 266-70; G. Goodman, *Court of King James I*, vol. I, 1839; H. Newland, *The Life and Contemporaneous Church History of A. De Dominis*, Oxford and London 1859.

[4] D. Cantimori, "Galileo e la crisi della Controriforma," post., edited by E. Garin in *Saggi su Galileo Galilei*, Florence 1967; reissued in D. Cantimori, *Storici e storia*, Turin 1971, pp. 657-74.

[5] D. Cantimori, "Avventuriero irenico," in *Prospettive di storia ereticale italiana del Cinquecento*, Bari 1960, pp. 97-110; id. "L'utopia ecclesiologica di M. A. de Dominis," in *Convegno di storia della Chiesa in Italia. Problemi di vita religiosa in Italia nel Cinquecento*, Padua 1960, pp. 103-123.

treatise, how an important trial—against an important defendant, who had actually compromised the summits of ecclesiastical power—was prepared. In short, this is a scandal exactly like the scandal that will have Galileo as its protagonist only eight years after the sentencing of De Dominis.

Two Abjurations, Three Condemnations

De Dominis had also been a scientist. He came from the Dalmatian dynasty at Spalato, which had given the Church a pope, and he had become a Jesuit and professor of mathematics, rhetoric, and philosophy at Padua and Brescia. His lessons in mathematics, published later in 1611, had attracted the interest of Paduan intellectual circles, although Galileo never declared his opinion on De Dominis's geometric optics.[6] De Dominis was friendly with Paolo Sarpi, Fulgenzio Micanzio, Giovanni Francesco Sagredo, and Traiano Boccalini before and after having removed his Jesuit habit to assume that of the Archbishop of Spalato. When the conflict between Venice and Rome, between Paolo Sarpi and Cardinal Bellarmino, began, De Dominis took the side of the interdicted man and returned to Venice to join the prelates faithful to the republic.

In Venice, De Dominis had entered the orbit of theological and political sedition of the British embassy and—as before him Bruno, Pucci, and Vanini—obtained the credentials for a secret passage to England. In exchange, De Dominis promised potent work against the pope's and the council's secular and religious power.

The English, however, were much more interested in getting their hands on something else: the manuscript of the *History of the Council of Trent*. In 1616, Paolo Sarpi had by now completed it, but he had not yet decided to send it abroad for publication. The principal object of the secret negotiations between De Dominis and Sir Henry Wotton (the English ambassador on the Venetian lagoon), through the mediation of His Majesty's chaplain William Bedell (the embassy's religious attaché), was the offer made by De Dominis to obtain from Sarpi and personally hand over in London a copy of the precious manuscript—which the Archbishop of Canterbury and King James were looking foward to as the most explosive book of the century. Paolo Sarpi's distrust of the ambitious archbishop was perhaps what defeated the plan.

With an English safe conduct in his pocket, De Dominis left Italy secretly through the Engadine and passed through the Spanish forts on the Rhine all

[6] M. A. de Dominis, *De radiis visus et lucis*, Venetiis 1611, to which was appended a theory (incomplete) of the optical functioning of the telescope on the basis of refraction. Galileo obtained copy of the book from G. F. Sagredo during the summer of 1612; see Sagredo's letter to Galileo, July 7, 1612, *Works*, XI, p. 355 ff., especially p. 356.

the way to Heidelberg, where he revealed in an open letter to the Catholic bishops the reason for his flight, which was inspired by the project of reunifying Christianity.[7] On December 3, 1617, in the Cathedral of St. Paul, and dressed in his archbishop's robes, De Dominis solemnly professed Calvinism: it was a kind of parody of the rites of abjuration performed in Rome.

Welcomed with great favor by James I, and by Abbot, the Archbishop of Canterbury, De Dominis obtained the prestigious and remunerative title of Deacon of Windsor and, by means of a series of propaganda sermons against the abuses of the Roman court, became the spokesman for all Italian exiles. His political success was sealed by the immediate publication of the manuscript he had brought from Venice, the *De republica ecclesiastica*, whose first parts appeared in London dedicated to James I. It was an interesting, though not too original, polemical work against Cardinal Bellarmino's theory of the ecclesiastical state and against the degeneration of the Roman regime. In Rome, the book was immediately put on the Index. The author, not having reported to the Holy Office to answer the accusation of heresy, was automatically excommunicated and deprived of ecclesiastical dignities by Pope Paul V. *De republica* caused a great fuss and greatly pleased, among others, Kepler.[8] But the most important parts—of a theological character and not merely political polemic—would not see the light until some years later. In the meantime, however, De Dominis had won immortal fame from a much more important book, with a great publishing success of which he was the protagonist.

In 1618, Paolo Sarpi had finally agreed to hand over his book to the English. A chain of diplomatic agents and Dutch booksellers carried the manuscript of the *Istoria del Concilio Tridentino* [History of the Council of Trent] across Europe and into the hands of the Archbishop Abbot. De Dominis saw the edition through the press, checking galleys and writing a long preface. In the summer of 1619, Sarpi's book was in the bookstores of Protestant Europe, where it sold like the Bible. For the Jesuits, it was a black beast; by November 22, it was already on the Index.

De Dominis had associated his name with one of the biggest publishing successes in the seventeenth century: the *History* had two editions in Italian, four in Latin, six in English, one in French, and another in German. The best publicity, which benefited the book for a long time, was the relentless persecution and denunciation that the Jesuits accorded it. Unquestionably, that book was dangerous for those who, like the Jesuits, had the task of defending the religious value of the decisions of the Council of Trent as a bulwark

[7] M. A. de Dominis, *Causae profectionis suae ex Italia*, n.p. 1616 (2nd ed., Amsterdam 1617).

[8] Id., *De Republica Ecclesiastica libri X*, 3 vols. Londini 1617-1622. On Kepler's evaluation of De Dominis's work, see J. Kepler, *Gesammelte Werke*, vol. VII, *Briefe* 1612-1620, pp. 283 (1618), 342, and 347.

of faith erected against heresy. The *History of the Council of Trent* robbed the council of its character of truthfulness: it recounted all the controversies, discussions, struggles, and internal and external dissensions which, until that moment, had been concealed in order to make the council's texts appear to the world as a law of truth and a mystery of religion.

Also, the quality of its truth concerning faith in regard to one of the Council of Trent's most important doctrinal decisions, the dogma of Eucharistic transubstantiation, was seriously impaired by revelation of the reservations and opportunistic motives that formed the background for the famous thirteenth session.

That dogma was in fact the crucial doctrinal point that rendered irreconcilable the schism with the Protestant world, as one diplomatic exponent had complained at the council. Sarpi told the story of the delicate behind-the-scenes politics in that decision and, above all, the insoluble and century-old dispute on the mode of real existence that set the Dominican group at the council in opposition to theology of a Franciscan stamp. Sarpi also recounted how, when confronted with disputes that would be incomprehensible to the layman, the council fathers had been forced to move cautiously—"not knowing themselves how to express their very own ideas"[9]—and ended up with the difficult compromise solution represented by the dogma's text. The edition of Sarpi's great book gave rise to thought. It must also have opened the mind of its editor to more profound and radical reflections of a theological nature, which went beyond the initial considerations of a Gallic political type. In the fifth and seventh volumes of his *De republica ecclesiastica*, De Dominis actually stepped onto the terrain of a high-level theological controversy, denouncing the council's authority when they claimed to assert *de fide* dogmas contrary to reason and absent from the Bible.

This polemic was the real theological root of De Dominis's so-called irenism; that is, his program of Christian reunification, or reunification on the basis of the common "fundamental" principles of faith. In the disputed teachings of Saint Irenaeus, a father of the Church, De Dominis had found his criterion: adherence solely to Scriptural revelation as a necessary and sufficient rule of faith. If the Catholic Church, De Dominis said, had maintained the Eucharistic mystery as it had been handed down by the devout spirituality of the fathers—an indefinable mystery—then "the major controversies and all the germs of reciprocal polemical intolerance would have been eliminated."[10]

The Eucharistic dogma, contrary to reason, absent from revelation—in the terms in which it had been imposed by Scholastic theology—was the fundamental obstacle to religious peace while, on the European continent,

[9] See P. Sarpi, *Istoria del Concilio Tridentino*, bk. IV, chap. II, Florence 1966, vol. I, p. 411.

[10] De Dominis, *De Republica Ecclesiastica*, II, p. 80.

the Thirty Years' War was going on. De Dominis, contesting the arbitrary decision to explain a sacramental mystery with the theory of transubstantiation and the permanence of accidents, appealed to the theory of Irenaeus: there are two realities in the Eucharist, an earthly and a celestial one.[11] That is the entire miracle, which does not go counter to common sense and natural philosophy, but leaves to bread and wine their values as natural substances. By saying this, De Dominis was repeating both Wyclif's thesis, condemned by another council, and the innumerable contentions of Protestant theologians from the first decades of the century against Bellarmino and Cardinal Du Perron.

Bellarmino and Du Perron had defended the Tridentine dogma of the Eucharist with the flawless argument of what is today called pragmatic linguistics: exegesis of the unequivocal meaning of the evangelical formula of the Consecration, "Hoc est corpus meum." The pronoun *hoc*, pronounced with respect to bread, and at a short distance from it, "signified" that the bread was no longer bread. To pronounce it meant to change the nature of things.

De Dominis, strong with the authority of Irenaeus and Justinian, rejected this interpretation. In short, his argument was this: the Consecration is an act of sanctification of the bread and wine by divine invocation. The crucial moment of the Consecration is the Benediction, not the succeeding formula "Hoc est corpus meum." If the *vis consacrandi* was in the act and not in the word, then the semantic proof of the truth of transubstantiation put forward by Bellarmino no longer had any demonstrative value, and the consecrated bread could remain bread. In point of fact, De Dominis said that it was sufficient to eliminate a pronoun from Catholic Eucharistic theology to open the way to Europe's religious reconciliation.

Like all intellectuals of his time, Marco Antonio de Dominis knew that the Eucharistic dogma was the focal point of great religious tension. Besides, it was not difficult to realize this: the liturgies, the theological controversies, even the artistic forms and architecture of the churches, everything stood in opposition around a center of tensions between intellect and faith. And that point had the physical dimensions of a Eucharistic particle. For one group, this was a fundamental basis of the liturgy as the celebration of a Eucharistic supper; for the other, it was the foundation of an opposed liturgy, centered on the value of the sacrifice of the altar. What for Catholics was the principal object of worship, celebrated solemnly by a fundamental feast day, Corpus Domini, venerated with the paroxysm of Baroque apotheosis and Eucharistic adoration, in the eyes of the Protestants was the source of an idolatrous cult, the tangible proof of the degeneration of Christian spirituality in Rome.

For an intellectual as restless and opportunistic as De Dominis, moreover,

[11] Ibid., p. 70 ff.

the political and juridical problems did not preclude taking a position on such a fundamental doctrinal point, which was felt to be dramatically lacerating. The conciliatory solution that he proposed consisted in reducing its dramatic quality, presenting the idea of transubstantiation sanctioned by the Tridentine Eucharistic dogma as not a fundamental principle of faith. The tendency was to reduce the role of dogmatic theology to its essentials, a tendency common to many philosophers and one that the institutions set up to safeguard the Council of Trent could not tolerate.

For De Dominis, as for many other errant heretics of his time, religious faith was a search interwoven inextricably with utopian expectations and with motives of opportunism—a constant flight from constituted religions and forms of hypocritical simulation. It was a journey without respite and an ever transitory compromise.

His *De republica* had barely been published when De Dominis was already thinking of a new defection, another *coup de théâtre* with himself in the role of protagonist. The political and religious expectations that he had cultivated, in the shadow of the most powerful Protestant monarch in Europe, had been disappointed. The hoped-for creation of an anti-Hapsburgian bloc among England, Savoy, the United Provinces of Holland, and Venice was caught off-balance by the victorious initiative of the Catholic emperor and Spain. The triumph of Catholicism isolated London and made it advisable to open a diplomatic dialogue with Rome.

It was the beginning of 1621, and De Dominis gave signs that he was having second thoughts. "Transubstantiation is an error for Philosophy but not in Theology,"[12] he wrote to the Deacon of Winchester, who implored him not to think of returning to Rome. But De Dominis had already made secret agreements with the Count of Gondomar, Spanish ambassador extraordinary at London. The Spanish ambassador also tried to dissuade him. "I go spontaneously, without any certainty or security, seeing in advance the danger to which I expose myself, disposed to meet it with an intrepid spirit, trusting only in divine protection,"[13] the incorrigible fugitive declared with these inspired words.

For, in 1621, both Pope Paul V and Cardinal Bellarmino were dead, and Gregory XV, an acquaintance of De Dominis, made new ambitions possible. At the beginning of 1622, King James received De Dominis's official request to go to Rome, still with the aim of Christian reconciliation.

Attempts were made to detain him, but in vain. "I believe that deception is the characteristic of this perfidious man raised in a Jesuit nest,"[14] an em-

[12] See R. Neyle, *Aliter Ecebolius M. A. De Dominis Pluribus Dominis inservire edoctus*, Londini 1624, p. 21.

[13] See De Dominis's letter to the Count of Gondomar, February 9, 1622, in M. A. de Dominis, *De pace religiosa Epistola*, Versuntione Sequanorum 1666, p. 16.

[14] Neyle, *Aliter Ecebolius*, p. 64 ff.

issary of the king remarked bitterly. From Venice came preoccupying echoes of the news of a new betrayal: the old friends Paolo Sarpi and Micanzio, indignant, were convinced that the price of impunity at Rome would be the secret revelation of what went on behind the scenes of the publication of the *History of the Council of Trent*. Indeed, even if the fear had been unjustified, De Dominis's defection represented a threat to relations between Venetian theologians and London.

Compromising letters and documents in De Dominis's hands were seized. Expelled as an undesirable element from England, De Dominis waited in Brussels for the necessary guarantees from Rome. He obtained the pope's pardon in exchange for a solemn abjuration.[15]

In Rome on November 24, 1622, De Dominis formally renounced and detested his heresies, rejected Sarpi's *History*, subscribed to the authority of the councils, and denounced his ideas about some dogmas being less fundamental than others. The abjuration had been spontaneous; thus, instead of being disgraceful, it opened a triumphal return to center stage for the Archbishop of Spalato. Restored to his dignity and to the benefits of his rank, De Dominis saw the doors of high Roman society open before him.

Cardinals and aristocrats competed over the former apostate and friend of King James and of his Lord Chancellor; and in the Barberinian pontificate's new climate of fashionable intellectuality, De Dominis became an associate even of the Cardinal-nephew Francesco Barberini. In fact, De Dominis, who had resumed his old scientific studies, dedicated a book to him on the tides, *Euripus seu de fluxu et refluxu maris*, published in Rome at the same time as *The Assayer*. The theological opinion for the imprimatur was signed by a mathematician we already know: Father Orazio Grassi.[16]

Having crossed the English Channel twice, De Dominis knew better than anyone in Rome what a real tide was. His explanation of the phenomenon was of the traditional type, on the basis of the "vis tractiva" [attractive power] of the sun and moon.

The Archbishop of Spalato's relations with important personages, his social success, preoccupied those who had not forgotten and who did not forgive him for the publication of Paolo Sarpi's book. His adversaries knew the ability at deception of that theological adventurer, that "servant of so many masters," as they had now contemptuously nicknamed him in England.

De Dominis spoke a bit too freely of his experiences and his convictions, which had not changed. Slander and delation, as always in such cases, ruined him: he was denounced to the Holy Office for having expressed dangerous opinions on the councils. He was protected in high places, however, and the

[15] See M. A. de Dominis, *Sui reditus ex Anglia Consilium exponit*, Romae 1623; on p. xix is the text of the abjuration pronounced on November 24, 1622.

[16] See M. A. de Dominis, *Euripus seu de fluxu et refluxu maris sententia*, Romae 1624 (the dedication is dated October 1623).

Holy Office moved with obligatory caution in such a circumstance, so as not to involve protectors of the suspected person in the scandal of a denunciation that might discredit their vigilance at the head of the Church.

A secret investigation confirmed the accusations. An incrimination for false oath was suggested. Unofficially, however, since it was not politic to circulate the rumor that De Dominis had tricked two pontiffs with heretical deception, a trial was based on a circumstantial and not too compromising juridical pretext. Given the accused's well-known propensity for running away, he was imprisoned not in the cells of the Holy Office's palace, but rather, as was fitting for an archbishop, at Castel Sant'Angelo, in a quarter where he could be attended by his own servants.

The denunciations hurled at De Dominis did make a trial necessary, but to find an incontestable ground for indictment that would not involve anyone but the accused was none too easy. For the simple reason that De Dominis had recently recanted, nothing he might have written previously could any longer be brought against him.

On this problem of incrimination, Father Bzowski said with great modesty that De Dominis's "writings had been rapidly examined and in them was found a proposition, on the sacrament of marriage, that contained various heretical statements."[17] This detail is confirmed by another direct source, the Jesuit Father Eudaemon-Johannes, who specifies that what had been found was a manuscript "in his hand" on the possibility of dissolving a consummated marriage, a manuscript that the accused recognized as having been written by him.[18] Archbishop De Dominis, known throughout Europe for his heresies on the councils, the pope, and the Eucharist, was thus officially incriminated for a manuscript on the lawfulness of divorce in the case of adultery.

In truth, the Commissary Cardinal Scaglia, after an interrogation "that kept after him for ten hours,"[19] wrung from the archbishop a confession to the actual imputation. De Dominis admitted that he still believed in the possibility that Catholics and Protestants could unite on the basis of common fundamental articles of Christian faith, the only ones necessary for salvation—the Trinity, the Incarnation, the Divinity, the Passion, and the Resurrection—while other articles of faith such as transubstantiation, predestination, and the primacy of St. Peter could still be discussed. "The Councils," he added, "and above all the Council of Trent, declare that many things are *de fide* which have nothing to do with the Faith."[20] The rejection of the principle of the councils' authority; the subversive incitement to re-

[17] Bzowski, *Annalium Ecclesiasticorum*, p. 172.

[18] See Eudaemon-Johannes, "Epistola de relapsu," p. 557.

[19] This detail was pointed out by the physician Johannes Faber, a member of the College of Prosecution Experts, in his letter to Galileo on September 14, 1624, *Works*, XIII, p. 207.

[20] Bzowski, *Annalium Ecclesiasticorum*, p. 172.

duce the number of essential dogmas through a liberal examination—these were the real accusations. At that point, only an exemplary condemnation could erase every trace of scandal.

One had to prevent, with a sentence of memorable severity, what "this innovator wanted; that is, the possibility of opening up, by means of a new examination, a study of matters not judged until now as well as of those regarded as established."[21] Father Bzowski, the historian, has thus left us his authoritative comment on the significance of the De Dominis affair. It was, in other words, a solemn condemnation of any criticism of the principle of theological authority. After the confession, the hearing of further proofs was not necessary. De Dominis was a pertinacious heretic. Yet, dissimulator as he was, Cardinal Scaglia, out of scruple, interrogated him again to find out whether the year before he had pretended to recant. He replied that he had not, that he was ready to recant again. And that is just what he did a few months later, in September 1624 on the point of death, before his inquisitor.

Was it the final feat of a virtuoso in dissimulation, or was it rather the feat of his jailors, pretending an official repentance at the last gasp? In either case, De Dominis had a death worthy of him and his aristocratic aloofness. He received the sacraments and allowed himself, or was credited with, a touch of final grandeur, sending a message of gratitude to Pope Urban VIII for the spiritual merits acquired during his retreat in Castel Sant'Angelo's prison.[22]

The Holy Office found itself with a dead man on its hands. The suspicion that De Dominis had been poisoned in prison could provoke international speculation around this delicate case. An autopsy was ordered and entrusted to an international group of experts: namely, the two illustrious doctors Giulio Mancini and (the Lyncean) Johannes Faber, summoned at dawn by functionaries of the Holy Office. Thus, despite the death of the accused, Faber received certain privileged information on the preparation for the trial and on the course the trial would have taken. One of the first to be informed, in minute detail, was Galileo.[23]

Tests of Pride

Galileo too, like Paolo Sarpi's ex-Jesuit friend, had defied authority by attacking the authoritativeness of the most important cultural institution of the Counter-Reformation Church: the Society of Jesus' Collegio Romano. And Galileo had also come to Rome benefiting from the reprehensible offi-

[21] Ibid.

[22] This detail is confirmed by Eudaemon-Johannes, "Epistola de relapsu," p. 560 ff., and noted again in the Protestant sources of the De Dominis affair. See P. van Limborch, *The History of Inquisition*, trans. by Samuel Chandler, London 1731, vol. II, pp. 282-87, especially p. 204.

[23] Ibid.

cial support given him by the highest authorities in the Church: the pope, his secretary of state, the cardinals, and the most influential men of the Curia—all forgetful of the serious theological charges hanging over the author of *The Assayer*.

Galileo had also benefited in Rome from frivolous demonstrations of high society and intellectual support, which had stretched official tolerance to the borders of mocking arrogance and subversive license. So now, under the indulgent protection of the pontifical political entourage as well as of innovative and libertine cultivated opinion, people were talking about freedom of examination in philosophy. The pleasure in provocation and the impertinent polemic of the innovative and licentious literati; the bold spiritual pretensions of aristocrats and high churchmen imbued with superficial philosophical and scientific curiosity—these betrayed, by their intellectual intolerance of institutions, the diffusion of an irreverent annoyance in the face of authority and tradition by the higher layers of Roman society, by refined, cultivated, and fashionable people. Whether an impatience with rules and dogmas or a reforming religious afflatus, both these aspects of tendencies among the Roman innovators were suspect in the eyes of the custodians of dogmatic orthodoxy.

The provocations, then, continued licentiously, tolerated by the highest authorities, amid jibes at Aristotelian education and petulant insults from the innovative and Galilean literati who knew nothing about science. These derisive provocations were like innocuous but insulting stones flung at the façade of the Collegio Romano's colossal and austere edifice (Fig. 3), the pride of the Jesuits. "The pride of the Jesuits," as Johannes Faber hoped, had been seriously vilified.

The Jesuits had much of which to be proud. They were the protagonists of their century. They were the intellectual and religious elite, molded by vigilance and prestigious expansion in the realm of politics and culture. The capillary organization of their pedagogy and their research covered Europe and was active on a world scale. The new humanists were Jesuits: new moralists, new scientists, new theologians, new philosophers. No one could give them lessons in being up-to-date, in having a desire for the new, and in valuing an openness to the world. With their down-to-earth religion, anti-intellectual and sentimental, with their fabulous missionary travels, their philological and historical inquiries, their philosophy founded on the concrete, their unquestionable, broad-minded competence in astronomy, their sensitivity to the most advanced aesthetic studies, the Jesuits certainly did not have to wait for Campanella, Galileo, or Cavalier Marino to know that the world must be discovered through the senses or that mathematics provides a rigorous knowledge of the facts.

Indeed, the Jesuits, with their will to homogeneous and disciplined action, possessed a further certainty with respect to the impertinent "worshippers of novelty." The Jesuits knew that considerations of a religious character must come before those of a narrowly philosophical, scientific, or artistic nature. Their religious motto was engraved in the Host radiating light, which was their symbol (Fig. 13); but their intellectual motto, their universal, cultural program, was inscribed at the beginning of their most authoritative manual of philosophy, Father Suárez's *Metaphysicarum disputationum*: "Our philosophy must be Christian and the handmaiden of Divine Theology. Metaphysical principles must be set forth and adapted in such a way as to confirm theological truths."[24]

It was a certainty without any shadows, at once ancient and very modern: a second Scholasticism. With the enthusiasm and firmness of this establishing will at the service of the Tridentine renovation of the Church, the culture of the Jesuits, welded to the principles of Catholic reform, was immune as an institution to the stealthy, tireless efforts at disruption and individualization on the part of those heretics and innovators who claimed, in the name of a new Scriptural faith, to be beyond the authority of doctrinal tradition.

The irresistible ascent of the Society of Jesus was the effect of an internal energy, which left behind (even to the point of humiliating them) religious orders both old and new. Benefiting from the Spanish influence on Rome, the Society's ascent culminated during the pontificate of Paul V, and this had profound repercussions on the next pontificate, that of Gregory XV. For it was in 1622 that the pride of the Jesuits triumphed with the magnificent celebration of two canonizations—those of St. Ignatius and St. Francis Xavier—which sanctioned the enormous importance the Society of Jesus then had in Rome, as it manned the helm of Catholic culture in that climate of Counter-Reformation crisis caused by the Thirty Years' War. For Galileo's contemporaries, the modern edifice of the Collegio Romano—where the *ratio studiorum* had been invented, with Bellarmino's decisive contribution—signified all of this.

The spirit we have described, that impassioned and total integration of Tridentine faith and culture, was not only contained in abstract programmatic statements, but was tangible, visible, available to all the senses, according to the religious mentality of its proponents, precisely in the structure of that colossal palace (Fig. 2). In fact, in order to give better material form to that vision of ideas and man, the Society of Jesus had decided during those years to enlarge the Collegio Romano, demolishing part of its original structure and adding a grandiose church dedicated to St. Ignatius. The church would be incorporated in the main edifice without any separation, in such a way as to create both an intellectual and a religious area, a single monumen-

[24] F. Suárez *Metaphysicarum disputationum in quibus et universa naturalis theologia ordinate traditur* . . . , Salamanticae 1597 (six editions until 1630).

tal space that could be used for cultural, research, and liturgical purposes. It was the architectonic materialization of a vision of the world and the celebration of a prestigious supremacy.

This grandiose and very expensive project was the Society of Jesus' largest and most spectacular church. It was a demonstration of pride and strength, at a point of success on the frontier of the Church and the world—and yet, right now, at a moment of isolation in Rome. Hence, considerations of prestige prevailed over all others.

The grandiose church was financed by Cardinal Ludovico Ludovisi, a leader of the pro-Spanish party who had accumulated enormous wealth as the cardinal-nephew of Gregory XV. The church would celebrate with an equal display of magnificence both the new saint and founder of the Jesuits and the Ludovisis who had canonized him. In 1626, with a ceremony of great solemnity, Cardinal Ludovisi blessed the first stone. But half a century would have to pass—amid difficulties, polemics, and second thoughts, as we shall see—before it took definitive form; and, in the end, the Church of St. Ignatius will remain forever unfinished, mutilated when compared to the initial pharaonic plans.[25]

Between 1623 and 1624, several designs had been submitted by fashionable architects. But the great realization must speak to the world of the excellence of the Jesuits, and must all of it be the work of the Jesuit fathers. Cardinal Ludovisi and the authorities at the Gesú chose a representative and authoritative exponent of the Collegio Romano, who gave proof of his scientific and religious value by making his debut in architecture with that astounding work. This was Father Orazio Grassi, the mathematician publicly satirized in *The Assayer*.

Tridentine Perspectives

It was a matter of building the largest and most magnificent church in Rome, inferior only to St. Peter's, with a cupola (which Father Grassi had designed) so vast as to approach that of Michelangelo's and to eclipse that which the new regime had commissioned Maderno to build for Sant'Andrea della Valle, the church representing the Barberinis (Fig. 4). Besides, the me-

[25] See R. Soprani, *Vite de' pittori, scultori e architetti genovesi*, 2nd ed., with notes by G. G. Ratti, Genoa 1768-1797, vol. II, pp. 9-11; L. Grillo, *Elogio di Liguri illustri*, Genoa 1846, vol. II, pp. 179-83. But, above all, see C. Bricarelli, "Il padre Orazio Grassi architetto della chiesa di Sant'Ignazio in Roma," *Civiltà Catholica* II (1922), pp. 13-25; id., "La chiesa di Sant'Ignazio e il suo architetto padre Orazio Grassi," in *L'Università gregoriana del Collegio romano nel primo secolo della restituzione*, Rome 1924, pp. 77-100. On the plans drawn by Father Grassi for the Church of St. Ignatius (Fig. 4) and to transform the church of the College of Siena (Grassi mission to Siena, 1626), see J. Vallery-Radot, *Le recueil de plans d'édifices de la Compagnie de Jésus . . . et de l'inventaire des plans des archives romaines de la Compagnie par Edmond Lamalle*, Paris 1960, p. 403 ff. and p. 16 (nn. 71 and 72 and Fig. 11). On the plans of Father Grassi for the College of Genoa and at the Church of Gesú in Genoa, ibid., pp. 101 and 103.

galomaniac cupola planned by Father Grassi would have blocked off the sunlight to the Dominican Library at nearby Minerva College. The Dominicans, already so overshadowed by their ambitious neighbors, succeeded in having their complaints against this act of architectural abuse heard.

Such were the requisites of prestige. The other requisites, those of an architectonic and aesthetic nature, were inspired by apologetic and religious considerations. In fact, the Church of St. Ignatius must exhibit, in an exemplary manner, the style and perspective of the great institution of education and research to which it was annexed as a grandiose chapel. Through the triumph of St. Ignatius, the work must testify to the vocation of Tridentine orthodoxy and apologetics, in every corner of the earth and in every sphere of human culture, and therefore also in architecture. Begun in 1629, Father Grassi's project gave perfect material form to this complex ensemble of requisites, and made of Father Grassi, until the end of the 1620s, one of the most representative and influential intellectual personalities—perhaps the most representative—of the Society of Jesus in Italy. This man was Sarsi. From a scientific, institutional, and religious standpoint, he was the most important opponent that Galileo ever encountered on his path, after the death of Bellarmino. Sarsi, Father Grassi's cover name, had been called a scorpion and a snake in *The Assayer*. In some succeeding personal notes, Galileo will describe his adversary with less poisonous, but certainly not more flattering epithets: "most foolish," "the perfect beast," "a complete dunce," and "a big buffalo."[26] But Galileo, so far as we know, never met face-to-face his adversary, of whom he knew only the Sarsi mask and some indirect opinions and images.

Who was Father Orazio Grassi, Galileo's most authoritative adversary? He is a man without a face for the historian, as well: not a single portrait has survived of the celebrated architect of St. Ignatius, the church in whose cellars—perhaps—he was buried, together with other more or less famous professors of the Collegio Romano.

Like Galileo, the historians know and remember of Father Grassi only the Sarsi mask and his polemics about the comets. Of the religious and intellectual personality of this adversary of Galileo there are only a few modest vestiges: some scientific letters to Baliani;[27] some elegant architectural

[26] See the footnote to the Galilean copy of the *Ratio*, in *Works*, VI.

[27] *Otto* [nove] *letere del Padre Orazio Grassi a Giovambattista Baliani* (1646-1653), ms. conserved among the Baliani papers at the National Library of Brera (Milan), ms. AF XIII, 13, n. 4. Cf. Fig. 8. These letters have been published (in a not always reliable French translation) in S. Moscovici, *L'expérience du mouvement. Jean-Baptiste Baliani disciple et critique de Galilée*, Paris 1967, and are described and discussed in C. Costantini, *Baliani e i gesuiti*, Florence 1969. Another letter of Father Grassi's is that of September 22, 1633 (State Archive, Rome) to Don Girolamo Bardi, ex-Jesuit and lecturer in philosophy at Pisa, published in A. Favaro, *Galileo Galilei e il P. Orazio Grassi. Nuovi studi galileani*, Venice 1891, pp. 203-220, especially p. 219.

drawings accompanied by his flowing calligraphy;[28] a thesis on optics published under the name of a student in the Collegio;[29] a solemn Easter sermon;[30] and the fame of a very successful sacred representation staged by him, *Apotheosis or Consecration of Saints Ignatius and Francesco Saverio*.[31] Yet these vestiges are more than enough to tell us that we have before us a rich, very varied, and complex intellectual personality. He was an official Jesuit scientist, a mathematician, an astronomer, and an architect—a great architect: this was his "profession." And yet, since the professionalism and scientific institutions of our day insist on distinguishing the figure of the architect from the figure of the scientist, historians of science have acted accordingly, never deigning to cast even the most cursory glance at Father Grassi's scientific work as an architect. Indeed, they have not even deigned to mention it. This omission is unpardonable for the history of science; for, when one is dealing with the seventeenth century, architecture is an integral part of science: "geometry, arithmetic, music, physics, medicine, architecture, and all the sciences that are subjected to experience and reasoning,"[32] as Pascal wrote in his famous *Preface to the Treatise on the Void*.

Father Orazio Grassi's most durable and important scientific work by far was his great church. It set the seal on his career and scientific fame, and in it he expressed himself freely, revealing the best of himself. Anyone who wishes to understand the personality and the qualities of Galileo's most authoritative scientific and institutional opponent must pay a visit to the Church of St. Ignatius at the Collegio Romano. And yet, when we enter St. Ignatius today, we are overwhelmed by a deceptive impression which can make us lose contact with Father Grassi's architectural project, as it was and as it should have been realized. The decoration has transfigured it. In fact, everything is illusory. The impression one gets is that of having entered an immense movie house or, as one might have put it in the seventeenth century, a "theater of marvels": a pavilion of optical illusions, of tricks, distorted

[28] Historical Archive of the Superior General Curia of the Society of Jesus (Rome), *Fondo gesuitico* 1245, 1/4; this is the memorandum (October 5, 1650) wit the title *Ricordi per la fabbrica della chiesa di Sant'Ignazio* (see Vallery-Radot, *Le recueil*, p. 421 ff.).

[29] See *Disputatio optica de iride proposita in Collegio Romano a Galeatio Mariscotto*, Romae 1618.

[30] *Divini Templi excisio, oratio habita in Vaticano sacello ad S.D.N. Urbanum VIII ispo parasceve die ab Horatio Grassio Savonensisi e Societate Iesu*, Romae 1631 (with dedication to Cardinal L. Ludovisi); republished in *Orationes quinquaginta de Christi Domini morte*, Romae 1641, p. 596 ff.

[31] See G. Nappi, *Annali del Seminario romano*, pt. III (ms. 2801 of the Archive of the Pontifical Gregorian University), folio published in C. Bricarelli, "Il padre Orazio Grassi," p. 21. On the festivals for the canonization of St. Ignatius, see Pastor, *Storia dei papi*, XIII, p. 94 ff. To this period, too, belongs a funeral eulogy written by Father Grassi as prefect of the Collegio Romano: "Relatione della beata morte del nostro fratello Giovanni Berchmans" (1621), published in French translation in P. Terwecoren, *Collection de précis historiques*, vol. 17, Brussels 1862, pp. 17-25, 36-44, 56-67. See Sommervogel, *Bibliothèque de la Compagnie de Jésus*, vol. III, cols. 1684-1686.

[32] B. Pascal, "Préface du Traité du vide" (1651), in *Œuvres completes*, Paris 1963.

images, *camere oscure*, and illusionistic perspectives that sent the people of that time into ecstasies.

The large opening of the cupola is covered by a canvas screen on which is projected the fictional perspective of a cupola. The vertical structures of the nave are prolonged in a new illusionistic effect as they gape before the infinite profundity of the heavens, which barely manage to contain the aerial and universal apotheosis of St. Ignatius and the Society of Jesus. "If one is to believe one's eyes," one reads in an account of the church's inauguration, "no matter how much one may know about the techniques of illusionistic painting, one must be deceived."[33] What is most striking in St. Ignatius are the illusory appearances of the visible structures. At the same time, the grandiose solemnity of the main altarspace, beneath the great quasi-cupola, makes one feel the invisible presence of a sacred religiosity.

The spectacular decoration in St. Ignatius is probably the most moving and famous of all of the Jesuits' churches. There is no other example of a Catholic church in which the illusionistic decoration exalts in such a persuasive manner the special liturgical and apologetic exigencies of its architecture, even to the point of abolishing it. The church had been conceived in order to serve the most solemn liturgical aim: "to display on the main altar the worship and prayers of the Eucharistic Sacrament." From this point of view, the scenery created for the Collegio Romano's church (by the Jesuit Father Andrea Pozzo in 1685) was insuperable. The emotion of transparent religious reality created by that altar—solemnly framed by illusory architectonic appearances, deceptive to the senses—makes St. Ignatius into an allegorical representation of the Tridentine dogma of the Eucharist, a great spectacle of the fundament of the Jesuits' vocation and apologetics in the seventeenth-century world. To our eyes, as to those of its contemporaries, the fascinating invention of Father Pozzo's brush and techniques of perspective is a daring study, a bold creation, which turns a pragmatic consideration (the enclosure of the sky by an unrealizable cupola) into the occasion for a provocative aesthetic solution.

Today, as on the day of its inauguration, this play of perspectives, an overwhelming profusion of marbles and precious reflections in the sumptuous chapels, this entire scenic apparatus, "grips one's gaze and spirit with a marvelous voluptuousness and voluptuous admiration."[34] This emotion of the senses must not distract us. The sumptuous and sensual delirium of the decoration refers to a spiritual climate already different from that in which Father Grassi had planned the church. However, Andrea Pozzo's inebriating decoration cannot, with its projections on a screen, make us forget the gran-

[33] See "Ragguagio dell'aprimmento solenne della nuova chiesa di Sant'Ignazio" (1650), in Bricarelli, "Il padre Orazio Grassi," p. 24 ff.

[34] Ibid., p. 25.

diose architectonic structures it had so brilliantly enhanced and transfigured. Father Grassi did the façade, the ground plan, the colossal row of pilasters, the rigorous proportion of chapels and ceiling vaults, the static perfection of the central nave, and the daring mathematical feat of having a very broad vault without even a single chain to betray three centuries later the illusionism of the fake architecture that covers it.

One cannot doubt Father Grassi's talent as a mathematician after having looked at that ground plan (Fig. 2), those proportions, and the buttresses visible on the outside of his church. Yet, Father Grassi's qualities as an architect were not those of a creator, of an avant-garde artist like Pozzo. The sensibility of his patrons in 1623 was still close to the ideals of a rigorous Counter-Reformation, in the climate of a victorious religious war.

Father Grassi's project revealed a scrupulous and intransigent fidelity to those ideals. At stake—in architecture as in natural philosophy, theology, and literature—were apologetic preoccupations concerning Protestantism and the resurrection of an individualistic, atheist or mystical, Renaissance rationalism.

It is not originality that is the particular merit of the architect of St. Ignatius, but rather his exemplary conformity to the style of the Society of Jesus and his obsessive orthodoxy in terms of the prescriptions of the Council of Trent for ecclesiastical structures, prescriptions that St. Carlo Borromeo had meticulously codified.[35] The façade of the Collegio Romano's church repeated that of the Gesú, the church of the order's curia. The doors had architraves as in the Christian basilicas, not curves as in pagan construction. Also, the floor plan was the one prescribed by Counter-Reformation architecture: a Latin cross, to be preferred to the Greek cross adopted in the Renaissance. Moreover, the interior repeated the lesson of the Church of Gesú. If envy had not prevented it, a dizzying cupola would have adequately crowned an effort at monumental emphasis; in any event, this was already attained by the enormous dimensions of the hall and the height of the naves, in conformity with Roman religious solemnity as against the bare simplicity of Protestant temples (Fig. 4).

Father Orazio Grassi was made of the same stuff as the walls of the Collegio Romano's church. He was not the sort of intellectual and scientist who would permit himself personal creative whims. He was what today would be called a prestigious intellectual tied organically to his party. Without those qualities, it would have been impossible to fill uninterruptedly for so many years one of the most difficult chairs in the Collegio Romano, exposed to the polemics of competitors and adversaries and to the presumptuous

[35] See St. Carlo Borromeo, *Instructiones Fabricae et suppellectibilibus Ecclesiasticae*, Mediolani 1557; reprinted in P. Barocchi, *Trattati d'arte del Cinquecento tra Manierismo e Controriforma*, Bari 1960-1962, vol. III, pp. 1-113.

judgments of ignorant folk. And Father Grassi lavished those qualities without stint: a few months as rector and architect at the College of Siena, in 1626; and then again at Rome, as prefect of the library, professor of mathematics, and in charge of the new project.

Proving himself as prefect of the construction of the Church of St. Ignatius (that is, as the person responsible for the first plan in the key sector of Jesuit architecture), Father Grassi, who remained in the chair of mathematics at the Collegio Romano without interruption until 1632, became the exponent and prestigious authority of the cultural policy of the Society of Jesus in Italy. From 1627 on, he had the delicate task of censoring the scientific works published in Italy by his fellow priests. His position as an authority was by now recognized with the title of pro-rector; that is, consultant to the rector of the Collegio Romano. For a Jesuit and an astronomer, it was the apogee of a career.[36]

Courses and Recourses

"A time will come in which your pride and your ambition will no longer have limits,"[37] the founder of the Collegio Romano, Father General Francesco Borgia, warned the fathers of the Society of Jesus in a letter of 1569. Two centuries later, Diderot will cite that premonition, which had proved true with the collapse of the Jesuits' glory.[38] But in 1628, when Father Grassi was rector of the Collegio Romano, this institution was the pride of the Jesuits (Fig. 2).

Ignatius of Loyola had wanted a college that would be a European university. Now, owing to its endowment of research tools, the quality of its teaching, the number and social standing of its students, it was one of the most important universities in Europe (Fig. 3). About two thousand students from all over the world had at their disposal the teaching of grammar, rhetoric, ethics, Aristotelian dialectics, and Scripture. There were courses in Hebrew and Greek as well as the fundamental ones of the six chairs of theology (three of which were Scholastic, positivist casuistic, and antiheretical).[39] At the Collegio Romano, the *cursus philosophicus* by tradition offered a particularly prestigious and selective teaching of mathematics.

[36] In the absence of a scientific monograph on Father Orazio Grassi, the information presented here is based on the biographical index card of the Historical Archive of the Superior General Curia of the Society of Jesus, kindly granted to me by its compiler, Father Edmond Lamalle, to whom I express my gratitude.

[37] See the epistle of Francesco Borgia, April 1569, "De mediis conservanda spiritum Societatis et vocationis nostrae," in S. Francesco Borgia, *Monumenta historiae Societatis Iesu*, tome IV, Matriti 1911, p. 71.

[38] See the article "Jésuite," in the Diderot-D'Alembert *Encyclopédie*, vol. VIII, p. 515.

[39] In the absence of a history of the culture and pedagogy of the Collegio Romano, see the essential information in R. G. Villoslada, S.J., *Storia dele Collegio romano dal suo inizio alla soppressione della Com-*

The Collegio Romano was not afflicted by the inveterate flaws of bureaucratic mumification and professorial gerontocracy. The professors alternated periodically in their courses, and changes were constant. Father Grassi's more than ten years' occupancy of the chair in mathematics—comparable to that of Father Clavius (fourteen years) and Father Lugo in the chair of theology—was an exception owing to reasons of "clear fame" and protection of the prestige that external polemics must not in the least endanger. Solidarity and vigilance—together with discipline, devotion to the order and the common life, and research—turned this large group of natural philosophers, mathematicians, and astronomers into a particularly united and competitive subgroup within the Collegio Romano's broad cultural community.

The seventeenth century saw the birth of the first "scientific communities," as historians of science called them: groups of scientists who know each other by their respective scientific titles and who form a professional order. Many historians of science have in fact seen in the Royal Academies, both in France and England, the model of the scientific communities' first institutions.

The word "community" has been taken by historians from sociology, but it derives from juridical and religious language, where it indicates a social patrimony and, in the second instance, an ensemble of persons called to live in common. From this point of view, the Collegio Romano, well before the creation of scientific academies, gathered within its walls in a close communion of life and study, a collective with a high degree of homogeneity, educational background, and solidarity which could boast of prestigious institutional recognition. Jesuit scientists in Italy and the world did not form a monolithic bloc. They often gave rise to demands for a return to orthodoxy because of their theoretical stands. But at the Collegio Romano, the cultural showcase of the Society of Jesus, full official orthodoxy was obligatory; the *ratio studiorum* would be faithfully applied, even at the sacrifice of personal scientific ambitions.

So, the Collegio Romano's group was an exceptional scientific community but was also extremely *sui generis*, for to be a member demanded certain requisites and aspirations that transcended what we today call "pure research" and that, in the seventeenth century, would have been called the "attempt to investigate some truth in nature." The scientific vocation of the Society's official scientists in Rome involved another kind of vocation: to conciliate and apply scientific preoccupations to doctrinal and apologetic needs.

---

pagnia di Gesú, Rome 1954. Much richer for the historian are the precious manuscripts of the Archive of the Pontifical Gregorian University: Anonymous, *Origine del Collegio romano e suoi progressi* (ms. 143) and Gerolamo Nappi's history of the Roman seminary (ms. cit.); the latter copy I consulted had gaps. I thank the conservator of the archive, Father Monachino for having kindly allowed me to consult them.

The quality of scientific instruction given at the Collegio Romano was not in any case an invention of apologetic propaganda. Father Christopher Clavius's teachings had created an important mathematical tradition. When *The Assayer* was published in Rome, there was a generation of mathematicians who had been Clavius's direct or indirect pupils.

The mathematics courses ranged from astronomy (the *Sphaera* and, subsequently, the *Tabulae Rudolphinae*) to geometric optics, from Euclidean geometry to arithmetic, from architecture to geography and chronology. In the 1620s, besides Father Grassi—or during his brief absences from Rome—such courses were conducted by his teacher, Father Cristopher Grienberger (Clavius's direct pupil) and by the excellent astronomer and physicist Father Niccolò Zucchi. But, even without holding official teaching positions, the mathematician Father Leone Santi, Father Paolo Guldin (very close to Father Grassi), and Father Christopher Scheiner—a specialist in astronomical observation capable of competing with Galileo—participated in the solid mathematical nucleus of the Collegio Romano's scientific community.

Besides the official course of mathematics in the second year of the Faculty of Philosophy, like that of Aristotelian physics, the Collegio held public lectures and debates by mathematicians on current astronomical themes. The new cultural orientations developed around astronomical ideas, not only literary and historical ones. Now, precisely that great endowment of astronomical and mathematical studies which the Collegio had at its disposition ensured it a deserved reputation for intellectual worth and up-to-dateness that even its philosophical detractors did not question.

But the teaching of natural philosophy was not up to that of mathematics. Indeed, in terms of natural philosophy, the Collegio Romano could not boast of any superiority over the other Italian universities. As in those other universities, here too the teaching of natural philosophy—that is, physics, cosmology, and psychology—dug in behind a tradition of engaging in polemic with the forms of anti-Aristotelianism circulating outside the university or in certain heterodox faculties. The Collegio's religious nature made of that subject, more than any other, the bastion of the inextricable alliance between physics and metaphysics. This inevitably, in the eyes of critics, produced a conservative cultural image, since natural philosophy, as in every university institution of that period, was a privileged scientific subject of study, more representative and important than the domain of the Faculty of Mathematics.

During those years, there succeeded each other in that teaching post Father Vincenzo Aranea, Father Fabio Ambrogio Spinola (a Ligurian like Father Grassi), Father Antonio Casiglio, and Father Gottifredi, whom we met at Virginio Cesarini's funeral. These are all rather obscure names, as we can see, men who have not made much of a mark on Jesuit philosophy; they, too, are rhetoricians and literati, like their Roman critics, even if eloquence here was employed in the work of religious devotion. The texts of some of these professors' lectures, which have been preserved in the archives of the present Gregorian University—Collegio Romano's heir—permit us to know in a rather exhaustive manner what the teaching of natural philosophy in those courses was like.[40]

Teaching meant to read Aristotle and comment on him, as in every other university. This meant an inexhaustible commentary on Aristotle's logic and natural philosophy: the *First and Second Analytics*; but above all *On the Soul*, the *Physics*, *On Generation and Corruption*, the *Meteors*, and *On Heaven and Earth*, with the help of the manuals of Father Suárez, Father Toleto, and Father Pereira for physics.[41] Nevertheless, one can find, beneath the instruction in Scholastic style, a vigilant openness to problems of experimental research: above all, in magnetism and optical physics. The commentary on Father Nicola Cabeo's *Meteors*, published later, was from this point of view a revelatory example.[42]

The Collegio Romano's fundamental didactic function imposed certain official rituals on this great institution of Roman cultural life. These served to strengthen and regularly renew the privileged relations between the institution and the chief figures of the political, ecclesiastical, and aristocratic worlds, who were largely former students of the Collegio.

Public and solemn manifestations of great resonance were, for example, the ceremonies accompanying discussions of the theses of some high-ranking student. Then, in the Collegio's main hall, adorned with sumptuous tapestries, all the professors, many prelates and cardinals, the entire aristocracy, and sometimes even the pope himself participated. There were scientific celebrations in which the candidate's "defense," at times prolonged for two or three sessions, was interspersed by music and choruses.[43]

Less solemn and sumptuous manifestations, though equally important for

[40] Archive of the Pontifical Gregorian University: F. A. Spinola, *In libros de generatione et corruptione* (1626); id., *In Aristotelis Organum quaestiones* (Collegio Romano, the so-called mss. "della Soffitta" [of the Attic—Trans.], m. 520; *In octo Aristotelis libros physicos disputationes*, ivi, n. 1218; *Disputationes in duos libros Aristotelis de genertione et corruptione*, ivi, n. 1229; N. Zucchi, *Tractatus philosophiae*, ivi, n. 1259; *Lectiones in primum librum de generationem*, ivi, n. 1188; Martino Escalente, *Commentarii in tres libros de anima*, ivi, n. 1546. To this manuscript documentation of the Gregorian archives (the"Attic mss."), which still waits to be studied in depth, should be added the printed didactic works of those years, among which are A. Casiglio, *Introductio in Aristotelis logicam . . .*, Romae 1629; V. Aranea, *De universa philosophia a Marchione Sfortia Pallavicino publice asserta*, ivi, 1625; and, besides the classic manuals of B. Pereira, *De communibus omnium rerum naturalium principiis*, Romae 1576, and F. Toleto's comments on the *Physics* and *Of Generation* (1585, 1579).

[41] On the manuals of the Jesuits' scientific instruction and their possible influence on Galileo's thought, see A. Crombie, "Sources of Galileo's Early Natural Philosophy," in M. L. Righini Bonelli and W. Shea, *Reason, Experience, and Mysticism in the Scientific Revolution*, New York 1975, pp. 157-75, especially p. 163 ff.

[42] N. Cabeo, *Inquatuor libros meteorologicarum Aristotelis commentaria et quaestiones*, Romae 1646.

[43] On the graduation celebrations at the Collegio Romano during this period, I advise looking at the anonymous ms. *Origine del Collegio romano e suoi progressi*, p. 49 ff.

the public relations of the Collegio and the Society, were the public lectures (*orationes*), such as Father Grassi's famous lecture on comets in 1619. But still more useful for reading the moods of the institution were the inaugural lessons at the beginning of the academic year (*prolusiones*).[44] Anyone who was anyone in Rome filled the Main Hall on these occasions. The professors presented their courses in a light particularly revelatory of the cultural preoccupations and tendencies favored by the Collegio Romano. These amounted to official messages.

When *The Assayer* came out in Rome, in November 1623, the tenor of its polemic had for some time already filtered through. Observers of Roman affairs, therefore, followed attentively the opening of the Collegio Romano's 1623/24 academic year, hoping to seize on some outward sign of the reactions of the Jesuits, famous for their impenetrability, both to the recent political shift caused by the election of Pope Barberini and to its cultural consequences.

Reactions were swift, from the first. We know only, however, that in the course of the inaugural lessons, there were vibrant invectives against "the discoverers of novelties in the sciences." According to Monsignor Virginio Cesarini's testimony, the Collegio Romano insisted that "outside of Aristotle there is nothing whatsoever but reproach and derision for anyone who dares rise above the servile yoke of authority."[45] This is all we know; but if there had been more, we probably would know it.

The initial rebuttal, on the spur of the moment, must have been impetuously worded, but prudent in its content. Monsignor Cesarini's brief and unalarmed testimony seems, in fact, to evoke the expected recourse to routine Aristotelian slogans (such as "Outside the Church there is no salvation"), but these are not aimed at anyone in particular.

For a more significant reaction, we must await a more knowledgeable evaluation of the situation's development. One year later, after *The Assayer* had been officially smiled upon by Pope Barberini—who had blessed the author—and after the intellectual innovators had used Galileo's insolence for their own ends, the Jesuits, in the inaugural lessons of the 1624/25 academic year, adjusted their line of fire.

Mario Guiducci, one of the Collegio's brilliant former students, could not pass up these ceremonies. At the end of September 1624, he informed Galileo that a very severe lecture (*oratio*) had been delivered "against the followers of the new opinions, or rather against those who do not follow Aristotle."[46] We do not know the content of this speech. But we do know the tenor of the inaugural lesson of the course in Holy Scripture given that

[44] See F. Strada, S.J., *Prolusiones academicae et paradigmata eloquentiae*, Romae 1617.
[45] *Works*, XIII, p. 107. [46] Ibid., p. 216.

year by Father Fabio Ambrogio Spinola. A Genoan, thirty years old, but already endowed with a literary man's eloquence, Father Spinola had taught natural philosophy the year before.[47] Therefore, confronted by the echo of the polemic between Galileo and Sarsi, he was well qualified to present the rebuttal and protest.

Father Orazio Grassi, obviously, could not reply: personally defending Sarsi would have meant dragging the Collegio into the debate, thus playing into Galileo's hands. Father Grassi had other things to do, and could not further compromise himself in the eyes of the Society of Jesus at this delicate moment, just when he had been chosen for the St. Ignatius project. That year, Father Grassi gave his course on Vitruvius; the texts of the lectures are still at the Brera Library in Milan.[48]

Father Zucchi, astronomer and formidable orator, would have been the perfect spokesman to voice the protest of the Collegio Romano. Father Zucchi, however, was absent, away on a special mission. For Father General Muzio Vitelleschi had seen to it that Father Zucchi was appointed confessor and theologian to Cardinal Orsini for the 1623 diplomatic mission to Prague.

The father general had not chosen this astronomer by chance, but rather for a very important mission which could be accomplished brilliantly in the course of that legation: to convert Kepler to Catholicism. Incidentally, the mission failed. In Prague, Father Zucchi had long discussions with the Palatine astronomer who had been seduced by De Dominis's ideas, but in vain. They spoke, of course, in Latin. At the beginning, Father Zucchi perhaps began to speak of his new theories on the telescope. We know, however, that at a certain point, he broached complex subjects: the Eucharist in its species and the pope. Finally Kepler realized that he faced not a colleague, but a Jesuit. He lost patience, "broke off the discussion," saying that he did not wish to talk about these things because he had his own ideas ("Ne argumenteris se suade"). Poor Father Zucchi will comment bitterly that this was the attitude "customary with modern heretics, to whom any convincing dialectic appears frightening."[49]

[47] On Father Spinola (1593-1671), besides the classic repertoire of the Society of Jesus (Sommervogel), see Grillo, *Elogio di Liguri illustri*, vol. II, pp. 295-98. Father Spinola followed closely in the footsteps of Father Grassi; after having taught philosophy and Holy Scripture at the Collegio Romano, he was rector of the Roman seminary and then succeeded Father Grassi as head of the Jesuit college in Genoa. He published works of Marian devotion, Jesuit martyrologies, and hagiographies.

[48] See Father Orazio Grassi, *In primum librum de architectura M. Vitruvii et in nonum eiusdem De horologiorum descriptione duo tractati*, Audiente (name shortened) 1624, ms. of the National Library of Brera, Milan, described in G. Rotondi, "Due trattatelli, inediti del padre O. Grassi," *Rendiconti del Reale Istituto Lombardo di Scienze e Lettere* 62 (1929), pp. 261-66.

[49] D. Bartoli, *Della vita del padre Nicolò Zucchi*, Rome 1683, p. 28 ff. Father Zucchi was a specialist in the field of telescopic techniques; his name will be linked with the discovery of spots on Jupiter's surface.

If the Kepler mission had attained its goal, a quite improbable matter, Galileo would perhaps have lost his most authoritative scientific ally, which was even less probable. Kepler's astronomic authority was favorable to Galileo, and it might be counterproductive for the Jesuits to question it by discrediting publicly, as they could have, the astronomical competence of *The Assayer*'s author on the subject of comets. Instead, one could and must attack Galileo and his Roman followers on the plane of anti-Aristotelian philosophical sedition, bringing them to trial before the court of the principle of authority.

Thus, that attack was advanced in the form of "a very vehement and violent invective against the followers of new opinions contrary to Peripatetic opinion,"[50] in Father Spinola's inaugural lesson on November 5, 1624. It was a lecture not to be missed.

Mario Guiducci perhaps witnessed that sensational declaration; in any case, he managed to get a copy of the text. He sent it to Galileo, with the addition of inserted polemical notes, perhaps with an eye to publishing a sequel to *The Assayer*.[51] That is how we know what the Jesuits' official reaction was to the controversy over *The Assayer*. Thus, a shaft of dark and menacing light leaked through what was otherwise a milieu of total reserve.

Without ever gratifying the adversary with an explicit citation, Father Spinola attacked in grand style "those who, being more avid for novelty than for truth," flout the fundamental rule of philosophy; that is, "the light of divine faith." Moreover, he continued:

> Faith must take first place among all the other laws of philosophy, so that what, by established authority, is the word of God may not be exposed to falsity. Philosophy worthy of the Christian man is therefore that placed at the service of Theology and in conformity with the principles of Faith. The only thing necessary to the Philosopher, in order to know the truth, which is one and simple, is to oppose whatever is contrary to Faith and to accept that which is contained in Faith.[52]

The spokesman for the Collegio Romano's philosophy had made a declaration in the grand style, reviving the great intellectual program of Father Suárez, who with his teaching of philosophy had set the tone of the Collegio's cultural atmosphere.

Confronted by a phenomenon or a problem, Father Spinola added, instead of accepting a new interpretation most likely taken from a non-Chris-

[50] *Works*, XIII, p. 226 ff.

[51] The text (Galilean mss., pt. VI, tome IV, cc. 3-38, National Library of Florence) is published as an appendix to A. Favaro, "Amici e corrispondenti di Galileo. Mario Guiducci," in *Atti del Regio Istituto Veneto di Scienze, Lettere ed Arti* 75 (1915-1916), II, pp. 1357-1418, especially p. 1395 ff.

[52] Ibid., p. 1402.

tian philosophy, one must "abandon the inventions of new opinions and embrace the view confirmed by the testimony of the authors."[53] Respect for the principle of authority as the rule of Christian philosophy must be absolute. Obviously, Father Spinola does not make direct reference to the book of nature as a slogan used by heretical and rash innovators. Instead, he says that, as in law so in the science of the laws of nature, one must base oneself on the authority of tradition, on the collection of opinions and decisions of the most famous official masters of jurisprudence. The book of the laws of nature is written as a digest. Aristotle and the approved commentators have compiled that book to guide research, and without that authority any and all subversive innovations would have free rein.[54]

Mario Guiducci, brilliant lay lawyer, does not let that metaphor of the digest get by and inserts an ironic comment. As a good lawyer, he thinks that Father Spinola intends to put the works of Aristotle on the same plane as Justinian's *Pandects*; that is, he has in mind the code of civil procedure.[55] The witticism is agreeable, but wide of the mark. Father Spinola certainly did not think of civil law as sanctioning the authority of Scholastic philosophy, but if anything canon law. The collection of laws to which the Jesuit philosopher alluded, as we shall see, comprised the theological sentences compiled by specialists of the Society of Jesus to aid students of canon law and the inquisitors. The civil codes, for a Jesuit of the seventeenth century, were transitory and modifiable texts. Not so the texts of theological tradition, summed up in a canonical code during the proceedings of the Council of Trent. Guiducci either did not realize this or pretended to be deaf in that ear.

Father Spinola's entire argument was conducted in the sibylline and metaphorical style that suited the ceremony. Instead of citing Galileo, he mentioned his metaphor (the discovery of a new world) in order to point out that it was the work of a modern man who had based himself on the stars of the ancients. The principle of authority in philosophy is paraphrased by means of considerations on military art and by polemical allusions to poetry and painting respectful of tradition. Father Spinola will reveal his intentions only at the close of his inaugural lesson, when he officially addresses the colleagues of the Collegio, lined up in the first rows to listen to him, and the students and onlookers who packed the great hall.

Father Spinola will then firmly condemn the attempts to build a "new structure of human knowledge."[56] The term "structure" (*structura*) was a term that made one think immediately of the architecture of sacred and profane edifices. Indeed, Father Spinola compared the foolish aspirations of the paganlike philosophy of the innovators to the construction of the Tower of Babel. This is no longer mere intellectual and scientific polemics, but a more

[53] Ibid., p. 1403. [54] Ibid., p. 1408. [55] Ibid. [56] Ibid., p. 1415.

serious matter. The innovators, with their libraries and encyclopedias (structures of human knowledge), want to scale the heavens. They are rebels against God and the faith; they want the Church to collapse.

The illustrious mathematicians who had planned the new Tower of Babel must be punished along with the workers building it. Against the Tower of Babel shall rise the Church of St. Ignatius at the Collegio Romano. And, turning to the author of that imminent "sacred building," who was seated among the Collegio's professors, Father Spinola evoked the nature of the peril: "The evil has insinuated itself for too long, Fathers, while those who would discuss divine things and human ones in the same manner envelop celestial knowledge in a dark mist and expunge the temple of the Christian religion, bringing it to ruination." Intentions were no longer veiled, and the final threat declared a repressive action as legitimate in a time of war against heretics:

> He who on the subject of philosophy dissents from the common opinion also separates himself, without much difficulty, from the common sentiment of the Fathers in Theology. Would that it were the will of Heaven that all the seeds of heresy that have derived therefrom should not bloody the Christian Republic with such a great ruination of souls. We certainly would not have to deplore the destruction of the most flourishing cities, the horrendous hatreds of unrestrainedly furibund nations; nor the horrible slaughters of peoples, the fields drenched in human blood, the cloisters of consecrated Virgins violated . . . or that alleged yoke of the pope's power; nor churches without priests; nor the rejection of God's word on the part of one about to die at the stake.[57]

Heresy. The most important word in the dictionary of seventeenth-century Rome has been uttered. It is an authoritative word in that place, shedding a sinister light, but finally revelatory like all the dark lights that illuminate the threatening premonition of this message: the fires of the sack of Prague, and of other Bohemian cities as soon as the explosions of the Battle of the White Mountain have subsided; the now imminent burning of the author of the *Republica ecclesiastica*, the new Wyclif; and, finally, the light of the flames surrounded by which the anti-Aristotelian atheist Bruno had rejected the Bible. He who speaks now, with such obvious references, speaks in the name of an army whose soldiers are in the vanguard, fighting heresy on the Bohemian front. On the home front, in Rome, the asserters of the new philosophy are the sly traitors of Catholic orthodoxy. The denunciation is precise: the new philosophy undermines the theological bases of doctrinal faith, and "literati of this ilk lead Religion to its defeat."[58]

[57] Ibid., p. 1417. [58] Ibid.

Challenged, the Jesuits of the Collegio Romano counterattacked on terrain where they knew they could neither be refuted nor mocked. The defense of the principle of authority contested by Galileo was shifting onto the plane of theological apologetics—which, in the atmosphere of the De Dominis trial, amounted to alerting the intransigent institutions of Tridentine Rome.

In fact, Mario Guiducci does not hide his fear. This time, faced by such a "tirade containing more ill will than truth,"[59] he has no desire to be ironic. Disoriented by that threat, whose significance he does not understand or wish to understand, in a comment destined for publication he takes comfort from the protections that have benefited his teacher:

> If it were true that heresies originate only from dissenting with the consensus of writers on the natural questions of Philosophy (which consensus in actuality comes down to the Peripatetic dogmas), the Shepherds watching over the human flocks would long ago with their supreme authority have taken measures and forbidden this doctrine to be contradicted. And yet, we see that it is permissible and free for everyone to follow Plato and the others who are contrary to it; hence your complaint and pathetic lament are not founded on the truth, and it becomes clear that, for lack of good arguments against the new opinions in Philosophy, you have put Aristotle's statements and interests on a par with those of Theology and Faith.[60]

We will not take Guiducci's simulated ingenuity literally. He had been a student at the Collegio Romano. But that harsh indictment must have produced a certain degree of astonishment; for, in his accompanying letter to Galileo, he wrote that he could not understand "how the Fathers of the Collegio permit their literati to say and write so falsely, with such great harm to their neighbor."[61] What counts is that, as even Guiducci has understood, the protest raised by the Collegio Romano does not have at all to do with Copernicus, but with natural philosophy. Both Father Spinola's speech and Mario Guiducci's parries completely exclude the problem of the Copernican ideas, officially condemned and officially repudiated by *The Assayer*. The problem evidently is another, and concerns the incompatibility between an anti-Aristotelian natural philosophy, daring but not yet officially prohibited, and Catholic faith.

But Mario Guiducci has not the slightest notion, so it would seem, of the nature of the problem. *The Assayer*'s theses on physics are heterodox with respect to official science; on the other hand, the present pope, ten years before, at the time of the controversy on the *Discourse on Floating Bodies*, had liked them enough to defend Galileo against the Aristotelian professors.

[59] Ibid. [60] Ibid. [61] *Works*, XIII, p. 244.

Not understanding what is at stake and frightened by those names and facts evoked so threateningly by Father Spinola, Guiducci, just to be on the safe side, reacted like a jurist. He had quickly gone to consult the most authoritative manual on heresiology: "having taken a quick look at the controversies of Cardinal Bellarmino, it is not said there that today's Heresies against Catholic faith originate from new opinions in Philosophy."[62]

Thus did Guiducci conclude his defense, with a clear conscience. Had he read Bellarmino's controversies more attentively, with a less juridical and more philosophic eye, he would have perhaps better understood why Father Spinola was so certain that the followers of the new natural philosophy were heretics.

Galileo proved to be more prudent than his pupil and did not follow up on Guiducci's plan to publish that sort of appendix to *The Assayer* which by now had degenerated into threats of forceful action. There was no point in circulating those threats when one could not defend oneself, since Guiducci's naive comments avoided the problems instead of confronting them.

The time of literary verve had passed, as had that of the political enthusiasms of the first months. The horizon was full of unfavorable signs, and the stand taken by the Jesuits against the heresies of the innovators at Rome was quickly turning into a reality. It was not only the hallucinatory demonstration of the institution's severity towards De Dominis; the Jesuits never officially pronounced the word heresy in vain. In fact, the inebriating season of success for the innovative literati was over. Since the summer (since June 11, 1624, to be precise), the Congregation of the Index had already condemned *L'Adone* and then—on July 17, 1625—Cavalier Marino's entire poetic work.[63] The bard of the innovative literati—despite the fact that he considered himself perfectly Catholic, despite the protection of Cardinal Maurizio of Savoy and the love of the collector Cardinal Francesco Barberini for his poetry—hurriedly had to leave both Rome and the Academy of the Humorists, of which he had been the ephemeral prince. On Father Lancillotti, too, the author of *Hoggidí*, fell the persecutions of his order which will later take the form of house arrest and trial at the Holy Office, despite the protection of Ciampoli.[64] And what about the other great literary event of 1623, *The Assayer*?

[62] In Favaro, "Amici e corrispondenti di Galileo. Mario Guiducci," n. 48, p. 1417.

[63] See *Index librorum prohibitorum* . . . , Romae 1948, p. 304. From this significant edition of the *Index* I take the exact dates of Marino's condemnation. I point out, in any case, from the *Elenchus librorum . . . ad annum 1640 prohibitorum*, 2nd ed., Romae 1640, that the decree of the condemnation of *L'Adone* was published at the beginning of 1627; see p. 208.

[64] See G. B. Vermiglioli, *Biografia degli scrittori perugini*, Perugia 1829, vol. II, pp. 51-57. After having taken refuge in France, Father Lancillotti will die in 1653. There exist autobiographical manuscripts on his persecutions at the Augusta Library in Perugia (mss. 1884 and 1885).

FIVE. THE HOLY OFFICE'S SECRECY

> . . . nulla notat formae contagia, nulla coloris, idem perstat odor, priscum tenuere enarem omnia. . . . Sic licet externa deceptus imagine sensus, terrenaque; oculi spevies pascantur inertes vivia vis animi velamine Numen adorat indepremsum, ingens, non nititur am alma colori pulsa nube fides, tantum tegamenta fatetur extima candentis Cereris manifesta.*

Sycophants and Informers

The Assayer was also denounced. It was denounced, moreover, at the Court of the Holy Office just when the favorable audience reaction to the book was at its peak among Vatican authorities. Neither the dedication to the pope, nor Father Riccardi's exultant authorization, nor the approval of the Master of the Sacred Palace, nor the succeeding certification of "religious fervor" released publicly to the author together with the papal embrace prevented the denunciation.

The denunciation is known to historians through the revelations in a letter sent to Galileo on April 18, 1625, by Mario Guiducci,[1] whom we left in Rome in the role of informer. But Mario Guiducci is not a character in a spy novel: we will be disappointed to learn that his revelation is late in coming, since the denunciation is many months old.

You needed the right stuff to be an informer in Rome. And Mario Guiducci, who had the right stuff for the law, furthered his own affairs by cultivating good social and political relations with persons highly placed in that "theater of marvels" which was the capital city and residence of the Church. Almost a year had passed since Galileo had left him in Rome. Guiducci had immediately gone to work, nosing around festivals and cardinals' palaces, spying out the capricious desires of the cardinals who asked of him, Galileo's

* . . . does not note contagion whatsoever, either in appearance or color, [but] the order persists and everything remains as before. . . . The senses can be deceived by the external and earthly appearance; let the eyes feed on inane appearances, but under that veil the vivified strength of the soul adores an elusive and great divine power. Faith does not make use of that color and, that sweet veil being removed, proclaims that in that guise are only the extreme appearances of white Ceres."—Benedetto Milani, *De rosantis Eucharistiae formis quae multis intactae in Compulutensi Societatis Iesu templo servantur, iorum carminum liber*, Romae 1624, p. 137

[1] Galilean mss. in the National Library of Florence, pt. VI, tome X, 91/D2, cc. 206-207, in *Works*, XIII, p. 265 ff. (first published in the Alberi edition, *Opere di Galileo*, Florence 1842-1856), vol. IX, p. 78 ff.

agent for Roman affairs, what they would never have dared ask of his master: a special fountain to embellish a courtyard, a portrait, a carriage with an ingenious system of seats, or (Cardinal Francesco Barberini's bright idea) a planetarium-cum-zodiac to be painted on the roof of his carriage.[2] Guiducci, though somewhat ashamed, repeated even these frivolities in his correspondence with Galileo.

In any event, the prestigious Galilean introduction had opened all doors to him, and had soon permitted him to become quite familiar with the circle of Cardinal Francesco Barberini—and even with the splendid, powerful cardinal himself, his "very special master," something of which he was proud. He was also proud of having entered the Academy of Lynceans, an exclusive club, even if, to tell the truth, after the lecture on comets in 1619 and a brief reply to the *Libra*, he had not published a single line on science.

Now, however, Guiducci began to think about returning to Florence. His letter of April 18 will be his penultimate, but in a previous letter he had already told Galileo of his nostalgia for home.[3] His great protector, Cardinal Francesco Barberini, had left at the beginning of spring for the famous legation in France. With him in the plethoric caravan of numerous intellectuals, political advisers, theologians, and Jesuit confessors—the most picturesque, costly, and noisy diplomatic mission ever seen—his friends had also left Rome: Cavalier Giorgio Coneo, Cavalier Cassiano dal Pozzo, Luigi Azzolini, Carlo and Cesare Magalotti, and Tommaso Rinuccini, Galileo's precious informant.[4]

Guiducci must have felt lonely; but, more probably, he was ill at ease because he felt exposed, especially since the previous winter when he had seen the stern face of Roman institutions. And it was in this mood of uncertainty and alarm that he wrote the letter of April 18, in which he wanted before all else to justify himself. Guiducci felt that he should not send Monsignor Francesco Ingoli the precious manuscript in which Galileo set down new arguments to launch again the old Copernican idea. Guiducci was preparing to leave, and prudence advised against running useless risks by circulating that compromising Copernican document.

Therefore, the first and chief piece of information in that April 18 letter was the prudential decision to put off once again the delivery of Galileo's reply on Copernicanism to Monsignor Francesco Ingoli. Galileo had formally charged him with this transmission, and to justify his reticence Mario

[2] See *Works*, XIII, pp. 193, 216 ff., 220 ff., and 251. See, on p. 266, Guiducci's accent of embarrassed justification, to which we allude.

[3] "Now I feel that it will be a thousand years before I return" (letter from M. Guiducci to Galileo on March 22, 1625), and he added, "about Sarsi I know nothing" (*Works*, XIII, p. 261). Cardinal Barberini's legation to France had departed from Rome on March 17.

[4] On Cardinal Francesco Barberini's diplomatic retinue, see Baillet, *La vie de Monsieur Des Cartes*, vol. II, p. 122; Pastor, *Storia dei papi*, vol. XIII, p. 285. On Rinuccini, see *Works*, XIII, p. 255.

Guiducci, who did not want to play the hero and run "the risk of a reprimand" for defending Copernicus, offered certain reasons for his fears. Guiducci, having met with Prince Federico Cesi a number of times, availed himself of Cesi's advice to let the question of Copernicus "sleep a bit."

The first argument of dissuasion presented in the letter had to do with a circumstance that had completely escaped Galileo's perspicacious friends (and, above all, Mario Guiducci) for many months before *The Assayer* was denounced. Guiducci, Galileo's Roman informer, was the last to hear it.

Guiducci's letter, the original of which we have seen among the Galilean manuscripts at the National Library of Florence, said:

> First, some months ago at the Congregation of the Holy Office, a pious person proposed to prohibit or correct *The Assayer*, charging that it praised the doctrine of Copernicus with respect to the earth's motion: as to which matter a cardinal assumed the task of informing himself about the situation and reporting it; and by good fortune he happened to hand over the case to Father Guevara, father general of some sort of Theatines, who I believe are called Minims, and this father then went to France with the Signor Cardinal Legate. He read the work diligently and, having enjoyed it very much, praised and celebrated it greatly to that cardinal, and besides put on paper certain defenses, according to which that doctrine of motion, even if it were held, did not seem to him to be condemnable; and so the matter quieted down for the moment. Now, not having this support from that cardinal who could help us, it does not seem that we should run the risk of a scolding; for, in the letter to Ingoli, Copernicus's opinion is defended *ex professo*, and although it states openly that with the assistance of a superior light it is proven false, all the same the malicious will not see it thus and will get into an uproar again; and as we lack the protection of His Excellency Cardinal Barberino, who is absent, and further having as an opponent in this matter another important gentleman, who was once one of the first to defend it, and furthermore N. S. being very annoyed by these bellicose intrigues so that one cannot talk to him about it, it would definitely remain up to the discretion and intelligence of the friars. For all these reasons, it seemed to me right, as I said, to let the matter sleep for a while, rather than keeping it alive with persecutions and having to ward off those who can deliver some sly blows. Meanwhile, time will help our cause.[5]

Time certainly could help Mario Guiducci's cause; for, a month later, he had returned to Florence safe from possible "persecutions" and "sly blows." Guiducci was really on tenterhooks: he concluded his letter by announcing

[5] Ibid., p. 265 ff.

his imminent return ("I hope to be down there before the middle of May") and showing clearly that he had not the slightest intention of carrying out the mission Galileo had requested of him. In fact, he had decided to hand over to someone else the hot potato of the Copernican letter addressed to Monsignor Ingoli ("when I depart, I will leave in Signor Filippo Magalotti's hands the letter written to Ingoli").[6]

He added a few words of information that he had already transmitted regularly until then; that is, that Sarsi was not publishing the book in answer to *The Assayer*. This information was simply repetition, and it was false; but Guiducci probably had not made the necessary inquiries to know that.

In a postscript, Mario Guiducci asked advice from Galileo concerning a fountain for Cardinal Alessandro Orsini. This cardinal—tied to the Jesuits and whom we have encountered, on the occasion of the Prague embassy of Ferdinand II, in Father Zucchi's company—was then financing the publication of *Rosa ursina* by Father Scheiner, one of Galileo's principal Jesuit opponents. But Guiducci, instead of finding out the contents of Father Scheiner's new book, had been discussing fountains with Cardinal Orsini.

This letter of Guiducci's is well known to modern specialists in Galilean biography. All of them have taken Guiducci's words at face value, recording the most interesting information it contains (that is, the mysterious denunciation against *The Assayer*) and expressing pleasure that in 1625 a cardinal and theologian had defended the Copernican ideas attributed to *The Assayer*. To our knowledge, Stillman Drake has been the only historian who at least was bold enough to remark in this connection that "the complaint that Galileo had praised Copernicanism in the book [*The Assayer*] was simply false."[7]

One does not need to be a specialist in scientific methodology to know that, when certain and irrefutable empirical proofs are lacking, it can be more difficult to demonstrate the falsity of a proposition than to demonstrate its truth. But even before establishing whether it is true or false, it must be comprehensible. Historians, too, often run into this difficulty. A basic problem of the historian's trade is, in fact, that of the reliability of one's sources. Usually, however, a series of other sources come to one's aid and allow one to accept or confute the reliability of a document containing a given piece of historical information. Things get complicated, obviously, when that document is the sole source of information. In this case, either one has faith in it or one must resign oneself to carrying out an internal and external analysis of the document.

Guiducci's communication is the only source that speaks explicitly of *The Assayer*'s denunciation. The information nevertheless was reassuring, and

[6] Ibid., p. 266.

[7] Stillman Drake, *Galileo at Work: His Scientific Biography*, Chicago and London 1978, p. 300.

since this denunciation will never be spoken of again, all this must have encouraged historians to take Guiducci's information literally.

For that matter, we have already mentioned that *The Assayer* never suffered any official persecution. But since our initial problem was the mysterious and belated remorse in Vincenzo Viviani's biography of Galileo, which held the polemic against Father Grassi of having been, inexplicably, at the origin of all his succeeding misfortunes, one naturally wants to reread this letter of Mario Guiducci's with care.

Opaque Information

If one reads this letter with care, without letting oneself be too distracted by the sensational information it reveals, one must inevitably take certain critical precautions, instead of accepting it literally. Even before its disquieting and at the same time reassuring revelation, it highlights (if examined closely) a series of questions of a logical and historical order: details that might pass unnoticed by a modern reader, but revelatory details that help us to appreciate the reliability of this important source of Galilean biography. Let us go about it in an orderly fashion, beginning this time at the end.

Only at the end of the letter, in fact, does Guiducci make up his mind to "come clean," as they say. In other words—and Galileo should not take it amiss—he is not going to be the one to expose himself and compromise his excellent Roman connections with the cardinals by officially putting into circulation a piece of writing in which Galileo defends *ex professo* a theory condemned by the Holy Office. Only a short while before, Guiducci had witnessed the intimidating spectacle of De Dominis's condemnation and the denunciations against innovators on the part of the Collegio Romano. We can understand his caution and his fears, since the only person who can guarantee him immunity from the inquisitorial institutions, the cardinal-nephew, is away from Rome, while Pope Barberini is thinking of the war between Genoa and Valtellina. We understand a little less easily, however, why Guiducci is so afraid to reveal to Galileo the name of the highly placed personage who is no longer willing to defend Galileo and that of the cardinal who, on the contrary, has taken his cause to heart with the Holy Office.

In regard to both these nameless persons, this can be explained by obvious motives of caution. These are the same motives of caution which, probably, prevented Federico Cesi from personally informing Galileo about what he had heard. For violation of the secret of a pretrial inquiry in cases like this, the penalty was excommunication. Prince Cesi could not compromise himself with a letter on such matters, which on top of everything else were not very clear and were gathered at second hand.

Indeed, this is the point. The arguments that Guiducci presented to justify himself were second- or third-hand bits of information: he says that he has spoken about the matter with Prince Cesi, but it is impossible to know whether Guiducci talked to the prince about the denunciation, or vice versa. The first hypothesis seems more likely, and it is also more likely that Guiducci would speak of conversations with Prince Cesi in order to justify himself, adducing the opinion of the head of the Academy of Lynceans regarding that which worried him the most; that is, the decision not to hand over the Copernican document.

In any case, the news of the denunciation has been "reported." What is most strange, however, is that it has not been even slightly evaluated by the man who reports it. Worse yet, we have glaring evidence that not only has Guiducci not evaluated the information, not only does he reveal it in an absolutely uncertain fashion, but when he does present a precise element, his fears and his concern to persuade Galileo bring into his mind certain entirely invented details, certain crude untruths.

Let me explain this better. One does not have to be a lawyer, like Mario Guiducci, to know that if one wants to inform a friend that someone has denounced him, the least that one can do for him, obviously, is to try to find out the date, the author, the development, and the aim of the denunciation. Now Guiducci has not made the slightest effort to find out anything more about the person who has vaguely informed him that, "some months before," a "pious person" had denounced *The Assayer*.

In the language of the seventeenth century, when one speaks of a "pious person" in cases like this, one means to say that the denunciation was anonymous and that it had been compiled by someone who wore a religious habit, and precisely the habit of an order in which this custom existed as a vocation. It is impossible to be more vague and generic.

One could perhaps object that the timorous Guiducci, even if he knew more, would not say more in writing. But that is probably not true, because the only precise thing which has been told to Guiducci, and which had to be kept secret in pretrial inquiry, is the name of the consultant who had suspended the continuation of the proceedings. Just so, Guiducci reports the sole identifiable element of this affair: the name of Father Guevara.

Someone must have whispered this name in Guiducci's ear, and that it was the name of the father general of a religious order. Guiducci did not at all know who the devil this Guevara was; he did not have the time to become a bit better informed, or he had forgotten part of the information, or he had not got it all. In any event, seized by the agitation of that preoccupied letter, he makes up things in order to reinforce the reliability of his information—a "father general of some sort of Theatines, who I believe are called Minims."

Father Giovanni Guevara was the authoritative provost of the Regular

Minor Clerks. A slight inexactness, an irrelevant oversight? Today the numbers of the Theatines, the Minims, and the Regular Clerks are so small that the difference appears trivial to our modern eyes.[8] A slight inexactness, in fact; but we must be grateful to Mario Guiducci, because it is a revelatory detail for anyone who would probe the contents of his letter.

In 1623, to mistake the father general of the Theatines for that of the Minims (Friars) or the Regular Clerks was, for an observer of Roman affairs, as crude an informational error as it would be today for a correspondent from the Italian capital to attribute a well-known political leader randomly to one or another of the many parties now enlivening the political scene. Despite their distant origins, the Minims (Mersenne was one of them), the Theatines of Sant'Andrea della Valle, and the Regular Minor Clerks could not be confused. The new College of Regular Minor Clerks at Sant'Agnese, on Piazza Navona, had a fine reputation in Rome. Its professors were the great linguistic specialists of the Propagation of the Faith. And who did not know Father Guevara, one of the political theorists favored at the Curia, held in great esteem for having contributed to the single brilliant diplomatic success of the new pontificate, so much so that Pope Barberini had entrusted him with instructing and advising the cardinal-nephew?

Why was it that Guiducci had not asked Ciampoli? Certainly, so long as Guiducci frequented the company of cardinals protective of the Jesuits, he had few chances of meeting Father Guevara. But why had he not asked Prince Cesi? Father Guevara had just published a valuable book on the theory of sensations, in which he praised Galileo and explained the function of the telescope. That book must certainly have reached the Cesi library.

Guiducci did not have the time to inform himself; his inexactitude is an indication of the agitation permeating the entire letter, of a lack of caution, of a superficial and justificatory self-persuasion. It casts a shadow on the reliability of this letter so full of indirect, generic information. And thus we may be forgiven for being perplexed by the peculiar reason adduced for the denunciation.

Guiducci reveals, or believes, that the object of the denunciation was *The Assayer*'s praise of the "doctrine of Copernicus with respect to the earth's motion." A little further on, he hastily reveals another fragment of rumor: "that doctrine of motion, even if it were held, did not seem to him condemnable." What doctrine of motion? Guiducci moves on quickly now to the preoccupying absence of the cardinal-nephew.

But let us go slowly. Copernicus had been on the Index for only nine

[8] See Clemente Piselli, *Notizia historica della religione dei PP. Chierici regolari minori*, Rome 1710 (dedicated to the Protector of the Order, Cardinal Francesco Barberini), p. 49 ff. On the present dimensions of this congregation, which flourished in the sixteenth and seventeenth centuries, see Abbé Omer Englobert, *La fleur des saints*, Paris 1946 (new rev. ed., Paris 1980), p. 439 ff.

years. In 1620 there had been proposals, perhaps at Cardinal Bellarmino's suggestion, for appropriate corrections to make the heliocentric system tolerable, as a pure mathematical hypothesis. *De revolutionibus* had been corrected, but not taken off the Index,[9] and Kepler had also been prohibited since 1619.

In 1616, all the theological advisers of the Holy Office had declared the doctrine of the earth's motion "foolish in philosophy and formally heretical." Before the condemnation, of course, some theologians had defended with exegetical arguments the rational legitimacy of that doctrine. But after that decree of condemnation, ratified in fact by the Holy Office and by the Congregation of the Index, there was only one person in the Catholic Church who could say that the doctrine of the earth's motion was not condemnable: the pope. And despite the best disposition and his assurances to Galileo the year before in Rome, the pope had not said it officially.[10]

How could Galileo believe that letter from the shaken Guiducci, who tried to tell him such sensational and incredible news, about Galileo's greatest concern, by letting it slip out so offhandedly? It would have been enough for Father Guevara to say that Copernicus was not praised in *The Assayer*: enough, and also correct. But no, according to Guiducci, Father Guevara, asked for a sympathetic opinion, apparently had exposed himself in writing, explaining to the Holy Office that all its theological consultants were in serious error. On what authority could he affirm such a thing? He was not a specialist in Scriptural theology, not even an astronomer. Besides, was it not perhaps written of Copernicus in *The Assayer* that "every idea must be revoked thanks to the recently damned hypothesis"?[11] Such a very strange and unrequested Copernican apology would have made the Holy Office quietly "calm down."

Do we need a further argument to demonstrate that it was impossible for Father Guevara officially to sanction the fact that the doctrine of the earth's motion was not damnable? If so, I will point out that in 1623 Father Guevara, as father general of an order, had approved the work of his fellow priest Father Aversa, *Philosophia metaphysicam physicamque complectens*. In the second volume, of 1627, that book described the Copernican theory at length and rejected it with all the more recent astronomical arguments and observations, especially the absence of stellar parallaxes. But, above all, Father Aversa remembered that besides being absurd that theory was against the

[9] See *Index librorum prohibitorum*, Romae 1681, p. 203.

[10] On Pope Urban VIII's offical tolerance for the Copernican doctrine, since "the Holy Church had not damned it nor was it to damn it as heretical, but only as reckless," reported by Cardinal E. F. Hoenzollern, see Galileo's well-known letter to Cesi on June 8, 1624, *Works*, XIII, p. 182.

[11] See *The Assayer*, p. 37 (*Works*, VI, p. 231).

faith and condemned by the decree of the Index of 1616.[12] It is very unlikely that in the same period Father Guevara would have decided to discredit himself to the point of declaring that the earth's movement was not a condemnable doctrine.

What doctrine of motion? That "of the earth," says Guiducci, thinking of the dangerous manuscript of the reply to Monsignor Ingoli which he did not want to make public. In reality, for Galileo the imputation was even more stupefying than the defense.

The Academy of Lynceans, seriously involved in the procedure of 1616 against Copernicus, had not wanted to run any risks with *The Assayer*. So that there would not be any blunders, the book had been read and reread by its most cultivated men in theology and letters.

Nobody had ever dreamed of using *The Assayer* to advance Copernican ideas on the sly, since Copernicus did not discuss comets. The manifest, desired intent had been to contest the Aristotelian philosophy of the Collegio Romano and its professor of mathematics. But there was no need for so much caution. For Galileo had put up a defense when the *Libra* tried to draw him into a Copernican trap: "Ptolemy and Copernicus never dealt with hypotheses relating to comets, and I do not see why they would."[13] In reality, as we have said, Galileo talked about comets in order to protect the Copernican system from possible falsification, but this did not in the least mean that he wanted to risk praising Copernicus. Galileo, suspected of Copernicanism, never let them catch him red-handed in *The Assayer*.

At the beginning of the book, Galileo had spoken out, denouncing his adversary, "who in order to find a way to blame me with I know not what concerning Copernicus, would have to find such things written down; and then, for lack of them, he decided to supply them himself."[14] From that moment on, whenever *The Assayer* pronounces the name of Copernicus (paragraphs 6 and 32) and even when it does not name him—when, for example, he accused Sarsi of not having "a perfect idea of the motions attributed to the earth"—a formal disavowal always follows: "As for the Copernican hypothesis, if for the benefit of us Catholics we had not been freed from error by more human knowledge and our sight restored, I do not believe that such grace and benefits could have been obtained by the reasons or events posited by Tycho." Thus the first time. As for the second: "if the movement attributed to the earth, which I as a pious person and a Catholic consider entirely

[12] Father Raffaello Aversa, *Philosophia metaphysicam physicamque complectens*, 2 vols., Romae 1625-1627, II, p. 4 ff. Father Aversa declared that, not withstanding corrections, the heliocentric Copernican hypothesis, physically incapable of being sustained, "cannot be tolerated with sure faith." He added that the most rigorous and modern publications of Tycho Brahe and Magini had rendered definitively vain the claims of Copernican astronomical doctrine.

[13] *The Assayer*, p. 33 (*Works*, VI, p. 229). [14] Ibid.

false and nil, lends itself to explain the many and varied appearances that one observes in the celestial bodies, I cannot be sure that he, so false, may not wrongly respond to the appearances of the comets."[15]

The only time when *The Assayer* speaks of the earth's motions according to Copernicus without adding this formal disavowal is on the occasion of the description of the Cesarini Palace experiment, which we mentioned before. But this is an illustrative event, a visual demonstration with which Galileo proved the "falsity" of Copernicus's doctrine of the earth's motion, since one of the terrestrial motions in that theory was shown to be merely apparent.[16]

Prudent Roman readers had also rightly considered it completely superfluous to insist here that the author of *The Assayer* was submitting to the condemnation of Copernicanism. Of course it was a dissimulation, a perfect fiction, this recourse to Copernican arguments *ex falsa suppositione*. But, formally, it was unexceptionable. On the basis of the letter, *The Assayer* could not be accused off Copernicanism.

What we have in *The Assayer* is an honest and rigorous dissimulation. Today we would call it hypocrisy and, as were certain illustrious interpreters, we too can be scandalized. In 1623, however, honest dissimulation was an intellectual virtue. And in any case, the Church had set up the Holy Office to persecute condemned opinions—formal heresies, not hypocrisy.

The Holy Office was a tribunal, and a denunciation required precise statements "in praise of" Copernicus. Where were obvious ones to be found in *The Assayer*? What might have been the upshot of denouncing as hypocritical a book enjoyed by the pope and dedicated to him?

Let us not forget that the times had passed when almost any friar could call Galileo to task. We are at the "marvelous conjuncture"; Galileo is the official Catholic scientist, not only the greatest scientist in Europe. Galileo was loved, respected, and feared in Rome. He was also hated, owing to *The Assayer*, but to denounce that book was not something that just anyone could do.

What "doctrine of motion"? If one wanted at all costs to be suspicious, the only doctrine of motion "praised" by *The Assayer*, as in a Lucretian poem, was that of the motion of atoms, a free Galilean version of the Aristotelian pronouncement that "motion is the cause of heat." But what had Copernicus to do with this?

The clarifications offered by Mario Guiducci's letter of April 18, 1625, are contradictory and incomprehensible in terms of either logic or history. They might represent a fascinating mystery, if the uncertainties and mistakes in

[15] Ibid., pp. 39 and 182 (*Works*, VI, pp. 233 and 311). [16] Ibid., p. 180 (*Works*, VI, p. 310).

this document did not cast doubt on its reliability. On the letter's single ascertainable point, we have seen, the information was real, fifty percent reliable. For the rest, it was the product of the author's agitation. Guiducci was frightened and he wanted to frighten Galileo, persuading him of the danger of an untimely profession of Copernicanism. Having received at the last moment the disquieting news that it was thanks only to the dismissive intervention of a certain Father Guevara that *The Assayer* had not been denounced for its questionable theory of motion, Mario Guiducci had perhaps unknowingly rounded out his information in the light of that danger which at the moment chiefly preoccupied him.

But this is only my hypothesis. As to Mario Guiducci's letter, only two things can be affirmed with full certainty. The first is that, without the shadow of a doubt, it tells us that Galileo was not a poor Christian like everybody else. Once again, thanks to his authoritative supporters in the Curia, Galileo evidently had the benefit of exceptionally favorable treatment.

Instead of following its normal course, the denunciation is withdrawn by a cardinal who arrogates the inquiry to himself, removes it from normal procedure, and quickly shelves it, recurring to the indulgent expertise of a theologian chosen specifically to ensure that there are no surprises. One thinks immediately of one of the two Barberini cardinals at the Congregation of the Holy Office, Francesco or Antonio, as the protagonist of this opportune and comprehensible stalling maneuver. Indeed, the maneuver was not only opportune and comprehensible, but absolutely indispensable. We have explained at length how the Galileo of 1624 was no longer simply a mathematician and philosopher from Florence, but rather the official scientist of the new regime in Rome. To denounce *The Assayer* in 1624 was tantamount to an underhanded political provocation, like so many others. It was not possible to threaten Galileo without automatically becoming, or already being, hostile to the new Barberinian pontificate, without trying to cast a shadow on the new regime's orthodoxy. To denounce *The Assayer* in 1624 was in fact equivalent to denouncing indirectly all who were directly involved in that book: the Academy of Lynceans, protected by the cardinal-nephew; the Curia's political and intellectual entourage; but, above all, the pope who had officially approved the book and sung the praises of its author for his faith as a Catholic, thus wiping out the entire past with one stroke.

Besides, everyone knew that Galileo had managed to obtain cautious but reassuring comprehension of Copernican doctrines. The pope—so the rumor ran all over Rome—was open-minded and tolerant in regard to Copernicus, even if officially Copernicus was condemned. The denunciation, whatever its object might be, had the flavor of a threatening reminder of Cardinal Bellarmino's decisions as well as a provocatory accusation of doctrinal laxity in the new pontificate's policy of cultural liberalization. For these reasons, it is clear, the futile scandal provoked by the general repercussions of that obscure denunciation must be snuffed out at birth.

The second absolutely certain thing, in regard to the revelations of Mario Guiducci's letter, is that the mysterious information—as incomprehensible as that in a coded diplomatic letter—has vanished altogether, like a soap bubble bursting in the air. Nobody since has made even the slightest allusion to the denunciation of *The Assayer*'s Copernicanism.

The official records of Galileo's trial, eight years later, do not reveal even the slightest mention of the circumstances recounted by Mario Guiducci. Neither the denunciation of Copernican doctrine nor its defense, which had calmed the Holy Office's anxiety, will be brought up again. Everything that Mario Guiducci writes to Galileo in April 1625 has vanished like a dream, and this is strange.

But the historian does not have the right to be satisfied with appearances. Nor does he have the right to discredit his sources too readily or to challenge (solely with logical arguments) witnesses who, three centuries later, cannot defend themselves. Before branding the testimony of Mario Guiducci's letter as unreliable, the historian must do everything possible to verify what it contains. Now, looked at carefully, Guiducci's letter also contains—amid so many reticences, mistakes, verbal and psychological justifications—a reference that the historian must make an effort to follow and track down.

Mario Guiducci offers a glimpse, albeit fleeting, of one ascertainable material element in that mysterious plot he recounts to Galileo. This is the sort of fragment that can leave something of itself behind in the archives of history, something that historians call objective: namely, written or printed pages, which were written to be read and to be preserved. Mario Guiducci writes that Father Guevara had "put on paper" something about the denunciation.

In order to understand that letter, confronted by its incomprehensible tangle, I could only do what Mario Guiducci did not have the time to do: try to find out more, to become a little better informed. After three hundred and fifty years, the only thing that might remain, if Guiducci had not invented everything, were those pages of Father Guevara addressed to the Holy Office.

Apparently, Mario Guiducci had not gone to ask for any further explanation from Father Ridolfi, Master of the Sacred Palace at the Holy Office. The furthest thing from his mind, probably, was the thought of knocking at the door of the Holy Office. And after him, nobody else thought of doing it. From where I now stood, I could only go and knock at that door myself.

The "Secret of the Holy Office"

I asked myself, and I asked the supreme authority of the Holy Office, the present Sacred Congregation for the Doctrine of the Faith, if those pages of Father Guevara had been preserved in the archives. To be able to establish the reliability of Mario Guiducci's revelations was of enormous historical interest. But even more exciting would be to find the arguments with which a theologian had officially illustrated, in the Rome of 1624, the noncondemnability of the Copernican system. For all these important historical reasons, I explained to the authorities of the Holy office, I was interested in seeing this document, if it still existed.

I was politely told that it did exist. In fact, my request had allowed them to find a document on *The Assayer*—contemporary, in manuscript, and unsigned, but sealed with the capital letter "G." Guevara's initial, probably. This allowed one to presume that the pages involved actually were Father Guevara's. It was added, however, that unlike other documents in the archives—such as official sentences and decisions—this type of documentation (i.e. theological advice and opinions) could not be consulted. In our case, however, various special circumstances seem to have surrounded that document, which had remained unknown until now. In the first place, we had here a judicial episode without a sequel. It was known that Father Guevara's opinion was completely dismissive. We had proof, through the trial published in Antonio Favaro's edition of Galileo's *Works*, with documents also taken from the Holy Office, that this episode had not had the slightest repercussion, negative or positive, on either the preliminary inquiry or the unfolding of the trial subsequently initiated against a much different book by Galileo: the *Dialogue*. Not even the shadow of *The Assayer*, we are certain, ever appeared on the horizon of the procedure or in the sentence, rendered in 1633 by the Holy Office's tribunal, condemning the *Dialogue*.

The sole outcome of Father Guevara's opinion was a decision for dismissal. This, evidently, diminished the judicial value of the document as such, but without (for this very reason) diminishing its interest for the history of ideas.

The second circumstance that made it possible for me to request authorization to consult the document was the kind of publicity it had received. Mario Guiducci's letter proved that knowledge of the fortunate judicial episode involving *The Assayer* had indeed circulated among his contemporaries. The circumstances, the name of the consultant, and the tenor of his opinion had been public knowledge. All of this obviously reduced—to the point of rendering it practically insignificant—the extent of the document's secrecy.

My request for consultation was therefore legitimate, motivated by a specific interest in the history of *The Assayer*'s fortunes in Roman ecclesiastical circles at the moment of the "marvelous conjuncture" of the 1620s. Furthermore, my reasons were authoritatively supported by the secretariat of the International Academy of the History of Sciences. In fact, it was through this organization, after a few months of waiting, that I was informed (in late May 1982) of the authorization to consult Father Guevara's opinion.

I shall take advantage here of this interim period to explain to the reader the nature of the Holy Office that had consulted Father Guevara. Since 1569, the sober edifice alongside St. Peter's had taken its name from the Supreme Tribunal and its adjacent prison on the ground floor of the Holy Office of the Congregation of the Supreme and Universal Inquisition. Founded in 1542 to centralize and guide the peripheral activities of the declining Inquisition, the Congregation of the Holy Office was the oldest and most important congregation of the Curia, so much so that the pope kept its presidency for himself. To one of the cardinals called to the noble work of this congregation fell the function of prefect (practically speaking, secretary), while a high prelate served as an assessor entrusted with extrajudicial affairs. The cases that had to be dealt with judicially were entrusted to an inquisitional commissary, assisted by two Dominican fathers. Then there was a fiscal promoter (or public accuser), a lawyer for the accused, and a notary. In charge of the assignment of imprimatur, and the palace's true "gray eminence," however, was the Master of the Sacred Palace, a post traditionally occupied by a member of the Dominican Order, who usually attained this post after having worked as a consultant.

The Master of the Sacred Palace, as also the cardinals of the congregation, availed themselves of the indispensable assistance of theological experts: "consultants," assigned to the preliminary examination or to the revision of a book, and "qualificators," a more precise and influential post because the person who filled it endorsed a written opinion requested in extraordinary cases for particularly complex doctrinal questions. We have seen the qualificators and consultants of the Holy Office gathered together on February 24, 1616, to endorse the expert opinion which declared that the Copernican doctrine was foolish and heretical. Such expert opinions had an advisory rather than a binding value, but were nevertheless influential.

The routine life of the Holy Office was carried on through four weekly meetings (ferials), at the palace or in the Vatican. First, consultants and qualificators examined the dossiers and the cases assigned to them. Then their votes were presented to the Cardinals' Commission. If the pope attended the final Thursday meeting, as was the usual practice, the decisions reached were finalized then and there.

For censorship of a book, the Holy Office not infrequently turned to temporary collaborators. We have already met this type of censor when Father

IL SAGGIATORE
Nel quale
Con bilancia eſquiſita e giuſta
ſi ponderano le coſe contenute
nella
LIBRA ASTRONOMICA E FILOSOFICA
DI LOTARIO SARSI SIGENSANO
Scritto in forma di lettera
All'Ill.mo et Reuer.mo Mons.re D.
VIRGINIO CESARINI
Acc.o Linceo M.o di Camera di N.S.
Dal Sig.re
GALILEO GALILEI
Acc.o Linceo Nobile Fiorentino
Filoſofo e Matematico Primario
del
Ser.mo Gran Duca di Toſcana.

FILOSOFIA NATVRALE

MATEMATICA

IN ROMA MDCXXIII.
Appreſſo Giacomo Maſcardi

F. Villamoena Fecit.

1. Frontispiece of original edition of *The Assayer*. "I have seen the frontispiece of *The Assayer*, sent me by Signor Stelluti, which pleases me very much, and if between the two words Astronomical and Philosophical is added a small 'and' above, this small printing error might be rectified." Galileo, in a letter to Federico Cesi, October 9, 1623 (*Works*, cit., XIII, p. 134).

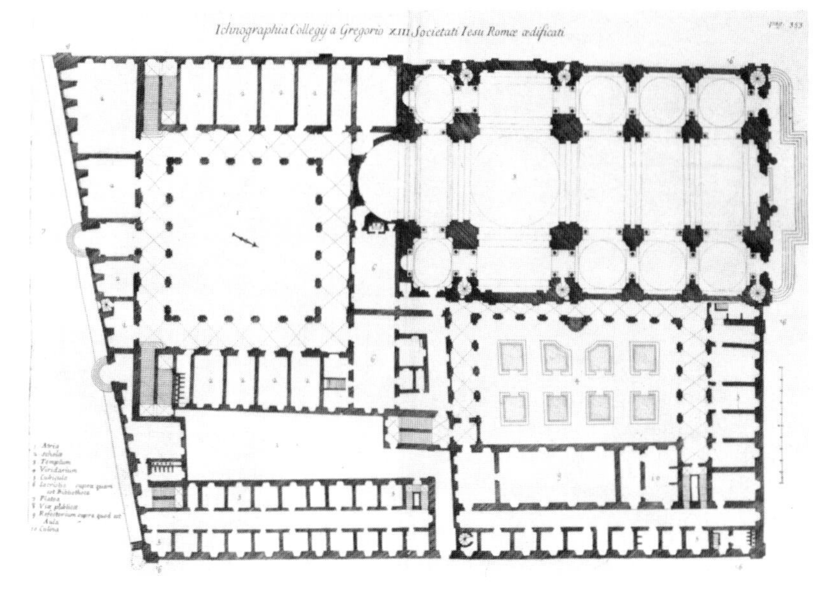

2. Floor plan of the Collegio Romano at end of the seventeenth century Erected at the order of Gregory XIII in 1582 by Father F. Valeriani, S.J., and Bartolomeo Ammannati, it was from 1584 the very modern and admired seat of the Collegium Nationum, created by Ignatius of Loyola (1551) to sanction a supranational Catholic education. Its functional arrangement imposed itself as the prototype of Jesuit university architecture (see the Louis Le Grand College in Paris). In fact, the arrangement around independent courtyards framed respectively by didactic and communitarian buildings realized a happy and ideal synthesis between the abbey or medieval convent and the Renaissance palace, between a studious community and public instruction. From the piazza (7) one entered the courtyard of the *scholae* (2). On the second floor, which could be reached by sumptuous twin staircases, are the congregations' halls and chapels, the Hall of Declamations or the Great Hall (in the southern corner) for theses and the theater, and the faculty libraries. The garden courtyard of the community (4) was surrounded by services: kitchen (10), refectory (9), offices, laundry, printshop, and (after 1631) a pharmacy where illustrious visitors were offered phials of balsam, cups of chocolate, and other refreshments. On the upper floors are the fathers' rooms (5), the Hall of Honor (*aula massima*) for grand ceremonies, the infirmary,

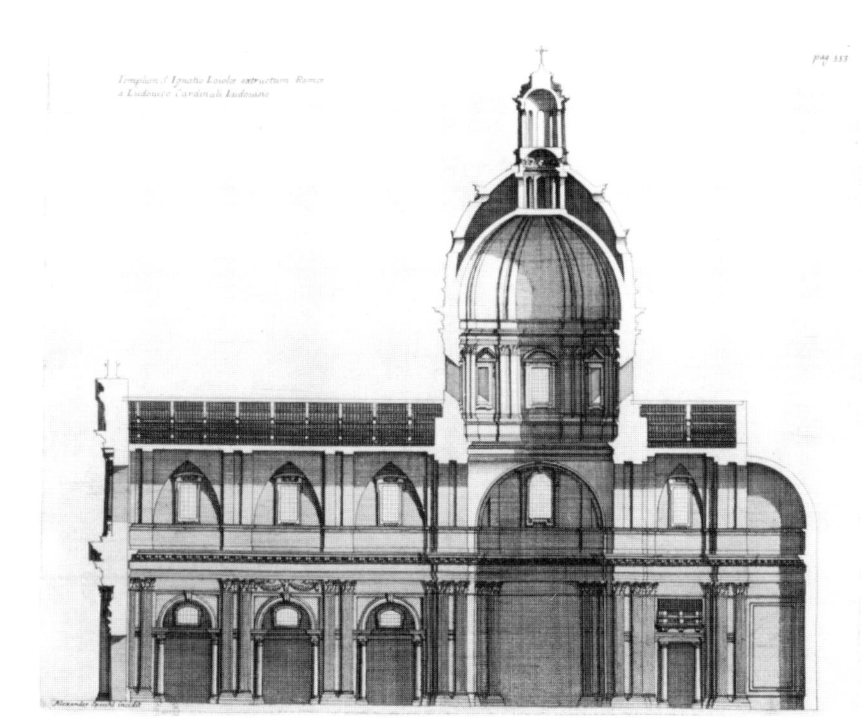

the "secret library" (the main library, reserved for the fathers, containing even heretical and forbidden books), and annexed to it the Kircheriano Naturalistic Museum (1651) and the loggia for astronomical observations. On the carriage courtyard (1) stood ordinary buildings (the stables, slaughterhouse, warehouses, granary, food storehouse, and workshops): the Collegio administered important lands and property left in trust. Connected to the *scholae* and the community through the sacristy (former Church of the Annunciation), the new Church of St. Ignatius (1626-1650), included in the edifice with the demolition of original sections, completed the rigorous harmony of the Collegio Romano's ensemble.

3. Bartolomeo Ammannati (?), façade of the Collegio Romano, 1584. At the center, beneath Pope Gregory XIII's coat of arms, is the motto of the Collegio: *Religioni ac bonis artibus*. Religious fervor and orthodox instruction at a high level went hand-in-hand, with identical methods set by the *consuetudines* and the *ratio studiorum*. The doctrinal lessons, the spiritual direction and meditation, the exercises, functions, and processions corresponded in method of instruction to the *praelectio*, the repetition, the disputes and *exercitationes*, the lectures, and the thesis ceremonies. Another didactic instrument was the Scholastic theater, auditorium of declamation for eloquence and sermons. Two thousand outside students came from thirty nations (from eight to eighteen years of age): novices, religious of every order, and laymen (who then pursued medicine or law in the universities). It was forbidden for the professors to administer corporal punishment. Only for the most turbulent of the small "grammar students" was a paid lay "corrector" resorted to. In conformity with Ignatian individual voluntarism, the pedagogic criterion was religious and Scholastic competition, stimulated by prizes and honors, with the selection of the best students in the academies and congregations. The docent system (thirty fathers plus assistants) maintained grammar schools and the humanities (rhetoric and history) and a Faculty of Theology (four years, five for novices) and of Philosophy (three years, courses of logic, Aristotelian physics, mathematics, and metaphysics).

4. Father Orazio Grassi, plan in cross-section of the Church of St. Ignatius at the Collegio Romano. There is a single grand nave, covered by a barrel vault, with ample openings over the chapels, spaced by pairs of pilasters. The plan displays, as does the façade, a scrupulous fidelity to the lesson of the interior of Vignola's Church of Gesú. The only variant here consists in the elegant free columns supporting the chapel's arches. But this is an episodic and decorative innovation attributable not to Father Grassi, but rather to the original designs of Domenichino, and was probably agreeable to the man who financed the church. Cardinal Ludovisi, known for his refined pictorial taste. Note the never-realized monumental cupola (more than 13 m in diameter). In the nineteenth century, however, on one of its enormous mutilated pillars will rise the Jesuit Father Angelo Secchi's famous astrophysical observatory.

RATIO
PONDERVM
LIBRÆ ET SIMBELLÆ:
IN QVA
QVID È LOTHARII SARSII
LIBRA ASTRONOMICA,
QVIDQVE E' GALILEI GALILEI
SIMBELLATORE,
De Cometis ſtatuendum ſit, collatis vtriuſque
rationum momentis, Philoſophorum
arbitrio proponitur.

Auctore eodem LOTHARIO SARSIO *Sigenſano.*

LVTETIÆ PARISIORVM,
Sumptibus SEBASTIANI CRAMOISY, viâ Iacobæâ,
ſub Ciconijs.

M. DC. XXVI.
CVM PRIVILEGIO REGIS.

5. Frontispiece of the original edition of *Ratio ponderum librae et simbellae*, Paris 1626, by Lotharius Sarsius Sigensanus, anagram of Horatius Grassius Salonensis. Like a herald sent to the enemy, the frontispiece exhibits a caduceus with the inscription "nunc omina pacis" (omens of peace now) above two entwined fighting serpents separated by an olive branch. Also, the book's epigraph repeats this spirit of reconciliation: "Compose the adverse spirits and extinguish these flames." In a handwritten postil is Galileo's comment in two lines of verse: "Simula il viso pace, ma vendetta / chiama il cor dentro, ed altro non attende" [The face simulates peace, but the heart within calls for vengeance, and lies in wait for nothing else].

diſputare cuius ignorem principia; nihil ideo de hac ſententia ſtatuo. Habeat hanc ille ſibi ſine riuali. illorum hac de re arbitrium eſto, qui rectè ſentiendi æquè ac loquendi Magiſtri, incorruptæ fidei tutelæ excubant. Nonnullus tamen qui me angit ſcrupulus aperiendus eſt. Hic mihi ex ijs naſcitur, quæ ex Patrum, Conciliorum, ac totius Eccleſiæ placitis, de Euchariſtiæ ſacramento apud nos indubitata cenſentur; qualia illa ſunt: Abſcedente, verborum potentiſſimorum vi, panis ac vini ſubſtantia, ſupereſſe nihilominus eorumdem ſenſibiles ſpecies, colorem ſcilicet, ſaporem, calorem aut frigus, has vero diuina tantùm vi, atque vt ipſi loquuntur, miraculosè ſuſtentari. Hæc illi. Galileus vero diſerte aſſerit calorem, colorem, ſaporemque ac reliqua huiuſmodi, extra ſentientem, ac proinde, in pane ac vino pura eſſe nomina; ergo abſcedente panis ac vini ſubſtantia, pura tantùm qualitatum nomina remanebunt. Quid ergo perpetuo opus miraculo eſt puris tantum nominibus ſuſtentandis? Videat ergo hic quàm longè ab ijs diſtet, qui tanto ſtudio harum ſpecierum veritatem ac duratione͂ firmare conati ſunt, vt etiam diuinam huic operi potentiam impenderint. Scio equidem lubricis ac verſutis ingenijs videri poſſe, patere hinc etiam effugium aliquod, ſi fas ſit ſanctiſſimorum fidei præſidum dicta ad libitum interpretari, eaque à vero & communi ſenſu aliò detorquere. Verum, quod in terræ motu ſuadendo non licuit, cuius tamen quies inter

6. Page 174 of "Examination XLVIII" (see Documents for translation) from Galileo's copy of *Ratio ponderum librae et simbellae*. In the margin, corresponding to the Eucharistic objection on this page, is an illegible note of reply, perhaps by Galileo.

G3. 292

Havendo ne' giorni passati scorso il libro del Sig. Galileo Galilei, intitolato Saggiatore, sono arrivato a considerare una dottrina insegnata già da alcuni antichi Filosofi, da Aristotele efficacemente rigettata, ma dal medesimo Sig. Galilei rinnovata: et havendola io voluto ragguagliare con la vera, et indubitata Regola delle dottrine revelate, ho trovato, che al Lume di quella Lucerna, quale per essercitio e merito della nostra Fede, riluce in luogo caliginoso, ma che però più sicura, e più certamente di qualsivoglia naturale evidenza ci illumina, apparisce falsa, ò pure (che io non giudico) molto difficile, e pericolosa, di sorte che chi per vera la riceve, non titubi poi nel discorso, e nel giuditio di cose più gravi; però ho pensato di proponerla a V.P. R.ma e pregarla, come faccio, a dirmene il suo senso, che servirà per mio avvertimento.

Dunque il pred. Auttore nel libro citato à fogli 196. lin. 29. volendo esplicare quella propositione spesso proferita da Aristotele in più luoghi: Il moto è causa di calore; et aggiustarla ad un suo proposito, prende a provare, che quelli accidenti, che communemente si chiamano colore, odore, sapore etc. dalla parte del soggetto, nel quale si stima communemente si trovino, non siano altro, che puri vocaboli, e solamente siano nel corpo sensitivo dell'animale che li sente. Và esplicando questo con l'essempio del Solletico, ò vogliam dire Titillatione, cagionata dal toccamento di qualche corpo in certe parti dell'animale, concludendo che come il solletico, quanto all'attione, tolto via il senso dell'animale, non è differente dal tatto, e movimento, che si farebbe sopra una statua di marmo, ma tutta è nostra affettione, così quegli accidenti, che si apprendono da nostri sensi, e si chiamano sapori, odori, colori etc. non sono, dice egli, ne' soggetti, ne' quali si stima volgarmente che siano, ma solamente ne' nostri sensi, sicome la titillatione, non è nella mano, nè nella penna, che tocca per essempio sotto le piante de' piedi, ma solamente nell'organo sensitivo dell'animale.

Ma questo discorso parmi che pecchi in assumere per provato quello, che doveria provare, cioè che in ogni caso l'oggetto, che si sente sia in noi, perche l'atto che è circa di quello è in noi; nè discorrerà bene chi dicesse: la vista, con la quale vedo la luce del sole, è in me, dunque la luce del sole è in me. Ma che che sia di sì fatto progresso, non mi fermo a essaminarlo.

Segue l'auttore ad esplicare questa sua dottrina, e s'ingegna di dimostrare, quello, che siano questi accidenti in ragione di oggetto, e termine delle nostre attioni; e come si vede a fogl. 198. lin. 12. comincia ad esplicarli con gl'atomi d'Anassagora, ò pure di Democrito, quali egli chiama minimi, ò particelle minime, et in questo, dice, continuamente si vanno risolvendo i corpi, che però applicati alli nostri sensi, penetrano la nostra sustanza, e secondo la

7. First page (292r) of manuscript G3, preserved in the Archive of the Sacred Congregation for the Doctrine of the Faith, former Holy Office, ser. AD EE. For a translation, see Documents.

8. Autograph letter from Father Orazio Grassi to Giovanni Battista Baliani, Genoa, April 17, 1648. The document is part of a confidential scientific correspondence between the two scientists, now both quite advanced in age, between 1647 and 1653. The content attests to the role played by Father Grassi in support of the Jesuit scientific initiative against a "proof of the vacuum." The letter was probably written in the Genoan Nuovo Collegio near Via Balbi, planned in 1635 by Father Grassi and inaugurated in 1642, or in the Casa Professa di Genoa at the Church of Sant'Ambrogio.

9. Raphael, *The Dispute Concerning the Holy Sacrament*, 1509. "Beneath the divine chorus on a blue background open the wings of a chorus of Fathers and Saints of the Old and New Testaments, seated in alternate order on the clouds, and they witness the great sacramental mystery. In the midst of four cherubim shines the Holy Ghost in the customary form of a white dove with opened wings, encircled by radiant light and suspended above the sacramental host on the Altar. The Altar is raised atop two tiers and a marble base opened up by two steps, from which one ascends to the upper level, on which are disposed the chief figures of this majestic holy action. On the Altar is displayed the golden monstrance with the particle of divine bread. At the sides are the four doctors of the Latin Church: Gregory, Jerome, Ambrose, and Augustine circumfused with the light of the Holy Ghost. . . . And here, with admirable industry, are varied the emotions and expressions of those who, standing behind the primary figures and who thus cannot see them, fix their eyes on the space between, thus managing to catch a glimpse of the sacramental host. . . ." (G. P. Bellori, *Descrizione delle immagini dipinte da Raffaello d'Urbino nelle camere del Palazzo Apostolico Vaticano*, Rome 1695, p. 8 ff.). "Saints, doctors, and Christians who in sixes, threes, and twos argue over history; one sees in their faces a certain curiosity and anxiety in trying to find certainty on that about which they are in doubt, giving evidence of this by

their arguing with the hands and by certain movements of their bodies, pricking up their ears, and knitting their brows. . . ." (G. Vasari, *Le Vite* [Lives of the Painters], ed. P. Barocchi, Florence 1976, vol. XIV, "Vita di Raffaello d'Urbino," p. 172). Among the volumes at the base of the altar, one can recognize the Bible and the Epistles. St. Gregory the Great (at the left, wearing the pope's tiara) is in a chair with zoomorphic decorations, arousing at the time of the Counter-Reformation suspicions of paganizing concessions. Bellori defended Raphael by citing a similar bishop's chair in the Church of Santa Maria in Cosmedin (Rome) as proof of the intended fidelity of the fresco to furnishings in use by the early Church.

10. Detail from a drawing for Raphael's *Dispute Concerning the Holy Sacrament*. This finished drawing, with all the figures already delineated, reveals the original elaboration of the goldsmith's object meant to be the protagonist of the fresco. Its design corresponds to traditional Eucharistic iconography connected with the ritual of the Mass: the chalice surmounted by the Host. In the fresco's realization, it will be replaced by a sunburst monstrance of more abstract and functional design for worship independent of the liturgy of the Mass, symbolic of a speculative theology.

11. Tintoretto, *The Last Supper*, 1594, Presbytery of the Church of St. Gregory the Great, Venice. "The night unfurled and all around made the ghosts rove among shadows and horrors" (G. Casoni, in C. Ridolfi, *Le meraviglie dell'arte*, Venice 1648, p. 66). Dark lights, like the nocturnal meteors of so many of his paintings, illuminate this institution of the Eucharist by Tintoretto—"in extravagant position, which is illuminated by a lantern at the center" (ibid., p. 51). With its modern setting in a Venetian tavern, anachronistic and not conforming to the Biblical text, this representation of Christ giving Communion to the disciples according to the Tridentine liturgy reflects an orthodox Counter-Reformation polemic at the moment of the publication of Bellarmino's *Controversiae* (Ingolstadt 1589-1593). We will have the proof of this a few years later: in 1599, the frontispiece of the Venetian edition of the *Controversiae* was decorated by a Eucharistic iconography inspired by this last work of Tintoretto. This iconography and these light effects reappear in 1608 in Rome in the *Institution of the Eucharist*, commissioned by Clement VIII from F. Barocci, and will glitter in the light of torches during the judicial ceremonies at the Church of Santa Maria sopra Minerva (Aldobrandini Chapel) near Filipino Lippi's *Triumph of St. Thomas* (Carafa Chapel).

12. Nicolas Poussin, *The Eucharist, The Seven Sacraments*, Dal Pozzo series, Rome 1634. "There are in this work three artificial lights, two of which come from an oil lamp with two wicks hanging above, casting light on the front of all the figures. The third is added by a candle located on a bench below, with the rays and shadows duplicated and triplicated, intersecting each other with larger and smaller, more or less apparent angles, depending on the distances, as can be seen in the bench itself and the leg of the couch" (G. P. Bellori, *Vite di pittori*, vol. I, 1672, p. 193). Contemporaries did not fail to notice the geometric control imposed by this "innovative" painting on the light effects used by Tintoretto and Barocci for the same theme. In fact, Bellori emphasized the head-on illumination of the scene and recalled that Poussin had studied the manuscripts on optics by the Theatine Matteo Zoccolini in the library of Cardinal Barberini. Nor did contemporaries fail to notice that this painting religiously reconciled such "natural and mathematical points" with fidelity to the Biblical text of the institution of the Eucharist by means of a subtle archaeological reconstruction: "the same figures in apostolic garb of the early Church" (ibid., p. 155) and the heterodox iconography of the couches, used before only by the Galilean painter L. Cardi da Cigoli in 1592. The "good and new disposition and expression" (ibid., p. 206) of Poussin's Eucharistic iconography inspired Champaigne, the painter of the Abbey of Port Royal des Champs, and was reflected in the Chantelou sacramental series.

13. Frontispiece of the book by Daniello Bartoli, S.J. *Historia della Compagnia di Gesú, Della vita e dell'Istituto di S. Ignatio* . . . , Rome 1650. "*Ite et inflammate omnia*. But since it is good that every fire and every celestial light come from the Father of Lights . . . Jesus communicates a ray of light to the heart of Ignatius, and from him it is transmitted to the four parts of the world . . . invested by so much light . . . fecundated by the light divine" (Father A. Pozzo, S.J., *Lettera al Principe Liechstenstein*, Rome 1924, p. 6, quoted in Villoslada, S.J., *Storia del Collegio Romano*, cit., p. 181). In the firm celestial hand of St. Ignatius, the identification of the Eucharistic dogma with a qualitative idea of light triumphs in the coat of arms of the Society of Jesus: a translucent Host giving off rays and decorated with the monogram I.H.S., or *Jesus Hominum Servator*. A constant iconographic element in all the Jesuit production

3

PRopensissimus fuit Galileus ut crederet, omnia corpora constare atomis, diversimodè figuratis & combinatis; ac non dari alias qualitates & accidentia, nisi motum localem atomorum. Ita disertè profitetur in opere inscripto *Il Saggiatore*, in quo, digressionem de hoc argumento, exorditur his verbis pagina 196. versu ult. *Dico che ben sento tirarmi dalla necessità, subito che concepisco una materia o sustanza corporea, a concepire insieme ch' ella è terminata, e figurata di questa o di quella figura; ch' ella in relazione ad altre, è grande o piccola, in questo o quel luogo o tempo: nè per veruna imaginazione, posso separarla da queste condizioni. ma che ella debba esser bianca o rossa, amara o dolce, sonora o muta, di grato o ingrato odore, non sento farmi forza alla mente, di* do-

A 3

(from the frontispieces of *Ratio studiorum* to the mathematical works of P. Clavius, from the frescoes of Baciccia to the fronts of the churches), this symbology is here the threshold image of an official history of the Society's activity and is emblematic of a universal palingenesis (the four female figures representing the four continents irradiated by penetration of the dogma's worship) to be performed under the guidance of St. Ignatius. In 1685, the subject and symbology of this frontispiece will be used by A. Pozzo in the *Universal Triumph of the Society of Jesus* depicted on the ceiling of the Church of St. Ignatius, according to the inspiration for his words quoted above.

14. First page of the anonymous *Exercitatio de formis substantialibus et de qualitatibus physicis*, Navesi, Florence 1677 (?). In 4°, 24 pp. bound in boards, it carries no printing information. From markings on the three quartos, one can assume the completeness of this book, of which no other copies are known. But it was intended to be bound together with the *Conclusiones philosophicae* with frontispiece and, at times, a copper engraving. The decorative border and the type faces are of a kind common in the second half of the seventeenth century, but typical of the Scholastic editions of the Jesuit College of San Giovannino in Florence by the printer Ippolito Navesi.

15. Agostino Ciampelli (1577-1642), *Adoration of the Eucharist*, fresco on ceiling of the Sacristy of the Church of Gesú, Rome. The liturgical exigencies imposed by Catholic reform inspired an exuberant architectonic production of tabernacles and baldaquins for the worship and display of the Eucharistic sacrament, such as, for example, C. Rainaldi's baldaquin-tabernacles in the Church of Santa Maria della Scala in Rome. The new devotional circumstances of the pious practice of Forty Hours led to the scenographic and theatrical experience of the Jesuits in putting up provisory Eucharistic settings with light effects of which this fresco gives us a partial notion.

16. G. Francisi, *Portrait of Saint Roberto Bellarmino*, oil on canvas, 1923. This unfaithful copy of the seventeenth-century painting is in the Palace of the Holy Office (Rome), reception hall, north side.

"Thus we shall have a new portrait of the celebrated personage, a portrait which the author is happy to present as more faithful because delineated by Bellarmino himself; that is, from his letters to relatives, from personal memories, from his very amnesias. . . . [L]ittle does it matter to him that the portrait has a physiognomy completely different from the traditional one, even that it might be a complete opposite" (P. Tacchi Venturi, S.J., *Il beato Roberto Bellarmino. Esame delle nuove accuse contro la sua santità*, Rome 1923, p. 3). Ambitious, insincere, nepotist, and above all affected by senile dementia, this is the new image of Cardinal Bellarmino proposed by the Catholic historian Monsignor P. M. Baumgarten in 1922, which Tacchi Venturi rejected as an insult to the traditional "most noble personage of the loftiest character" (ibid.). Hence the contemporary apologetic interest in the picture of a senile Bellarmino. And this is proved by the selection of an unfaithful seventeenth-century etching (1644) taken from the original painting (see fig. 17) to decorate Tacchi Venturi's above-mentioned apology. He was the official historian of the Society, and important cultural and political personality (head of the *Italian Encyclopedia*'s ecclesiastical subjects section), from 1919 rector of the Church of Gesú, responsible for the artistic patrimony of the church and the Casa Professa, which are indebted to him for reconstructions and replacements of "dignified rearrangement," with pavements, altars, and balustrades "in perfect Baroque style" (see A. Dionisi, S.J., *Le stanze di S. Ignazio*, Rome 1982, pp. 50-52). His predecessor, the Society's historian P. Daniello Bartoli, wrote in the seventeenth century about Bellarmino's image: "As regards that truer and more worthy portrait of him which will be that of his soul, it is of little importance to know about his body. . . ." (*Vita*, cit., p. 256). There existed in Rome another seventeenth-century portrait of Bellarmino (now in the Church of St. Ignatius of Loyola) which was tampered with by painting in a halo at the moment of his canonization. Is it possible that the original and now lost "senile portrait" of the cardinal-inquisitor is hidden beneath the deceptive appearance of this painted beatitude?

17. Portrait of Cardinal Bellarmino, location unknown. The photograph reproduced in the Jesuit G. Castellani's article "Bellarmino R.," in the *Italian Encyclopedia*, vol. VI, Rome 1930, p. 548, came from the Superior General Curia of the Society of Jesus. "Chosen by the eternal Argonaut to steer the great ship of Ignatius's squad against the furor of the icy wind which from the north brought cruel and horrid tempests and the fierce deceits of British assaults, I pulled the powerful oars of holiness. I hurled the flaming darts of truth" (G. B. Marino, in "Ritratto del cardinale Bellarmino," in *La galleria del cavalier Marino*, Venice 1620, p. 123). "He was of a somewhat less than average stature, had a broad forehead, lively and pleasant eyes, a rather hooked nose, large ears, a decent mouth, a thick head of hair, at first black and then white, a sparse beard, a pink-and-white skin, a rather smooth complexion, only somewhat choleric and sanguine; he had good health and a temperate nature and was very well suited to bear the long labors of studies and negotiations" (G. Fuligatti, S.J., *Vita del cardinale Bellarmino*, Rome 1624, p. 344). This first biography of Bellarmino was decorated by an etching (touched up for hagiographic reasons) of the original portrait, which provides the only clue for a hypothetical dating.

18. Cristiano Banti, *Galileo before the Inquisition* (106 × 140.5 cm), Florence 1857. "Banti could not have chosen a subject of greater importance for any heart that feels love of country. . . . Galileo, the bold and indefatigable investigator of the secrets of science, for having affirmed an innocent astronomical truth . . . is called before that tenebrous tribunal and, almost as if he were a mischievous student, is questioned and sharply reprimanded. . . . On a small table the Bible is open. On the face of the friar . . . one sees expressed all the ferocity of ignorance. . . . The figure of Galileo seems to us too vulgar. . . . According to us, the scene is too monotonous and homely; the painter could have introduced other figures, some of which might have alluded to the consquences of a sentence pronounced by the Holy Office" (*Rivista di Firenze* 1 [1837], p. 464 ff., quoted in Matteucci, *Banti*, cit., p. 58). The painting reflected the then strong desire for "historical truth" apropos of the 1633 trial. It reconstructs the first audience as described in 1852 in the grand-ducal Alberi edition of *Opere di Galileo*, cit., vol. IX, p. 457: "in the presence of Father Maculano and the assistants, Fiscal Procurator Carlo Sincero and another who is not named and who shows him a book with the title *Dialogue* . . . he goes on to speak of the precept imposed on him by Cardinal Bellarmino." Note the anonymous assistant who is depicted with his face obscured, the *Dialogue* and the very evident unsealed protocol sheet, and the indictment revealed in Monsignor M. Marini's *Galileo e l'Inquisizione*, Rome 1850. Details and title lead one to believe that the painter was inspired by contemporary historical documentation, emphasizing Galileo's innocence with edifying features. However, the severe criticism quoted above expressed reservations that can no longer be ours. Where we see a forced light, it found the dramatic and didactic rendering of the scene insufficient. The superficial reading of it, as if the inquisitor were fanatically quoting the Bible, simplified the liberal lesson of the painting and gave it the aspect of melodramatic legend.

Grassi authorized De Dominis's book on the tides and Father Riccardi authorized *The Assayer*. True enough, Father Riccardi was a consultant *in pectore*; but Father Grassi, so far as we know, never set foot in the Holy Office's palace.

Canon law gave the Tribunal of the Holy Office very broad judiciary prerogatives. Indeed, this tribunal was, and still is today, competent on anything that might directly or indirectly involve Catholic doctrine or endanger orthodoxy. In other words, this was a tribunal which had criminal jurisdiction over accusations of heresy and related matters—magic; superstition; doubts in matters of faith; above all, the dogmatic doctrine of the sacraments, the reading of forbidden books, and conversation with heretics—and which passed sentences, made declarations, or gave dispensations on all subjects. It was a very important tribunal of canon law. However, it differed from almost any other kind of judicial administration, before or after it on the face of the earth, in two exceptional respects. In contrast to any other tribunal, the Holy Office knew no limits to its supreme jurisdiction—no limits, in terms either of territory or of persons. Having jurisdiction over every land inhabited by man, this tribunal of the faith could send its emissaries to all corners of the earth and persecute anyone: not only Catholics, but also heretics, schismatics, pagans, and even Jews. Also included were bishops and apostolic nuncios; from the point of view of the Holy Office, only cardinals and the pope enjoyed juridical immunity.

But the second special characteristic was, and is, the "secret of the Holy Office." This was a special and very strict bond, the most rigorous form of judicial secrecy. It bound lawyers, witnesses, experts, and all parties to the case—these were reconfirmed by periodic oaths—to the most complete silence on the documents, discussions, and votes relating to criminal as well as inquisitional trials.[17]

The guilty party (we are thinking, for example, of Galileo after his condemnation in 1633) must also observe the "secret of the Holy Office" with respect to the possible behind-the-scenes events of his trial, or incur the maximum punishment and the declaration of ecclesiastical unfitness which is meted out to violators. Superior to every other secrecy, that of the Holy Office was equal to "the sacramental seal of confession," and was equally protected.

On the other hand, as we have seen, the widest publicity was accorded to the sentence and decrees. This was rather like the publicity accorded political treaties, with the difference that here the secrecy on the behind-the-scenes events was more rigorous, the revelations rarer and more distorted, and espionage practically impossible. That is why, although in Paris, Prague, and

[17] See *Cod. Iur. can.*, can. 1623.

Madrid there were just as many spies as in Rome, only in Rome was the profession of spy an invaluable art.

Manuscript, Anonymous, Undated

It now appears clear that the Galilean document whose existence I discovered was not protected by any trial or pretrial secrecy, since there had been no trial or pretrial procedure, but only a *nolle prosequi* concerning *The Assayer*. This was and is still today an absolute historical certainty.

The documentation preserved in this connection by the Holy Office had a biographical and intellectual interest for the historian, but it did not have any judicial connection with the famous "Galileo case"; that is, with the dossier, complete or partially complete, on the trial of the *Dialogue* relating to the 1616 procedure against Copernicus. This is probably why, on the morning of June 11, 1982, I was able to consult at the Holy Office, beneath Cardinal Bellarmino's benevolent gaze, that mysterious document: Father Guevara's papers, which had cleared *The Assayer* of all accusations of Copernicanism.

The document I had discovered was a manuscript, unquestionably written in the calligraphy of the early-seventeenth century (Fig. 7). The first thing that struck me was the handwriting: a flowing, tidy hand, regular, and with a uniform slant. The style was very personal, inimitable, and could not be confused with automatic, thoughtless handwriting, with the exactly uniform laying out of letters by the hand of a copyist from the period. It was, then, an autograph, but the person whose hand had written it had eschewed a signature; nor was it dated.

The manuscript consisted of two sheets of ordinary paper (20 × 27 cm), the first sheet written on both recto and verso, the second only on recto. The two sheets had been countersigned, but by a different hand—probably at the end of the seventeenth century—with the numbers 292 and 293 at the top-right. These numbers corresponded to the progressive pagination of the bound volume that contained our document. This was an old binding, identified with the initials "EE" of the series "AD," where AD stands for "Acta et Documenta." If, at this point, surprised as I was by the absence of a signature at the foot of sheet 293, you instinctively turn the page to see whether some other sign might be found on the other side, you will be rewarded with a new surprise and a sharp disappointment: page 294 is missing.

Even if the next page had once contained something related to the two preceding pages, we will probably never know; for it has been torn out, a not infrequent occurrence in the archives of that period. What was written on page 294? A postscript? A decision? A notation? Was it blank? Any hypothesis is both legitimate and perfectly futile. In order not to discourage the reader too much, though, I will try at least to give him a few further details on the binding that contained our document; that is, the "Acta et Documenta" series, of which I had volume EE open before me. But, I must point out, I had been authorized to consult only those two pages, not any of the others in that thick binding.

Even if, with the suffering felt by a historian in such cases, I obeyed the restriction of consulting only those two pages, it was not difficult to see that no rigorous filing system had presided over the gathering of this mass of papers (quite normal for the epoch). A criterion of rigorous filing would have allowed me to advance a provisory hypothesis—if not on the moment when it was written, at least on the date at which, by and large, those pages had come to the Holy Office's chancellery. The volume contained petitions and opinions, not sentences, decrees, or, so far as I could tell, reports of interrogations. Otherwise, the series would probably have been entitled "Decreta" and not "Acta et Documenta."

This, however, must have been the only criterion of classification. Leafing quickly through the pages, I could see documents, dated and undated alike, which seemed to confirm my first impression (i.e. that this was a collection of documents from the first half of the seventeenth century) and which also made it clear that they had been put together pell-mell, without concern for their chronological sequence. Thus, the pagination provided only an arbitrary ordering of the sheets in that volume—and this only to make possible a convenient table of contents at the beginning of the volume when it was bound.

The same hand that had numbered those pages—again, quite likely at the end of the seventeenth century—also wrote the table of contents for the volume. At this point, another surprise awaits us: the table of contents, made purposely to enable one to find the documents contained in the bound volume, neglects to cite our document.

But let us not be too surprised, for omissions of this sort are not infrequent in seventeenth-century archives. Still, there is no doubt that a professor of paleography who might want to give a lesson on the shortcomings of the period's archivists could easily use this document for a case study. Men and time seem to have conspired to keep it secret through the centuries.

Trial Records on File

If we are to understand how in the world, after more than three centuries, the fascination of that mysterious document has only now fallen into our hands, we must take a step back (though, to tell the truth, not too long a step). Anyone who has even leafed through the famous national edition of Galileo's voluminous *Works* knows very well that at the beginning of our century Antonio Favaro, the impassioned editor of that great edition, had gained access to all the Galilean documentation existing in the Vatican's secret archives as well as in the more secret Archive of the Holy Office, "opened exceptionally owing to the lofty and enlightened wisdom of Pope Leo XIII,"[18] as he recalled with emotion. In this way, the great work was worthily crowned by the first and still exemplary integral publication of the trial documents relating to the proceedings against Galileo in 1616 and 1633.

This new document confirmed that all the official documentation relating to Galileo's trials was contained in archival material already identified, but published in an incomplete manner. The Holy Office, as we know, did not preserve the files that are most pertinent to the events of the 1633 trial. And that was simply because the archives had been repeatedly flung open, without requests for permission, during the course of a recent history that deserves to be briefly told.

Like those of any trial, the records of Galileo's trial were collected in two series in the Holy Office's archives: "Decreta," which gathers together very brief trial records, decisions, and chronologies; and "Processus," containing summaries, notes taken during interrogations, witnesses' testimony, defense statements, and sentences, i.e. the real and proper contents of the trials. In 1810, by order of Napoleon, the Holy Office's archives went to Paris, together with three thousand cases of pontifical documents. In a special package addressed to the minister of cults, there arrived at Versailles a volume of documents from the "Processus" series: the documents concerning Galileo.[19] But the passage from the Holy Office to the Versailles Palace did not increase public knowledge of the contents of those Galilean folders. For France, in any event, this was historical loot and an enviable instrument of political pressure on the Holy See. More ephemeral than the latter, Napoleon was forced to exit the scene before beginning the proposed publication, of which we will find only paltry signs during the Restoration.[20] The Bour-

[18] A. Favaro, *Galileo e l'Inquisizione. Documenti sul processo galileiano esistenti nell'Archivio del Sant'Uffizio e nell'Archivio segreto Vaticano*, Florence 1907, p. 8 (*Works*, XIX, pp. 272-74).

[19] See the paleographic description and history of vol. 1181 of the Secret Archive of the Vatican, in A. Favaro, "I documenti del processo di Galileo," *Atti del Regio Istituto Veneto di Scienze, Lettere ed Arti* 61 (1901-1902), pt. II, pp. 757-806. Favaro supported J. B. Biot's version, in the *Journal des Savants* (1858), p. 398 concerning the return to Rome of the trial documents from Paris, through the offices of Pellegrino Rossi. This version has since been corrected on the basis of new Viennese documents that came to the Vatican Archive, published in Monsignor A. Mercati, "Come e quando ritornò a Roma il codice del processo di Galileo," *Atti della Pontificia Accademia delle Scienze, Nuovi Lincei* 8 (1926), pp. 58-63. Besides the preceding sources, see D. Berti, *Il processo originale di Galileo Galilei*, Rome 1878 (2nd ed., augmented), p. 3 ff. See, also, M. Marini, *Galileo e l'Inquisizione*, Rome 1850, for the first partial publication; and L. Firpo, "Il processo di Galileo," in *Nel quarto centenario della nascita di Galileo Galilei* (Pubblicazioni dell'Università cattolica del S.C.), Milan 1966, pp. 83-101. The critical edition of the trial dossier in *Works*, vol. XIX, pp. 272-421, is more modern and complete.

[20] See B. Venturi, *Memorie e lettere finora inedite o disperse di Galileo Galilei*, Modena 1821, pt. II, pp. 197-99. For Lalande's letter on these trial documents and their translations, harking back to the projected Napoleonic edition, see *Opere di Galileo*, ed. E. Alberi, vol. XVI, p. 305 ff.

bons, too, exit the scene: enter Louis Philippe, "roi de tous les Français."[21]

Its archives having been returned, Rome will bring pressure to obtain the restitution of that sensitive trial dossier. But this is to no avail, for it cannot be found in Paris. The volume, in fact, does not seem to bring luck to its new, illegitimate owners and, at the fall of Charles X, is carried into exile by the Court of the Duke of Blacas. Only in 1843, without fuss, will the volume be returned to Rome through the nunciature of Vienna, apparently in the same condition in which it had left. In Paris, rumors spread concerning secret diplomatic agreements and a Roman promise to publish the trial.

Yet, new events were at Rome's gates. "Long live Pius IX," the Italian liberals wrote on the walls, but when revolution came to Rome, in November 1848, the new pope fled precipitously to Gaeta. Just as precipitously, there arrived in Rome Professor Silvestro Gherardi, exiled scientist, deputy of the Constituent Assembly of the Roman Republic, and future minister of public instruction during that short-lived Risorgimental adventure. The clarion voice of positivist anticlericalism and a scholar of the history of Italian science, Professor Gherardi rushes to the Palace of the Holy Office in search of the famous Galilean dossier. The palace is guarded by republican militiamen, and there is a great coming and going of patriots: "too many people had free access." The dossier is not there. In truth, it had never been returned there: it was in the pope's personal library. A fleeing Pius IX had entrusted it to Monsignor Marino Marini, prefect of the Vatican Archives, who would deposit it (together with other documents of exceptional importance) in capsule X, where Antonio Favaro will find it and later publish it in its entirety. Professor Gherardi, "bitterly disappointed," had to content himself with seeing the Galilean documents of the "Decreta" series.[22]

The Holy Office's "Decreta" had meanwhile been transferred, in April of 1849. On order of the triumvir Giuseppe Mazzini the entire Archive of the Holy Office, for reasons of security, was moved to the Church of Sant'Apollinare, as the palace had become a kind of court of miracles, a refuge for the people of the Trastevere. The archives, perhaps "already stripped hastily by their masters before running away," must not be put at further risk. But a final adventure awaited it.

When General Oudinot restored order and the pope to Rome, the archives also returned to its illustrious seat. In the confusion, a French officer, perhaps more out of intimate republican conviction than personal interest, saw to it that an astounding quantity of those documents were once and for all removed from the protection of that palace: seventy-seven volumes of the

[21] On the Parisian stay and the vicissitudes of the Holy Office's documents after the negotiations carried out by Monsignors Gaetano Manni and Marino Marini and later by Cardinal Consalvi, see Favaro, "I documenti;" and Firpo, "Il processo di Galileo."

[22] See S. Gherardi, *Il processo di Galileo riveduto sopra documenti di nuova fonte*, Florence 1870, p. 5.

Inquisition were stolen away. They will end up in the library of the Protestant Trinity College in Dublin.[23]

But this time the "Decreta" documents had not left Rome. They were still there, as the same Professor Gherardi verified, braving the French patrols and slipping quietly into the Holy Office's palace. The anticlerical Gherardi could have taken them; but, trusting in Independence and a Roman capital of Italy, he had done nothing more than copy those documents, with the intention of publishing them. And this he did in 1870, when he saw that Rome the capital was confined to one side of the Tiber. Antonio Favaro will republish them in their entirety in 1907.

All this is brought up by way of explaining how in the Archives of the Holy Office there could be so very little about Galileo's trials. This small amount was published in the national edition of the complete *Works*: the report on a request for a confidential investigation of contacts between Galileo and Cremonini in 1611; about thirty reports and decisions of meetings of the Congregation concerning the two procedures. Before publishing this documentation, however, Antonio Favaro remarked that one was still far from certain of having everything concerning the reasons for Galileo's incrimination. "Other documents," Favaro recognized, "were not originally included among the trial documents themselves, or might be dispersed in other series, or, at least in part, might have been destroyed because, according to the ancient practices of the Holy Office, not everything is preserved."[24] Besides "ancient practices" there were more recent vicissitudes, even if Favaro discreetly neglected to mention the latter.

Our discovery has confirmed Antonio Favaro's prediction, even if the document did not pertain to any trial. And yet, not even Favaro, despite his indefatigable tenacity as a pursuer of Galilean documents, would have been able, at the time of his search at the Holy Office, to find what I sought and found there several decades after him. Indeed, we must remember that the table of contents of volume EE of the series in which it is found does not list the document. Thus, with only a cursory examination of the archives, without a systematic page-by-page search, the document was invisible.

It was only later, according to the Holy Office—after Antonio Favaro's foray and between the two world wars—that a catalogue of the various series of documents, injunctions, instances, petitions, denunciations, and opinions was compiled. The catalogue was twofold: by name of incriminated author and by title of examined book. We are not in a position to know whether it was an exhaustive catalogue, but we do know that it was exhaustive for volume EE.

The person entrusted with the task of cataloguing did a careful job: he did

[23] See H. Gaidoz, "De quelques registres de l'Inquisition soustraits aux Archives romaines," *Revue e l'Instruction Publique, de la Littérature et de Sciences* (May 1867), pp. 102-104 and 114-17.

[24] *Works*, XIX, p. 274.

not confine himself to sifting through the tables of contents, but (at least in the case of volume EE) systematically checked the documents contained. In this way, our document became visible and also acquired (probably, since I did not see them personally) its two file cards with the key words "Galileo" and "*The Assayer*." Thanks to this double-card index, my request made it possible to ascertain quickly that the Holy Office's archives preserved under series AD what I was looking for.

Hence, although we must be grateful for the diligent work of those who put together that precious card index, we may rightly ask why they limited themselves simply to filing our Galilean document, apparently without remarking on its existence. We can say only that this document seems to have gathered around itself a series of truly inexplicable circumstances, deliberate or not, but at all times serving to ensure maximum discretion. This constitutes its most surprising enigma—but not the only one. At the very moment when it at long last came to light for the historian, it disclosed a series of mysterious questions.

"G3"

The reader will find the text of the document reproduced at the back of this book (see Documents). The manuscript is in Italian, a language not exactly attuned to the dignity of an officious or official theological opinion. The text is set out in seven paragraphs, and there is no sign of intercalation between them. There is a single reference (in the left margin of the verso of sheet 292) to the *Metaphysicarum disputationum*, the famous manual of Catholic theology and philosophy by the Jesuit Father Francisco Suárez. Within the text, another reference: to the proceedings of the Council of Trent. As was to be expected, one finds numerous quotations, very accurate, from *The Assayer*.

Before examining the contents of the document, it must be pointed out that in the upper left-hand margin of the first page, a figure or word has been deleted. Instead, at the top of this page, one reads another code, "G3," in a hand different from that of the text. But it seems to be the same hand that has numbered the pages. Most probably, this is a registration reference, an archival signature whose meaning is hidden to us: Galileo? Guevara? Three copies made or, much more likely, three pages?

Having reached this point, the reader will by now be impatient to know what Father Guevara had "put on paper" on those pages:

> Having in past days [thus begins the mysterious document] perused Signor Galileo Galilei's book entitled *The Assayer*, I have come to consider a doctrine already taught by certain ancient philosophers and effectively rejected by Aristotle, but renewed by the same Signor Galilei. . . . [T]his doctrine appears false, or even (which I do not judge) very difficult and dangerous. . . . I have therefore thought to propose it to

you, Very Reverend Father, and beg you, as I am doing, to tell me its meaning, which will serve as my warning.

The reader will once again be disappointed, and again surprised, to discover that, despite everyone's good will and good faith, the manuscript which was brought to me and which everyone expected would be Father Guevara's dismissing opinion was not the opinion of Father Guevara.

As one sees from the end of the first paragraph, this anonymous document is indeed addressed to a responsible authority of the Holy Office, almost certainly to the Master of the Sacred Palace (that is, Father Niccolò Ridolfi, who had granted the imprimatur to *The Assayer*). But it is not at all an expert theological opinion given at the request of the Holy Office, such as that which Father Guevara gave or was supposed to have given. Here no authoritative, officially absolving opinion is expressed; rather, an opinion is requested on private and denunciatory grounds. This is a denunciation.

The exordium is of the kind prescribed in cases of denunciation, even if this one was officially anonymous. It served to turn aside all suspicion of premeditation, of personal rivalry with the accused, of a presumption of judgment.

A denunciation to the Holy Office was deposited in the sole interest of the faith. This is why someone declares preventively that the reading of Galileo's book—an almost casual reading, without preconceived intentions—has aroused serious religious scruples, which the denouncer humbly requests the tribunal to dispel by pronouncing itself authoritatively concerning these doubts of a Christian conscience.

In short, this is a denunciation in good and proper form, as will also be apparent from the document's structure and contents. Indeed, it is a model denunciation, perfectly in line with the prescriptions of Martin del Rio's famous book[25] as well as with the more recent manual of penal procedure, the *De haeresi*, by the Roman jurist Prospero Farinacci;[26] the latter work was also certainly consulted and reliable, for it had been dedicated to the cardinals of the Congregation of the Holy Office and scrupulously checked by Cardinal Bellarmino.

An anonymous denunciation, to be effective, was not available to just any fanatical friar. In cases of anonymous denunciation, in fact, the preliminary judicial procedure was entirely documentary; as the sycophant did not wish to appear, it was thus impossible to proceed to the examination of witnesses.

If the denunciation was to be effective, moreover, it had to be stated with both mathematical and theological-juridical rigor. One had to point out the precise incriminating passages in the denounced book: passages, statements, precise and relevant words. Not only that, but one had to indicate with pre-

[25] See M. del Rio, *Disquisitionum magicarum libri sex*, Moguntiae 1624, p. 755; P. Farinacci, *Tractatus de haeresi*, Romae 1616.

[26] Father B. Pereira, *De communibus omnium rerum naturalium principiis*, Romae 1576.

cision on what canonical texts and what doctrinal points these passages and affirmations by the author aroused legitimate suspicions of heresy. All this had to be confected according to an elegant protocol and with a sincere Christian spirit, in order not to discredit the seriousness of those doubts by suggesting motives of personal hostility or to appear to be assuming the prerogatives and authority of the Holy Office.

Our denunciation too, after its exordium, was followed by the rigorous selection of incriminating quotations from *The Assayer*. The pages and lines quoted were from the original edition of 1623. Immediately after this, the denouncer described the philosophical and doctrinal difficulties which those statements produced with respect to orthodoxy, as in every text of controversial dialectics worthy of the name. *In fine*, it having been explained why these propositions of *The Assayer* were foolish in philosophy and having been made clear that they were also formally heretical, came the conclusion, repeating the denouncer's submission to the authority of the tribunal.

From the point of view of its formal redaction, and given the situation, the denunciation could not have been improved upon. We shall see, too, that the substance was no less perfect. But, before going into that, we shall first present some elements for a legitimate hypothesis as to the dating of the document.

A first element of the dating is offered by the quotations taken from *The Assayer*'s first edition (1623). A second element—much weightier than the first—is the absence of any reference to Galileo's official condemnation in 1633. Now it should be observed that it would not have made any sense to denounce the author of *The Assayer* after he had already been condemned for heresy because of the *Dialogue*. Above all, it would have made no sense to denounce *The Assayer* after 1633, without some allusion to the author's condemnation for heresy. Even if the incriminating book or point of discussion were different, in a denunciation one could not overlook the condemnation already inflicted on the author. About this it is not reasonable to have any doubts.

Thus, it becomes necessary to suppose that the document was drawn up sometime before 1633. Moreover, the initial expressions used by the anonymous writer leads one to think that "Signor Galileo" was alive and *The Assayer* only recently published. Internal analysis offers other elements.

Exegetical Cunning

The denouncer knew a very sure test for measuring the truth of a theory such as that of *The Assayer*: "compare it" with Catholic doctrinal truth. Should that comparison or translation be possible, good; if, however, the concordance were not exact, the translation contradictory, then it would be clear that the theory is false. That is why he turned to the Holy Office instead of remaining within the normal confines of university philosophical controversies.

It is quite clear from the very first sentences of the exordium that the denouncer was not a fanatical follower of Aristotle, since what interested him more than the honor of Aristotle was the concordance between philosophy and faith. That is to say, the author of the denunciation declared himself quite emphatically to be more than an Aristotelian—to be a follower of "theological philosophy," a very modern and respected current of thought.

To be a follower of theological philosophy meant not to be a succubus at any cost of Aristotle, even less of St. Thomas Aquinas, but to aim constantly at "linking philosophy and the sacred words" in order to guard "against the modern cunning of modern heretics."[27] These principles were to be found in the famous manual of natural philosophy by Father Pereira, *De communibus omnium rerum naturalium principiis et affectionibus*, but were also in Suárez and in all the Jesuit books on philosophy. In fact, "theological philosophy" was the official philosophy of the Society of Jesus.

To put it so prominently on view from the very start, our denouncer must have studied at a Jesuit college. Moreover, the author of the document must have attended university for a long time, because the text shows us that he had full command of the golden rules of argumentative method typical of Jesuits in theology, and had the assurance of a professor schooled in all the techniques of *disputationes*:

> You shall reduce the adversary to his principles [the first golden rule]. When you demonstrate the truth you will adopt the descending method (that is, argument from general to specific); but when you impugn the false, you must use the ascending method, and when confronted by a pertinacious adversary, you can reduce him to the absurd.[28]

The text of the denunciation was articulated in accordance with these instructions. Thus, the plan of the text is as follows. First of all, two suspicious opinions are selected. The first is that of the subjective nature of the perception of color, odor, and taste, together with the subjectivity of the sense of touch illustrated by Galileo with the example of tickling. The second is the explanation of the perception of such phenomena by means of "the atoms of Anaxagoras or of Democritus, which [Galileo] calls minims or minimal particles . . . so that . . . the aforesaid accidents are not distinguished from atoms except in name."

The identification of incriminating opinions, accompanied by verbatim or

[27] Ibid.

[28] Petrus Mosnerius, *Philosophia tomus primus*, ed. Father H. Fabri, S.J., Lugduni 1646, art. VI ("De Methodo concertationis et exercitationis" and art. VI ("De Methodo exercitationis pro respondente"), pp. 25-28.

paraphrased quotations from *The Assayer*, is followed by the objection. This second part contains a first criticism, only hinted at, tending to demonstrate that the first incriminating opinion was false in philosophy. The second objection, more articulated and serious, suggests instead how the second incriminating opinion is formally heretical, being irreconcilable with the Catholic faith and contrary to the authority of theological tradition.

Let us comment on the first criticism, concerning philosophical character. This had not demanded a great effort of the imagination: it opposed to Galileo's philosophy the classic principles of Scholastic realism. "I deny what is assumed," the denouncer seems to say. And in any event, Galileo could not demonstrate his "assumption." Colors are perceived by the sense of sight, but can one thus exclude the possibility that they might exist outside the perceiving organ? When I feel a tickle, I can say that I am tickled; but when I see red, can I therefore think of myself as being red? Obviously not.

The denouncer made the same dialectical argument in another way, demonstrating by *reductio ad absurdum* the falsity of Galileo's opinion: although there is in me the sight with which I see the sun, I cannot therefore say that the light of the sun is in me. Nor was this argument original; but for us it is very valuable, because it constitutes an element of internal analysis which identifies the point in time of our manuscript, and the state of mind behind it.

This controversial criticism, centered on the image of the "sun in me," was in fact fashionable in 1624. Provoking objections of this sort in Rome was a book by the noted Protestant philosopher Jean Mestrezat, who was consecrated as pastor by the infamous Du Moulin and who continued the latter's polemic against Cardinal Bellarmino. The book, entitled *De la communion à Jésus Christ au Sacrement de l'Eucharistie*,[29] had led the Jesuits to designate as the order of the day and the order of rebuttal the problem of the sense perception of tickling, of sound, and of the light of the sun. As the light of the sun is present to our sight, the French Protestant had written, even though the sun is distant from us, so in the Eucharist we perceive the divine presence, even though it is not a local and substantial presence as Cardinal Bellarmino and the Council of Trent had claimed.

Official Catholic reactions during those same months, in any event, were no less active and influential. Cardinal Jacques-David Du Perron, Primate of

[29] J. Mestrezat, *De la communion à Jésus Christ au Sacrement de l'Eucharistie contre les cardinaux Bellarmin e Du Perron*, Sedan 1624, p. 84 ff. On the heated Eucharistic polemic that procedes publication of *The Assayer*, I cite A. Chrastovius, *Triumphus Iesuiticus, hoc est, Redargutio Iesuiticarum contradictionum quas propugnat Bellarminus in doctrina de Eucharestiae Mysterio*, Lipsiae 1608, 2nd ed. 1620; Father Pierre Coton, S.J., *Du traes sainct et très auguste Sacrement et Sacrifice de la Messe*, Avignon 1600; P. Du Moulin, *Apologie pour la sainte cène du Seigneur contre la présence corporelle et transubstantiation*, La Rochelle, 1609; Father F. Veron, S.J., *La manducation fantastique des prétendus réformez en leur cène*, Paris 1621. For the most important Catholic position, see R. Bellarmino, *Disputationum de controversiis Christianae fidei*, Ingolstadt 1607, tome II, p. 747 ff. and J.-D. Du Perron, *Traité du Saint-Sacrament de l'Eucharistie*, Paris 1622.

France and protagonist of the Eucharistic controversy from the beginning of the century, was dead, but in 1624 his fundamental *Réfutation de toutes les objections tirées des passages de Saint Augustin alléguées par les hérétiques contre le Saint Sacrement de l'Eucharistie* appeared posthumously in Paris. The philosophical and theological Augustinianism of the new sacramentalist heretics—in particular, the old nominalist idea that "the shape is the truth"—was denounced *en bloc*.[30]

Thus, the crucial importance of the Eucharistic apology for Catholic faith was once again the order of the day—more than ever in 1624—and it is at this moment and by this logic that we must now consider the second criticism concerning *The Assayer*'s atomist materialism. This is, as we shall see, the real heart of the denunciation, forming its doctrinal nucleus.

There was, however, a difficulty. *The Assayer* had managed to be none too explicit—speaking now of minims, like the Aristotelians; now of atoms; now of particles of natural Aristotelian elements; now of fire corpuscles and shaped corpuscles. Our Tartuffe, who was an Aristotelian but certainly not a professional philosopher, preferred to advance with leaden feet, with scientific rigor, so as not to find himself in a poor position and leave the adversary possible escape routes.

Hence, the denouncer approached *The Assayer*, making use of his instruments (which were Aristotelian instruments) to perform a rigorous exegesis of Galilean atomism and to demonstrate that, whatever else it might be, it was certainly a heretical doctrine. Basing himself on Aristotle, *Of the Heavens* and the *Physics*, our anonymous person suggests that *The Assayer*'s doctrine is derived from Anaxagoras's substantial (that is, materialist) atomism or—alternatively—from Democritus's theory of shaped atoms.[31] The denunciation evaluated the consequences in the case of the first attribution, which was based entirely on the quotation of a single line from *The Assayer* (p. 200, l. 28), where Galileo had actually mentioned the "matter of these bodies," speaking of their particles.

Before reporting the consequences, we must first add a brief comment here. Why, indeed, Anaxagoras and not simply Democritus, traditionally relied on to stigmatize atomism? In reality, this attribution is an internal element revelatory of the anonymous person's knowledge and preoccupations, and perhaps also of his identity.

It is quite obvious why apropos of *The Assayer* its denouncer tried to drag in Anaxagoras. Was it not Anaxagoras, the successful Athenian philosopher of recognized impiety, whom Aristotle presented in the *Meteors*[32] as the author of the theory of celestial lights through the clouds (by way of explaining

[30] See J.-D. Du Perron, *Réfutation* . . . , Paris 1624, p. 1.

[31] See Aristotle, *Of the Heavens*, 302a 10-20, 302b; *Physics* 187a 26 and 187b 8 ff.

[32] Aristotle, *Meteors* 345a 27.

lightning flashes), just as Galileo had explained the optical effects of the comets as lights through the clouds?[33]

In other words, the person who denounced *The Assayer*'s atomism as a return to Anaxagoras must have immediately identified Galileo with Anaxagoras in light of Galileo's theory of the optical nature of comets. We shall see that the astronomical origin of this identification will soon be officially brought forward as another objection to *The Assayer*.

But let us turn now to the denunciation itself. It declares that if Galileo's atoms are substantial, like Anaxagoras's homœomeries, then Galileo's doctrine is not compatible with the existence of the Eucharistic accidents sanctioned by the second canon of Session 13 of the Council of Trent.

A great "experimental" principle of philosophic and theological value was the miraculous permanence of heat, color, taste, smell, and the other sensible accidents of the bread and wine after the Consecration, which transformed their entire substance into the body and blood of Christ. If we interpret those accidents as *The Assayer* wished (that is, in terms of "minimal particles" of substance), then even after the Consecration there will be particles of the substance of the Eucharistic bread which produce those sensations. Thus, if we adopt the ideas of *The Assayer* in physics, there would remain particles of the bread's substance in the Consecrated Host; but this is an error anathematized by the Council of Trent. And it is with this anathema that the first part of the denunciation of Galilean atomism ends.

It should be said in this connection that the proceedings of the Council of Trent had just been republished in a "Roman" edition in 1618. So our anonymous person had before his eyes the fundamental council decision on the dogma of the Eucharist. It read:

> If someone says that in the sacrosanct sacrament of the Eucharist there remains the substance of the bread and wine along with the body and blood of Our Lord Jesus Christ, and denies this marvelous and unique conversion of the entire substance of bread into body and wine into blood, which allows only [*dumtaxat*] the species of bread and wine to subsist, a conversion that the Church very appropriately calls transubstantiation, on him be anathema.[34]

Need we recall how often this great formula of Tridentine Catholic faith was repeated in catechisms and theological discussion? But the denunciation

[33] Ibid. 369b 15 ff. For Anaxagoras, the lightning flashes were fires descending from the sky through the clouds. On comets, ibid. 342b 27. Both in the *Discourse on the Comets* (1619) and in *The Assayer*, Galileo had mentioned Anaxagoras's astronomical and meteorological theories.

[34] See *Decreta Sacrosanti Concili Tridentini . . . auctoritate apostolica edita*, Vallisolati 1618, sess. XIII, "Decretum de SS. Eucharistia," can. 2; Denzinger, *Enchiridion 884, Concilium Tridentium Diariorum actoruum, epistularum*, tome VII, pt. IV, vol. II, Freiburg 1976, p. 216.

continued with an examination of the other characteristics of *The Assayer*'s atomism, those which made one think of Democritus's shaped atoms.

Perhaps *The Assayer* had an escape route. Galileo had written that something of the sensible qualities still remained: the shaped quantity; that is, shape and size. Is such a quantity enough as an objectively knowable datum to save the existence of the Eucharistic accidents?

No, it is not enough, replies the author of the denunciation. And he goes on to explain. Even though he gets mixed up, stumbles, and creates difficulties—he is obviously not a philosopher by profession—whenever his pen runs into philosophical terminology (quantity, mode), he nonetheless manages to make himself clear. Galileo had identified quantity with substance. Thus, the Eucharistic accidents must be "triangular shapes, acute or obtuse"; that is, forms of matter. The denouncer does not impugn this opinion, since he is not a philosopher; rather, he appeals to his teachers, who belonged to the dominant theological current and whose teaching it was to keep quantity and substance distinct so as not to fall into the nominalist error and make transubstantiation contradictory.

The authority cited by the anonymous person is Father Suárez's philosophy and his "modal" theory of quantity as separate from substance. The Suárez passage cited is from the treatise *Metaphysicarum disputationum*, a textbook in the major Jesuit colleges. In that disputation, Father Suárez had rejected the ideas of Ockham and other nominalists as regards their erroneous opinion on the identification of quantity and substance.[35]

The denouncer avoided repeating that destructive analysis. For the moment, we shall do the same and content ourselves with this observation: the "philosophy" of accidents in *The Assayer* is accused of challenging the accredited theological philosophy and of contradicting the dogma of transubstantiation by affirming the real permanence of quantity as a shaped substance.

Let us, therefore, summarize the two arguments of the accusation. To translate the Eucharistic dogma into the grammar of *The Assayer*'s physics implies a violation of the Tridentine formulation on two counts. Indeed, *si primum*, the permanence of substantial particles by way of explicating sensible accidents implies the permanence of the bread and wine even after the Consecration. Thus, there would not be any miracle. *Si secundum*, there would be no miracle even to make the quantity persist, because one is still talking about a shaped substance.

The third argument, in the case of the adversary's pertinacity, was, as noted, *reductio ad absurdum*: one grants, absurdly, that Galileo accepts the

[35] See F. Suárez, *Metaphysicarum disputationum*, Parisiis 1619, tome II, disputation 40 ("De quantitate continua," pp. 365-406), sec. 2, p. 368 ff.: "Utrum quantitas molis fit res distincta a substantia materiali et qualitatibus eius."

dogma of the Eucharist. That is, one admits Galileo is disposed to believe that the entire substance of the bread and wine, shaped or not, disappears. From this it follows, on the basis of his theory, that all the accidents must also disappear. But the Council of Trent (Canon 2) says infallibly that they exist; all persist and are real. On this grave doubt of conscience the denouncer has asked the authoritative judgment of the Holy Office. It is clear that if he submitted that doubt to the Tribunal of the Holy Office, rather than to his own confessor, he did so in order to solicit the intervention of the judicial authority.

In 1624, moreover, the accusation was certainly the most serious of all, owing to the renewed challenges from heretics at that moment. It should suffice to recall that the argument of transubstantiation had been one of those which had permitted Cardinal Bellarmino to clear the path in the long trial against Giordano Bruno and, quite recently, allowed Cardinal Scaglia to go all the way in the De Dominis trial.

Protection of the Scholastic interpretation of the Tridentine dogma was the crucial point of official Catholic theology. From Cardinal Bellarmino to Cardinal Du Perron, from Father Suárez to Father Vazquez, the absolute value of sensible qualities had been, as we shall see, the impregnable defensive bastion of Counter-Reformation faith.

Imbroglio

How can this document discovered in the Holy Office be reconciled with Guiducci's contradictory revelations? If we take Mario Guiducci's statement as true, the hypothesis spontaneously arises that *The Assayer* had been subject to two denunciations, almost at the same time. One of these was a strange and mysterious imputation of Copernicanism; such an imputation was false and Father Guevara had headed it off, albeit in an incomprehensible manner. The second denunciation was the one we have before us.

However, a second hypothesis fits better with the reliability of Guiducci's letter and goes as follows: *The Assayer* has not been incriminated by its apology of Copernicanism, but for its atomist doctrine. The true denunciation, which the preoccupied Guiducci referred to, perhaps without knowing its contents, was this one. Father Guevara had been asked to head off this denunciation with a favorable opinion. The first hypothesis, until contrary evidence is available, does not stand up. The second was coherent, and I could argue for it by answering a series of questions.

Why—first question—had I discovered this denunciation instead of Father Guevara's famous sheets, which I had looked for and thought I had found? Because the only thing I could find at the Holy Office was what had been deposited at the Holy Office. The denunciation, obviously, must necessarily have been deposited. Father Guevara's papers must not necessarily have been sent to the Holy Office.

Looked at carefully, Mario Guiducci's letter did not say that that defense had been deposited with the tribunal. We must remember that, for reasons of political convenience, the proceedings opened by the denunciation against *The Assayer* did not follow normal procedure. There was the impending intervention, on the part of a friendly cardinal, which violated the procedure and had recourse to the expert opinion of a theologian having no official standing with the Holy Office. Father Guevara was neither a consultant nor a qualificator.

In an exceptional procedure of this kind, there are two possibilities: either the opinion written by Father Guevara had been drawn up in the form of a memorandum for the private use of the cardinal who had arrogated to himself the denunciation, or these papers had been brought in as proof that there was no case on which to proceed. In the first instance, there was no point in looking for it in the Holy Office; in the second, they either were lost or I had not found them.

Was Father Guevara's opinion written on the torn out sheet no. 294? Was the decision not to proceed with the case on that missing sheet? We shall never know.

Was Father Guevara's opinion deposited at the Holy Office, and is it still there—in another series, not yet catalogued; or perhaps bound in the same volume EE in which I found the denunciation, among those documents that I was not authorized to study, presented in such a way as not to be easily identifiable either through the volume's table of contents or through the card catalogue of the archive? In this case, a historian will find it one day.

The second question is as follows: Why Father Guevara? In other words, why, on the basis of the denunciation we now have, can we finally understand Guiducci's allusion to Father Guevara as a theologian truly able to ward off the danger facing *The Assayer*?

Who was Father Guevara? Historians know only that he was a prominent personage inasmuch as he was father general of the Order of Regular Minor Clerks, a religious authority in the Rome of that period. To understand how matters stood, it was necessary to track down this little known personage of the Galilean experience, find out what he did and what he thought, study what he wrote, and perhaps even find among his papers letters or drafts relating to the famous "sheets."

Father Giovanni di Guevara

In 1624, the order of which Guevara was father general was protected by Urban VIII and in great ascendancy. Today it is in complete decline, reduced to a few dozen members.

One must go to the eastern periphery of Rome in the Monte Sacro quarter to find, near the parish of the Church of Santi Angeli Custodi, the very modest seat of what was one of the liveliest religious congregations in seventeenth-century Italy and Spain. What remains of the archives of the Regular Minor Clerks no longer contains any trace, so it would seem, of the former father general.

Objective history is not only made up of archives accumulated over time, still unpublished, but also of dispersions and destructions. The destructions of the last world conflict have erased any traces of Father Guevara where it was probable to find them: in the archives of the archiepiscopal curia of Teano, a southern Italian town near Caserta, where Guevara was named bishop by Urban VIII in 1627, and where a tablet in the cathedral records the political and intellectual qualities of its ancient head. But it is precisely these qualities which have left behind indirect and direct clues that permit us to know Father Guevara's personality.[36]

Giovanni di Guevara was a Neapolitan patrician attracted to the intellectual vocation that distinguished the Congregation of Regular Minor Clerks—also called *caracciolini* from the name of their founder—which had arisen in Naples at the end of the sixteenth century. From the College of San Guiseppi in Naples, where Guevara had studied, the Regular Minor Clerks had fanned out into Spain and Rome. In Rome they had a college for Neapolitan novices, first on the Piazza Giudea and later on Piazza Navona at the Church of Sant'Agnese.

During the first decades of the seventeenth century, the order made an effort to spread across central Italy, and Father Guevara gave evidence of political talents by establishing privileged relations with the various Italian states. In 1607 he was sent to Pesaro, where he found a college, taught logic, and ensured the fathers of the Regular Minor Clerks the enviable position of an order protected by the grand duke. He was recalled to Rome for a new diplomatic mission to the Republic of Genoa, where he could count on the support of an aristocratic Genoan family with which he was related on his mother's side. He stayed in Genoa until 1610 and, during that period, founded the Collegio di San Fedele.

He probably continued to teach logic, philosophy, and theology, but he also participated in the effort of his order to compile, under the guidance of the Neapolitan Father Gennaro Campana, tables of astronomical observations of Mars, Venus, and Jupiter. Thus it was that, like other Regular Minor Clerks, Guevara also carried out an occasional observation of the new

[36] See L. Allacci, *Apes Urbanae*, Romae 1683, p. 158; B. Chioccarelli, *De illustribus scriptoribus qui in civitate et Regno Neapolis floruerunt*, Neapoli 1780, p. 335; N. Toppi, *Biblioteca Napoletana*, ivi, 1678, p. 119; and V. F. Uguelli, *Italia sacra*, Venetiis 1720, tome VI, p. 575. On the role played by Father Guevara during the first phase of the Barberini pontificate and within the Clerks Regular Minor, see Piselli, *Notizia historica della religione dei C.R.M.*, p. 105 ff.

star that appeared in 1604. He imparted this information to Professor Antonio Magini, of the University of Bologna, who was then the greatest lay authority on astronomy in Italy.[37]

Though Father Guevara did not pursue his initial astronomical interests, he remained a passionate dilettante in astronomy. Yet, even his sporadic experience with celestial observations permitted him immediately to become a great admirer of the author of *Starry Messenger*.

Father Guevara again left Rome, this time to go to the Duchy of Urbino; for Duke Francesco Maria had now decided to found a house of Regular Minor Clerks. At Urbino the Regular Minor Clerks had become the duke's spiritual advisers and librarians. The special merits he obtained in Urbino earned Guevara his election in 1619 as father general of the order. The following year, the pope bestowed on the order the honor of the official oration in the pontifical chapel on the Day of the Circumcision.

In 1623, the Duchy of Urbino lost its sole heir. The duchy was in decline and was fought over by Florence and Rome, both desirous of obtaining its annexation. Father Guevara's privileged connections in the Court of Urbino must have played a determining role in helping the Church's interest to prevail. Francesco Maria abdicated in favor of the pope and retired to the Regular Minor Clerks' house of Castel Durante to pray and study philosophy.

Father Guevara, on the other hand, in order to ratify the duchy's annexation to the States of the Church in 1623, at the beginning of the new pontificate, went on a diplomatic mission to Florence. The result of that visit was the dedication to the new young grand duke of Florence, Ferdinando II, of a book of political morality, *Horologio spirituale dei principi*, which Father Guevara published in Rome in 1624 and which partly consoled the Medicis for renunciation of their claims to Urbino.

This fortunate annexation of the Duchy of Urbino was the initial diplomatic success (it would prove to be the only one) of the new Barberini pontificate; and Father Guevara was involved in it. Urban VIII therefore had reason to keep him in the Curia as a political consultant, "for his knowledge as well as for his skill and prudence in handling great affairs,"[38] and to put his experiences and diplomatic abilities at the service of the Church. The opportunity presented itself immediately, because the pope wanted to settle the Valtellina conflict diplomatically, by means of an apostolic legation to Paris and Madrid entrusted to Cardinal Francesco Barberini. The cardinal, as his uncle knew very well, understood nothing about politics. Urban VIII put by his side Father Guevara, who, with the "great talent he had for polit-

[37] See A. Favaro, *Carteggio inedito di Ticone Brahe, Giovanni Keplero e di altri celebri astronomi e matematici dei secoli XVI e XVII*, Bologna 1886, p. 287 ff.

[38] Ibid.

ical maneuvers,"[39] was the consultant needed to deal with Cardinal Richelieu and the Count of Olivares.

The independence of mind, if we can call it that, of Richelieu and the diplomatic clumsiness of the cardinal-legate were superior to Father Guevara's best efforts. All the same, the political crisis was settled in the sense desired by the pope; and perhaps some merit must go after all to Father Guevara, because on his return from Spain in 1627 the pope rewarded him with the title of Bishop of Teano. This was the first step to the purple, which was not difficult to foresee.

In any case, Cardinal Barberini's absence from Rome had not been useless for Father Guevara. On the ship that brought him, together with Cardinal Barberini, from Civitavecchia to Marseilles, then to Lisbon and finally to Rome, these two friends and defenders of Galileo had many philosophical discussions. Father Guevara had even found the time to begin writing a commentary on Aristotle's *Mechanics*. He will finish and publish the book on his return, availing himself of Galileo's advice.[40]

In the midst of his political maneuvers, Father Guevara had not in fact neglected to cultivate philosophical studies, and he was already well known in Rome for his important treatise *De interiori sensu*, a very interesting book, unjustly forgotten by Galileo's historians.[41] Father Guevara was also—like the more authoritative philosopher of his order, Father Aversa—one of those Scholastics and religious who had lined up with Galileo, disrupting the front of official culture in Rome, which was only apparently compact.

The philosophical formation of the Regular Minor Clerks was molded by the most traditional Thomism. This order could not boast, like the Jesuits, of a prestigious tradition of mathematical studies, an aggressive philosophy, or even a marked vocation for cultural controversy. They were a cultivated religious order, traditionalists in philosophy, but neither bold nor combative. Their intellectual vocation was wedded to their religious one, which was above all contemplative and devoted to good works. The mystical contemplation of the Eucharistic sacrament was, as with the Capuchins, one of the mystical dispositions of their faith and of their religious attitude toward this great seventeenth-century nucleus of culture and spirituality.

In *De interiori sensu*, the book that brought notice to Father Guevara as a philosopher, the problem of sensory perceptions was discussed by means of

[39] See Piselli, *Notizia historica della religione dei C.R.M.*, p. 165.

[40] Father G. de Guevara, *In Aristotelis Mechanica Commentariis*, Romae 1627, dedicated to Cardinal Francesco Barberini. On his return from the Vatican legation to France and Spain in the autumn of 1626, Father Guevara came to know Galileo personally (*Works*, XIII, p. 341 ff.) and until 1628 (ibid., p. 389 ff.) had a scientific correspondence with him. Galileo in his turn cited favorably the comment of Guevara's *Meccaniche* in the *Discourses* (*Works*, VIII, pp. 68 and 165).

[41] Father G. di Guevara, *De interiori sensu libri tres*, Mascardi, Rome 1622. The copy I studied is the one in the Vittorio Emanuele II National Library, in Rome.

a commentary concerning *On the Soul*. Father Guevara developed at length the old Thomist theory of intentional forms or species. The senses, according to this theory now fallen into disuse, perceive the formal aspects of things because they are "impressed" by images—by simulcra or "species" transmitted by the objects to the sense organs. The subject knows things because it assimilates in itself these species, as wax receives the imprint of a die and reproduces it.

The theory that Father Guevara defended had already served Aristotle's Arab commentators as a way of explaining the phenomenon of vision and the permanence of images on the retina. Now Father Guevara used it to demonstrate the efficacy of the "telescope or *perspicillum*" of the "very illustrious Galileo"; indeed, the concavity of the telescope's lens permits it to condense and render sharper the visual species feebly emitted by the most distant stars.[42]

In 1622, such a theory, put forward by a dilettante astronomer, was sure to produce a certain discomfort in the Collegio Romano's scientific circles, where for many years now, with the researches of Father De Dominis, Father Grassi, and Father Zucchi, the telescope had been studied on the basis of a geometrical optics much more weighty than such metaphysical beliefs.

The doctrine of "species," which St. Thomas had codified and Father Guevara harked back to, was no longer in current use and, in fact, will soon disappear completely from the history of philosophy. But the terminology survived and survives even today; the term "species" continued and still continues to be used in theological language. Indeed, we already know that it had played a central role in the Council of Trent's definition of the Eucharistic dogma, where the sensible appearances of consecrated bread and wine were defined as "species."

Of course, like all philosophy books in the seventeenth century, Father Guevara's *De interiori sensu* also assigned a central place to the dogma of the Eucharist. This was all the more to be expected in a book on sense perception, where it was essential to deal with the problem of the physical sensation of color, odor, taste, and the other accidents in a manner compatible with the Eucharistic ones. The Council of Trent had not gone into this sophisticated problem of perceptive psychology; it had said only that one still dealt with permanent "species" after the conversion of all the original substance of bread and wine.

The dominant theological interpretation, throughout the whole presti-

[42] Ibid., pp. 64-68. The theory of the optical functioning of the telescope by "the multiplication of luminous species" was based on "the argument of refractions" of which Galileo himself made use; see, for example, the *Discourse on the Comets* (*Works*, VI, p. 75). This theory harks back to the beginning of G. B. della Porta's *De refractione*, on the basis of which luminous species, traversing a dense, transparent medium, multiply.

gious genealogy of great Jesuit theologians that we mentioned before, was for a realistic theory: real accidents without substance. The Jesuits, masters of controversial and dogmatic theology, would not even deign to use the old Scholastic word in their treatises and their disputes with heretics, and emphasized their interpretation by speaking of "accidents" and of "qualitates reales sine subiecto" (real qualities without subject).

But as a good, belated Thomist, Father Guevara did not want to subscribe to such a draconian philosophical interpretation of the Tridentine Eucharistic dogma; nor did he want to humiliate his order by being hitched to the chariot of a triumphant theology of Jesuit stamp. Father Guevara and his order had a religious aspiration and an intellectual identity not to be confused with others. And when he spoke of the Eucharist, he did it with perfect fidelity to the letter of the Tridentine Council and to the Thomist philosophical tradition of subjective perception. That is to say, he spoke of species subjectively impressed on the sense organs by Eucharistic appearances.

The new spiritual substance—Father Guevara's book observes in this connection—is a spiritual substance inaccessible to the external senses. The Eucharistic species, on the other hand, are perceptible.[43] And anyone who had read his book knew also that those formal appearances were, like the others, perceptible through the emission of simulcra, or images, to the intention of the sense organs.[44]

Father Guevara was not at all obsessed by the need to offer a rational and controversial theory of the Eucharistic mystery. But neither did he hesitate to believe in an idea of the perception of the Eucharistic species which, although a minority view and out of date in his own time, nonetheless did not violate, until the contrary was proved, the dictates of Tridentine dogma, even though it could not prevail over an authoritative theological tradition.

And Father Guevara was right to think that way. Still in the middle of the century, even such a bitter fighter in Eucharistic controversies as Father Thomas Compton Carleton, professor at the College of Jesuits at Liège, will recognize, albeit against his will, that the interpretation of Eucharistic appearances not in realist terms of absolute accidents, but in subjective terms of the "impression" of species, was not formally heretical, though it was by now discredited, heterodox, and perhaps rash.[45] In reality, as Father Compton Carleton will admit, the reconciling term "species" was "not too clear," and therefore it was not possible to condemn a theory inclined to consider

[43] See Di Guevara, *De interiori sensu libri tres*, n. 37, p. 50 ff., on the philosophical problem of quantity in relation to the eucharist.

[44] Ibid.

[45] Father Th. Compton Carleton, S.J., *Philosophia universa*, Antverpiae 1649, disputation XII, p. 246. This book was approved by Father A. Gottifredi, Father General Vincenzo Carafa's emissary in Belgium.

the Eucharistic appearances not as objective reality, accidents without substance, but impressions of a subjective nature.

At this point, nothing more need be said in order to understand why Father Guevara was the person most suited to silencing the accusation against *The Assayer*. Galileo—this pupil of St. Thomas and admirer of Galileo could well respond—does not support the doctrine of the subjective nature of Eucharistic appearances, since *The Assayer* does not speak of this theological question directly and explicitly. But even if *The Assayer* had held it explicitly, that doctrine was not to be damned, because it was not contrary to the dictates of the Council of Trent to declare that the sensible appearances of the Eucharist were "names," corresponding only to individual sensory perceptions.

The expert opinion would have been correct and, as a defense, vastly sufficient to silence the most heavily documented and frightening part of the denunciation. From this point of view, and only from this, the defensive argument reported by Guiducci's letter becomes acceptable and comprehensible. Father Guevara was authorized to remind the cardinals of the Holy Office that a realist interpretation of Eucharistic accidents could not be based on anything but its privilege as a greatly accredited theological theory, certainly not on a restrictive doctrinal definition. As for atomism, *De interiori sensu* had also cited the ideas of "the very ancient philosophers Leucippus and Democritus" as among the philosophical theories of sensation discussed in Aristotle's *On the Soul*.

An opinion of this kind from an authoritative Scholastic philosopher, author of a book on the problem of sense perception, exculpated *The Assayer* and made unnecessary any further documentary inquiry. Thus, the denunciation could be filed away with a *nolle prosequi*. The reasons we have described permit us perhaps to solve the mysterious and contradictory questions that Guiducci's letter until now left dangling in midair.

Therefore, having taken apart and interpreted Guiducci's document, we lean toward the hypothesis that the denunciation revealed in it, as well as the defense on the part of Father Guevara which that document lets us glimpse, referred not to an accusation of Copernicanism, but to the more recently uncovered denunciation concerning the incompatibility of *The Assayer*'s philosophy and Eucharistic dogma. The rest of our story will confirm this hypothesis.

But, before reliving it, we must ask ourselves whether Galileo had unwittingly fallen into a theological trap or been taken in by a slanderous quibble, without having the slightest inkling of it, or whether indeed Galileo and his Roman friends were even aware of the potential problem? The historian cannot attest to any reply on Galileo's part. So far as we know, he never pronounced himself on the question, at least in writing, by offering any speculation concerning a resolution of Eucharistic theology. Besides, he could not do otherwise. Officially warned against making proposals for the rational exegesis of certain Biblical passages, Galileo certainly could not venture to present personal interpretative solutions that differed from the dominant theological tradition entrenched behind the bulwark of the Council of Trent's words concerning the most important Catholic dogma.

On the contrary, as though to discourage historians from the temptation of superficial insinuations, Galileo will always explicitly insist, even with his assured friends, on his profound religious fidelity and devotion to the Church—a devotion that has been offended by the calumnies of his adversaries. This, in a period of widespread intellectual irreligiosity, was an answer, even if its significance is evidently too intimate to be easily deciphered.

However, we can exclude with absolute certainty the notion that Galileo and the men of the Church who supported him could overlook the fact that behind the words "heat, smell, and taste" lay dramatic centuries of Eucharistic debate which had again become topical. Nor could they be unaware that Peripatetic philosophy was welded to Scholastic theology precisely through the (until that moment) dominant interpretation of those words. It was the most difficult and disquieting link between "the material of faith" and "natural terms"; the new philosophy must perforce pass that way.

The Eucharist was the daily bread not only of religious life, but also of Catholic philosophical disputes. Nobody knew this better than Galileo, who in one of his manuscripts of Aristotelian notes called *Juvenilia*, influenced by the Jesuit philosophy manuals, had recourse to that argument. Disputing with Democritus, Ockham, Averroës, and Achillini on substantial forms and the nature of weight, Galileo had written in his own hand: "In the Very Blessed Sacrament there exists weight and lightness, and yet there is no substance whatsoever."[46]

We cannot bestow on the author of *The Assayer* a presumption of too naive an innocence, therefore, when his raising the question of atomism and Ockham's theory again ran into the Eucharistic objection. We can only surmise that he avoided confronting the question directly, since in the Catholic Church it was not permissible for a layman to discuss dogmatic or Scriptural theology. But his Roman friends also knew that it was a time of renewal and that new theologies aspired to supplant the Scholastic one.

With the justification of speaking about "pure natural terms," *The Assayer* proposed a materialistic theory of sensible phenomena to sever the knot: to separate natural philosophy from Scholastic theology; perhaps, from another point of view, to reconcile the book of the universe with the Bible concerning that great mystery of the faith. From such a perspective, too, our

[46] See "Tractatus de elementis, Quaestio tertia," in *Juvenilia* (*Works*, I, p. 132).

history will confirm this hypothesis. The rest of the story was all written and published, even if until now barely visible, in Galileo's *Works*. Before continuing it, the most important question still demands to be asked: who was the author of the autograph?

It is certain that it must have been an enemy of Galileo's who wrote those pages, but which one? At this point, Galileo had had three kinds of enemies. First came the Dominicans, who had publicly wanted to defend, in their own name, the literal interpretation of the Bible against an exegetical use of the Copernican hypothesis.

A second group comprised laymen, philosophers, and university professors; that is, Peripatetics and Aristotelians influenced by Averroës. These were the pupils of Achillini, Nifo, and Pomponazzi, who looked on Aristotle's scientific hypotheses as metaphysical truths: among them Giulio Libri, Flaminio Papazzoni, Giovanni Antonio Magini, Fortunio Liceti, Antonio Rocco, Scipione Chiaramonti, Antonio Lorenzini, Francesco Sizi, and Ludovico della Colombe.

The third group Galileo had acquired recently: the Jesuits. Initially great friends of Galileo, as they too were students of celestial novelties, they had turned adversaries when Galileo betrayed the trust of the Collegio Romano by being anti-Aristotelian and criticizing the modern astronomical theories of Tycho Brahe.

Now we must probably exclude the first group. The Dominican Order had officially dissociated itself when one of its members had taken the initiative of denouncing Galileo. Later, in 1624, the Dominicans were officially implicated in *The Assayer*.

The second group should also be excluded. Galileo's atomism had encountered strong criticisms in secular university circles. But the Eucharistic argument had never been raised at all, insofar as it was not within a layman's competence to deal with such subjects. Neither Vincenzo di Grazia nor Claude Bérigard, not even the monk Antonio Rocco, will resort to such a lethal argument. Nor was it the style of these secular university men to turn to the Holy Office to have an adversary condemned—for the simple reason that they were in no position to do so, since very often they had to be on guard against the Holy Office. Try to imagine, for example, even the most Peripatetic and least secular of these Aristotelian adversaries of Galileo, Antonio Rocco—a libertine, a declared unbeliever despite his cloth, author of the most famous pornographic novel in seventeenth-century Italy, he too denounced by a "pious soul" to the Inquisition of Venice—wanting to turn to the Holy Office with a doubt of conscience on the Eucharistic orthodoxy of *The Assayer*.[47] No—in fact, the "pious soul" who had denounced Galileo must wear a habit. But which one?

[47] Antonio Rocco—to whom is attributed the *Alcibiade fanciullo a scuola* (1630), published clandes-

Who, indeed, could be interested in butting into a dispute between such feared, powerful, and intractable personages as Galileo and Father Orazio Grassi? What religious could declare himself explicitly to be a pupil of Suárez's philosophy, if not someone of the third group, who felt duty bound to be Cardinal Bellarmino's heir and pupil in Rome?

---

tinely in Venice around 1650 and a classic of licentious Italian literature in the seventeenth century—will be subjected to ecclesiastical censure and will be put on the Index for his Averroistic book *Animae rationalis immortalitas*, Frankfurt 1644. On the personality of this anti-Galilean disbeliever and libertine, see G. Spini, *Ricerca dei libertini. La teoria dell'impostura delle religioni nel Seicento italiano*, Rome 1950, p. 152 ff.

SIX. IDENTIFICATION OF AN ANONYMOUS PERSON

Tu pulchra coeli decora, tu coeli iubar;
omnia tu minimo
Miracula puncto colligis;
Coelum, quod uno lucet intus sydere,
quod rotat Angelica
Motor Sacerdos dextera.
. . . .
Te lucis autor adversus dedit
Occiduus nebulas
Arram triumphatae necis
Redivivus in te, victor in te non semel
Vincis io populos
Divini amoris Telifer.*

A Missed Vocation

Cardinal Bellarmino—everyone knew—had always nurtured in his heart a passionate predilection for science. His many and onerous official duties, however, prevented him from cultivating this vocation. Having reached the age of seventy in 1612, and having brought to a victorious close the last serious theological and political controversies, the old statesman no longer had any desire to continue working as "the Congregation's porter," as he had been nicknamed.

His health was getting worse and worse; his legs were swollen, and he was beginning to go deaf. He decided to continue only routine activity at the Holy Office and the Congregation of Rites, which for a man like him was tantamount to going into retirement. He still attended the consistories and Paul V's audiences. But only on the occasion of the Conclave of 1612 or for some special tasks—such as the injunction to Galileo, the Campanella trial, or the defense of the Immaculate Conception—did Bellarmino return to center stage.

He spent less and less time at his official residence in the Vatican Palace.

* "Splendid ornament of the sky, light of the sky, / you gather in a minimal point all miracles: / Sky made radiant by a single star / which its Motor the Sacerdote causes to rotate with an angelic hand . . . / . . . The author of light made you / against the vapors of the sunset / pledge triumph over death, / in you victorious and resurrected many times; / oh, you conquer the peoples, / bearer of the arrows of divine love."—P. Guglielmo de Waha, S.J., "Ad Eucharisticam Panis speciem Ode prior," in P. Theophile P. Raynaud, S.J., *Exuviae panis et vini in Eucharistia, qua ostenditur esse veras qualitates, Opera*, vol. VI, Lugduni 1665, p. 426

He preferred, as though he were an ordinary Jesuit priest, to make long ascetic retreats at the College of Novices of Sant'Andrea, on the Quirinale, or to spend long days at his private residence in order to work quietly on his devotional books. Finally he had been able to finish work on his "favorite child," the ascetic treatise entitled *De ascensione mentis in Deum per scalas rerum creatarum* (1614), over which he had lingered through his entire life.

He enjoyed finishing that exciting book because he had been able to slip into it many scientific references on astronomy and physics which had reminded him of the naive scientific fervor of his youth. Once, Bellarmino wrote, he had decided to estimate the velocity of the sun's rotation around the earth, to get an approximate idea of the infinite power of the Creator. He had waited on a beach (perhaps on the road to Capua, when he was bishop) for the sunset. Though he had no instruments to measure the velocity of the sun, he did have at his disposal a standard of time whose constancy was guaranteed by long habit: as the sun disappeared on the horizon in a symphony of colors, he recited two miserere, truly a very brief fraction of time.[1]

Now, in 1612, the old cardinal still had many things to learn. Only the year before, in April 1611, he had anxiously looked through a telescope and had seen "some very marvelous things around the Moon and Venus"[2] announced by Galileo. He had been able to discuss them with his scientist friend Father Clavius, who assured him—with the competence of his mathematical calculations—that Galileo had actually discovered the moons of Jupiter.

Now, in deferential homage, Galileo had sent him a copy of his *Discourse on Floating Bodies*. Bellarmino had been quick to thank him, even before reading it, flattered as he was by the homage of Europe's most famous scientist.[3] But with whom could he talk about it, now that Father Clavius was dead?[4]

During those years of study, Cardinal Bellarmino did not resign himself to a lonely old age. He wanted continually to keep abreast of the latest scientific ideas. Later on, he had even tried to establish a relationship with the nephew of his friend Cardinal Cesi, Federico Cesi, also a dilettante scientist, but perhaps now a trifle presumptuous after have become Galileo's great

[1] See R. Bellarmino, *De ascensione mentis in Deum per scalas rerum creatarum*, Antverpiae 1614 (2nd ed., Antverpiae 1634), p. 110 ff. On p. 113, however, we have an example of Cardinal Bellarmino's rigid Aristotelian convictions on the subject of theories of light and color. On this very widely circulated book of contemplative cosmology, see A. O. Lovejoy, *The Great Chain of Being*, 1933 (6th ed., Cambridge [Mass.] 1957), p. 90 ff.

[2] See Cardinal Bellarmino's letter to Father Clavius and the mathematicians of the Collegio Romano, April 19, 1611, *Works*, XI, p. 87.

[3] See Bellarmino's note to Galileo, June 23, 1622, *Works*, XI, p. 337 ff.

[4] Father Clavius died on Fabruary 6, 1612. See ms. Urb. Lat., published in J.A.F. Orbaan, *Documenti sul Barocco in Roma* (Memorie della Società Romana di Storia Patria), Rome 1920, p. 285.

Roman friend. In fact, when Bellarmino had him asked as a peer for an opinion on the mutation of the North Star, Federico Cesi had replied with a long dissertation full of pretentious purple prose in an inelegant and murky Latin, explaining to him, as to a student burdened with religious scruples, that the sky was composed of fluid matter.

Bellarmino did not like that reply: too many Biblical quotations, as if to remind him that he was the notary of the anti-Copernican injunction; too many confused ideas on the infinity of the universe, forgetting that the recipient had been Bruno's inquisitor; and, finally, too much assurance for a dilettante scientist.

Bellarmino, always very affable and ironic, replied in Italian, with great good humor, thanking and praising. He confined himself to correcting with a single, erudite Biblical quotation Prince Cesi's unfounded ideas on the infinity of the universe. In short, he responded with a paternal rebuke, advising his correspondent to continue in the fear of God in order to earn heaven, from which he would be able to contemplate the motions of the stars and planets with greater assurance.[5] Their scientific relationship ended there.

Fortunately, Cardinal Bellarmino had finally been able to establish his private residence in a house right next to the Collegio Romano, on the side facing the Minerva, with windows that looked out on the Cloister of Dominican Novices. And often the young monks would try to catch a glimpse through the window panes of the famous Cardinal Bellarmino studying or reciting the rosary.[6] This house was convenient for Cardinal Bellarmino, as he could make daily use of the books in the Collegio's nearby library, attend the doctoral festivals and the lectures, and chat with the fathers between one lesson and another. He felt strongly about having ties with the Collegio's professors and scholars, so much so that when the swellings on his legs began to make him feel that the large twin staircase of the Collegio was too steep a climb, he actually thought of building a sort of bridge between his house and the adjoining building.

Before devoting himself to theology, Bellarmino had in fact been a brilliant university philosopher at the Collegio Romano. He had been the model pupil of Father Pietro Parra's philosophy course, not to mention the commentaries of Father Toleto, and had the satisfaction of competing in the public debates at the Collegio, defending Peripatetic conclusions. He was very good in "Aristotelian physics" and had been chosen, the only one in his

[5] F. Cesi, "De caeli unitate, tenuitate fusaque et pervia stellarum motibus natur" (August 14, 1618), published in C. Scheiner, *Rosa ursina*, Bracciani 1630, p. 775 ff. Cardinal Bellarmino's response (August 25, 1618) to the receipt of this Cesi memoir is also published in the volume, on p. 775.

[6] G. Fuligatti, *Vita del cardinale Bellarmino*, Milan 1624, p. 345. For the exact location of Cardinal Bellarmino's private residence alongside the Collegio, it should be added that it was an area purchased *en bloc* by the Society of Jesus in 1611; see Orbaan, *Documenti sul Barocco*, p. 86. Cardinal Bellarmino must have moved as was the custom, in the autumn of that year.

course, for the end-of-the-year lecture on the occasion of the solemn ceremony of the awarding of diplomas. The subject of his brilliant philosophical exhibition had been *On the Soul*, and he still remembered it very well.

Later on, as a famous theologian, he would insist particularly that the *ratio studiorum* of the Collegio Romano be based on absolute fidelity to Aristotle in its philosophical instruction. Bellarmino had then returned to the Collegio Romano as professor of theological debate; and now, if he could not actually reside there, he at least wanted to have influential relations with the professors of the new generation, and to participate in the intellectual life of the university.[7]

At his house, the day's schedule was marked off by the sound of the Collegio bell. But this perfect syntony of life was threatened more and more by the cardinal's growing deafness. It was perhaps this preoccupation, or perhaps his old passion for astronomy, that induced Cardinal Bellarmino to repair the damaged sundial at the front of his house. He therefore politely asked Father Grienberger, Father Clavius's successor in the chair of mathematics and astronomy at the Collegio Romano, to come to his house and see what should be done. Repairing a sundial was a bricklayer's job, and his thus inconveniencing a professor of astronomy was excessive; but perhaps Cardinal Bellarmino wished to take advantage of this moment to talk about scientific matters with the new docent. And since one could not say no to Cardinal Bellarmino, Father Grienberger went up to his house accompanied by his thirty-year-old assistant Orazio Grassi, who was not yet a Jesuit priest but had already revealed remarkable talents as a mathematician.[8]

Thus it was that Cardinal Bellarmino personally met the young Collegio Romano scientist. It was on an unspecified June day in 1612, and Cardinal Bellarmino must by now have had plenty of time to read the book which Galileo had sent him in homage, and which had set off so many discussions over the atomist theory of heat it upheld against the Aristotelians.

Father Orazio Grassi, Jesuit

Orazio Grassi had been in Rome for more than twelve years. An eighteen-year-old patrician from Savona, he had come to the College of Novices of the Society of Jesus at Sant'Andrea in 1600 and, at the Collegio Romano,

[7] See Bartoli, *Della vita di Roberto cardinale Bellarmino*, p. 428.

[8] There could not fail to have been other opportunities for meetings between Cardinal Bellarmino and the mathematician Grassi. But this first personal contact, with Galileo's new book on atoms open on Bellarmino's desk, seems to me worth remembering. The circumstance was mentioned by Grassi himself to his friend Father Eudaemon-Johannes, and afterward the latter referred to it in *Summarium*, n. 29, p. 101, a lesson taken up again by Brodrick, *The Life and Work of Blessed Robert F. Cardinal Bellarmino, S.J.*, vol. II, n. 2, p. 346. Bartoli, *Della vita di Roberto cardinale Bellarmino*, p. 76, incorrectly places the episode during Bellarmino's previous residence in the Trastevere.

had attended the three-year philosophy course of logic, natural philosophy, and metaphysics. In 1605, he had been the only student to take the complementary course of mathematics, practically on a private basis. After that, he took the long, difficult four-year course in theology.

He took a sojourn at home and then, before the start of the academic year 1616/17, he was named father general to take over the prestigious teaching of mathematics at the Collegio Romano. Until 1624, and then again from 1627 to 1632, he taught uninterruptedly. Yet, by 1618, when he took his vows, he had already published a *Disputatio optica de iride*,[9] under the name of one of his students. And again the following year he confirmed his uncommon qualities as a scientist with the famous *De tribus cometis anni MDCXIII disputatio astronomica*, which the Collegio Romano decided to publish officially, anonymously, and under its auspices. From that brilliant lecture on the most modern theories of astronomy, as we know, sprang the polemic with Galileo which went on until the publication of *The Assayer* in late October 1623.

Well, when *The Assayer* appeared in the bookstores, destiny seemed to will that the first copy of the book be purchased, in the Sun Bookstore, by none other than Father Grassi. Nobody more than he was burning with the desire to get his hands on that book, although information about its contents had been circulating for some time now. Hence, as a professor of the Collegio Romano had heard that Galileo's book was on sale in that bookstore, frequented by virtuosi, innovators, and libertines, he "immediately rushed there,"[10] as might be expected.

He arrived out of breath. Right off he saw the frontispiece; the sarcastic title came immediately below the pope's coat of arms and that of the Academy of Lynceans (Fig. 1). "He changed color"[11] and, with that irascible and impulsive character known by all, could not restrain himself from scolding the bookstore owner, as if he had anything to do with it. Although Galileo had kept him waiting three years for a reply, he said, it would take him only three months to reciprocate ("in three months he would relieve his anxiety").[12] He slipped the book under his arm and left as he had come.

As foreseen, the lure had worked: Father Grassi had betrayed himself before the Sun bookseller. Rather than a surprising coincidence, the sale of the first copy of *The Assayer* to Father Grassi himself (i.e. Sarsi) had much more probably been a trap set for the book's buyer-victim thanks to the complicity of a friendly book dealer. In fact, while Father Grassi was keeping an eye out for *The Assayer*'s appearance in the bookstores, Galileo's friends were "keep-

[9] *Disputatio optica de iride, proposita in Collegio Romano a Galeatio Mariscotto*, Romae 1618.
[10] See F. Stelluti's letter to Galileo, November 4, 1623, *Works*, XIII, p. 147.
[11] Ibid.
[12] See T. Rinuccini's letter to Galileo, November 3, 1623, ibid., p. 145.

ing an eye out to discover with what tolerance of spirit it is seen and read by those for whom it is specifically written;[13] they were to inform Galileo in detail.

Spy Games

But to spy out Father Grassi's "state of mind" was not so easy. Monsignors Cesarini and Ciampoli—who among Galileo's Roman friends knew the Collegio Romano professor more personally?—knew that "Sarsi," when he spoke, always concealed his intentions behind many compliments for Galileo. But they also knew that Father Grassi had a temper. The only way to discover Sarsi's recondite intentions was to put Father Grassi in a rage. The ideal way would be to observe his first spontaneous reaction at the sight of *The Assayer*. To do so, however, one must make sure that a chance buyer—an assistant, a student or colleague—did not bring him the book or warn him, thus spoiling the surprise effect.

The owner of the Sun Bookstore, a friend of the Roman virtuosi, lent himself to the plot. Well before distribution by the printer, the book dealer had obtained, through the indulgence of Father Niccolò Ridolfi, Master of the Sacred Palace, an advance copy from among those that had been deposited at the Holy Office for the imprimatur.[14]

Grassi took the bait. The owner immediately passed on what Grassi had said to two Galileans, Tommaso Rinuccini and Francesco Stelluti. They in turn informed Galileo of the important news: Sarsi was preparing to give a lightninglike reply in three months.

The news, including the time and the predictable tenor of the forthcoming reply, was essential for Galileo. In fact, he was preparing to come to Rome within a few months, with the idea of opening negotiations for a resumption of the Copernican campaign. Knowing in advance the adversary's moves and time schedule was a decisive advantage.

From Florence, Galileo insisted on a confirmation of what he had been told; and one month later, the information was confirmed by Rome. This confirmation, too, was rendered possible by an angry outburst on the part of Grassi—this time in front of an acquaintance of Tommaso Rinuccini. Father Grassi was skeptical about the possibility of preparing a reply in print (that is to say, a book) before the end of 1625 and complained that he did not have a patron disposed to finance it. Grassi had also said that his book would not contain any sardonic remarks, and he expressed the hope of a reconciliation with Galileo on his approaching visit to Rome.[15] The person speaking, however, was the ceremonious Father Grassi. Instead, a few days

[13] Ibid. [14] Ibid., p. 147.

[15] See T. Rinuccini's letter to Galileo, December 2, 1623, ibid., p. 153.

later, Rinuccini's acquaintance found Grassi "in a furor" at having heard that in Florence it was thought *The Assayer* "must have shut the mouths of all the Jesuits, who will not know how to reply." At that, Father Grassi had again betrayed himself: "if the Jesuits can know by the end of the year how to respond to a hundred heretics, they should be able to do the same with a Catholic."[16]

This threat confirmed and made more specific the first one. Father Grassi held to his promise of a swift reaction to Galileo's book, in the spirit of struggle against heresy typical of Jesuit priests.

Tommaso Rinuccini, unable to understand how one could accuse *The Assayer* of heresy, commented on the information as if it were a piece of nonsense and reassured Galileo that among the Jesuits "there is a strict commandment not to discuss such doctrines."[17] This view was plausible. Father Grassi was deeply involved in the project for a new Collegio Romano church. He did not have the time, nor was it opportune for him, to publish a book of polemics that ran the risk of compromising the Society of Jesus at a politically delicate moment. The Society could not commit itself too openly against an adversary who was about to be officially and triumphantly welcomed by the pope and the Curia.

So, for the moment, the book in reply to *The Assayer* is postponed. Nonetheless, Father Grassi has the firm intention of giving Galileo his just deserts without delay. For obvious reasons, Galileo must first arrive in Rome; and so for the time being all is quiet, at least apparently. Only some unverifiable information filters through.

At the end of February, on the eve of Galileo's arrival, Johannes Faber writes to Prince Cesi that the rumor is that "Father Grassi has answered but does not want to publish.[18] During Galileo's stay in Rome, what Sarsi had begun to think of saying, without publishing, becomes known.

In April and May of 1624, Galileo stayed in Rome with Mario Guiducci. The great success of his visit did not seem to be disturbed by annoying rumors. But this is only an appearance, owing to the absence of letters. As soon as Galileo returns to Florence and Guiducci's informative reports begin to reach him, we shall be certain that a very thorny problem was being discussed in Rome. Indeed, the muffled rumors had now become a threatening drum beat: "I hear from all sides the growing noise of battle that Sarsi promises us with his reply," a frightened Mario Guiducci writes by the month of June, 1624, "so much so that I am almost led to believe he has made it."[19]

Guiducci had maintained good relations with the professors of the Collegio Romano, and his information came from reliable sources. In fact, he

[16] Ibid., p. 154. [17] Ibid. [18] Ibid., p. 167.

[19] See M. Guiducci's letter to Galileo, June 21, 1624, ibid., p. 186 ff.

has added the possible theme of this threatening reply, which Galileo must have been aware of since his sojourn in Rome the month before, for Guiducci had only to touch on it in passing. "But, on the other side," Guiducci also wrote,

> I do not see what [Sarsi] could latch on to, since Count Malvezzi as much as guaranteed to me that on those opinions about heat, tastes and smells, etc., he will not reply at all, because, he says, it is quite clear that you, Sir [i.e. Galileo], have included them to start a new battle, for which you must be very well prepared and armed; and the aforesaid Count and a certain Marchese Pallavicino are dissuading Sarsi from getting into the controversy.[20]

This information is very important for us. The lightning reaction that Father Grassi meant to hurl at *The Assayer*, as he had promised, as though he already knew what to "latch on to," hinged on the doctrine of heat, tastes, and smells; that is, on atomism and the accidents implicated in the Eucharist.

Not only that, but this intention was probably already known a month before. Besides, Roman Galilean circles were doing their utmost to discourage an initiative on Father Grassi's part, trying to convince him that Galileo had arguments available in his favor, and that it was not advisable to attack him.

We are already quite familiar with Count Malvezzi and Marchese Pietro Sforza Pallavicino. But Guiducci knew them only by name. As usual, Guiducci is mainly trying to reassure himself; hearing talk of a marchese, he thinks of some grand personality, and that is enough to make him feel at ease. In fact, Marchese Pallavicino is only a seventeen-year-old student at the Collegio Romano, a brilliant student working for his diploma under Father Aranea. His defense of Galileo is certainly useful, but his powers of dissuasion should not be overestimated. We shall hear about this at the proper time.

Signals of Alarm

Meanwhile, though, the situation had not yet come to a head. On July 6, Guiducci still had not got a whiff of it, and confined himself to relaying to Galileo the good news that Father Riccardi had become a consultant to the Holy Office.[21] But during that very same period, in Florence, Galileo had become alarmed. Someone had told him that Father Grassi has carried out his promised threats.

Something must have happened in Rome, because Guiducci does not an-

[20] Ibid. [21] Ibid., p. 192 ff.

swer. Nor does Father Riccardi—to whom Galileo has written to get first-hand information on what is happening at the Holy Office—give any sign of life. Galileo turns to Tommaso Rinuccini with an urgent request to find out whether what is being said is true; that is, whether Father Grassi has made his move.

We do not have the letters sent to Galileo, but we know the tone of alarm and the pressing questions from the reply that Tommaso Rinuccini immediately sends him. Rinuccini is caught by surprise, because by now it was Guiducci's task, and no longer his, to spy out Father Grassi's moves. Indeed, the information network previously set up by Guiducci collapses just then, at the most delicate moment of Father Grassi's reaction; perhaps Grassi has taken advantage of this moment to act undisturbed, unknown to anyone.

In truth, an unpredictable and disastrous thing had occurred. Just at the crucial moment, in July 1624, Guiducci has come down with malarial fever; for a month and a half, during the muggy Roman summer, he is in bed, completely out of commission.

Tommaso Rinuccini, the standby informer to whom Galileo has turned, is not familiar with Guiducci's information network and has to improvise another one quickly in order to collect details on the alarm signals that have reached Florence. For the moment, the first superficial pieces of information he can collect from his confidants are negative, permitting a cautious optimism. On July 20, 1624, he writes to Galileo:

> In order to be able to tell you definitely, Dear Sir, that it is not true that Father Grassi has answered, I believe that all I would have to do is have Signor Lodovico Serristori, as if on his own account, ask the said father about it. For by other means up until now I have no proof of it; nor do the others know anything about it, nor do they believe it, Dear Sir, so that I think the rumor is unfounded.[22]

However, probably in deference to Galileo's palpable concern, Tomasso Rinuccini adds the promise of further inquiries

> All the same, this coming week I will meet Signor Lodovico [Serristori] and will make further inquiries, which owing to the short time available, I have not yet been able to do. And I will by the first post give you, Dear Sir, reliable information, while I thank you for the favor you do me in turning to me, but please remember that I wish to serve you in greater matters.[23]

Tommaso Rinuccini was most likely not up-to-date on Father Grassi's hinted-at threat of attacking *The Assayer*'s atomism on the plane of doctrinal

[22] T. Rinuccini's letter to Galileo, July 20, 1624, ibid., p. 194 ff. [23] Ibid.

theology. Therefore, he does not understand how important it is for Galileo to obtain reliable information from him, now that Guiducci is out of action and in bed with a fever amid medicines and bloodlettings. The following week, on July 27, as promised, Tommaso Rinuccini again hastened to write to Galileo about this problem, but with nothing achieved and a somewhat too reassuring optimism:

> Signor Lodovico Serristori has not yet been able to find out anything about Sarsi, so that I believe I can assure you, Dear Sir, that the rumor was unfounded; but I shall not fail to keep you posted, as necessary, about everything that I may come to know.[24]

Actually, it turns out, the "rumor" that had reached Galileo from Rome, concerning a mysterious effort at a reply made by Father Grassi against *The Assayer*, cannot be either denied or confirmed. It should be noted that this is not at all a matter of a printed reply (that is to say, a book), for it is known that the veto of the Jesuits and Grassi's activities have caused him to postpone such a book until at least the following year. The fact is that Galileo's informants at the beginning of the summer of 1624 have lost contact with Father Grassi, and are moiling aimlessly around the scent of a mysterious rebuttal about which people in Florence were better informed than those in Rome.

Father Grassi, having covered his tracks and ahead of the game, now moves with assurance and even takes the initiative, as if he no longer has anything to hide—so much so that he can reverse roles and change from a man being spied on to a man spying on the would-be spy. On August 10, Father Grassi's surprising reappearance in this new garb was reported in Tommaso Rinuccini's information bulletin to Galileo:

> Signor Mario [Guiducci] did not have a fever as of yesterday evening, but I am afraid that he will have a long convalescence. During his illness he was visited by Father Grassi, so peace is made, and the friendship will serve perhaps as an excuse not to reply. Signor Lodovico Serristori, moreover, never verified the warning that you, Dear Sir, received; indeed, just the opposite.[25]

So the trail leading to something more about Galileo's "warning" peters out. In recompense, Father Grassi had made a courtesy visit to the feverish Mario Guiducci. The game is now in Grassi's hands. Galileo must not have been at all reassured by that unexpected visit and, through Rinuccini, asks Guiducci for an explanation.

The explanation will arrive on September 6, when Guiducci feels well

[24] Ibid., p. 196. [25] Ibid., p. 198 ff.

again and ready to "resume serving friends and masters."[26] But Galileo wanted to know why he had agreed to meet Grassi without first discussing it with him, as well as what Father Grassi had asked him. Guiducci could only reply in an embarrassed fashion, with a series of excuses: his illness, the intercession of an anonymous "very important prelate, and my particular master," the insistence of his old professor of rhetoric at the Collegio Romano, Father Tarquinio Galluzzi.

The scene described can be easily imagined: a gentleman in sickbed, and around that bed a helter-skelter of visiting Jesuit spiritual directors; whispered advice and pressures, until the goal is attained. Thus it was that by dint of the insistences of "several Jesuit fathers" who came to visit during his illness, Guiducci, who was supposed to keep an eye on Father Grassi's intentions, was officially visited in his house by no less a person than Father Grassi himself. And so, the latter was able to worm out of him certain information about Galileo. The polite scene of that turnabout in espionage roles is described by Guiducci in terms so naively true that one cannot abstain from quoting it:

> I was visited by Father Grassi with much courtesy and affability, as if we had known each other before this for a long time. We did not go at all into past matters; rather, a large part of our conversation consisted in praising your writings, Dear Sir, and the introduction to this discourse was as follows. Speaking of many printed works of philosophy and other subjects, and of the objections that are sometimes made by the censors, Father Grassi, either because his conscience bothered him, or because he thought I might speak on his behalf, told me that over the past few days he had again seen and appreciated that fine work of the Archbishop of Spalato [Marco Antonio De Dominis] on flux and reflux, and although there was nothing valid in it proved by reason he could not help but approve of it, as he did. And, he and I praising in accord the aforesaid writing, he added: "But we do, however, have the writings of Signor Galileo, which are very ingenious on the same subject."[27]

Whatever Guiducci may have thought, nothing was troubling Father Grassi's conscience. As for the De Dominis trial, of such great interest in those days, the good father had spontaneously exculpated himself: he had approved the unhappy archbishop's theory of the tides, it was true, since that theory was perfectly orthodox, even if unfounded. Galileo's theory of tides, however, turned out to be much more "interesting" and new.

[26] M. Guiducci's letter to Galileo, ibid., p. 202 ff. [27] Ibid.

Remember that in 1616 Galileo had presented a manuscript, proving the earth's motion and based on the argument of the tides, to the young Jesuitophile Cardinal Alessandro Orsini. Now Galileo once again announced that Copernican proof at the end of his reply to Monsignor Ingoli.[28] Father Grassi must have known the first version and was now trying to obtain information on the second.

So he had begun, as usual, showering Galileo with praise and letting it be understood that he was a scientist without prejudices of any kind, so much so that he had obtained the approval of the book on tides by that pertinacious heretic De Dominis. And Galileo, what new things did he have in mind concerning the tides? The naive Guiducci, who had a fever besides, told him that in Galileo's mind tides were now to be demonstrated with the earth's motion, and he explained to Father Grassi in what way. Indeed, Guiducci added, Galileo had other arguments to perfect his system. Father Grassi did not miss a syllable.

"And here we fell to talking about the earth's motion"; but Guiducci had not failed to specify that Galileo, as a good Catholic, resorted to that idea "as a hypothesis and not as a principle established as true." This was a useless clarification—there was no point in having so many scruples with Father Grassi, who knew even better than Galileo himself how Cardinal Bellarmino had put the matter to Galileo. Now Father Grassi, who perhaps had a chance to discuss the point personally with Bellarmino, declared his complete tolerance for Copernicus's theory: "if a demonstration for the said motion were found," Father Grassi declared, "it would be a good idea to interpret Holy Scripture in another manner than has been done"; and this point of view Father Grassi knew "ex sententia Bellarmini," at least this is what Guiducci in his fever remembered having understood, forgetting that in the *Libra* his interlocutor's opinion on Copernicus had been different. And with this reassuring statement of solidarity, "and with polite expressions, the aforesaid meeting ended" between Guiducci and Father Grassi.[29]

Father Grassi already had some further clues as to the new book that Galileo, with the agreement of the ecclesiastical authorities, was preparing on Copernicus, but he still had many things to extract from Guiducci. On September 13, a week after his recovery, Guiducci had gone to hear a lesson of Father Galluzzi at the Collegio Romano. Again he bumped into Father Grassi, who by now treated him with familiarity and who hung onto him in order to find out something about this new physical proof of the earth's motion that Galileo, as Father Eudaemon-Johannes had reported, intended to

[28] See G. Galilei, *Discourse on the Flux and Reflux of the Sea* (1616), *Works*, V, pp. 378-401; and *Reply to Monsignor Francesco Ingoli* (1624), *Works*, VI, pp. 510, 561.

[29] *Works*, XIII, p. 202 ff.

develop: "do not be amazed," wrote a happy Guiducci to Galileo, "if sooner or later he comes over to you completely, because he shows a great desire to understand your opinions and praises you to the skies, although this could well be a sham."[30]

All these compliments and polite attentions are beginning to fill Guiducci with suspicion, as his eyes were gradually being opened. At the same time, moreover, Galileo was still certain that something had taken place at the Holy Office at the beginning of the summer. Father Riccardi's silence worried him. Mario Guiducci promises to meet with the Holy Office's consultant, far from indiscreet eyes, after the usual sermon at San Luigi dei Fiorentini and to ask him "what he has to say about not having replied."[31] We shall learn from Guiducci's next letter, on September 28, that actually Father Riccardi had received Galileo's July request for information and that Galileo had sent him another through Guiducci.[32]

But certainly Father Riccardi could not compromise himself with a letter containing revelations on questions covered by the secrecy of the Holy Office. He replied by word of mouth. Nothing remains either of those letters or of that word-of-mouth communication.

As for Grassi, Guiducci could no longer say anything to him now that he realized his demonstrations of friendship had devious motives. Now he made efforts to avoid him whenever he went to the Collegio Romano: "the aforesaid Father [Grassi] showers me with all sorts of courtesies when I go there, which have begun to annoy me, because if he is with some other person he leaves his company to come over to me, and if I do not slip away he will not let go of me, accompanying me right to the door."[33] Guiducci at least hoped that Grassi would not publish any work in reply to *The Assayer*, "although about this I can only conjecture."[34]

By now it was too late to go on playing at spies; now there was nothing to hide. At the beginning of the month of November, the academic year of the Collegio Romano opened with violent attacks against the followers of the new philosophy, thus exposing the Jesuits' intransigent and bellicose intentions.

Mario Guiducci had been the victim of a deception that repeated the one carried out in 1615 by the Archbishop of Pisa against Father Castelli. Then, too, an interest in Copernicanism was simulated in order to obtain more precise information on the dangerousness of Galileo's ideas. Father Grassi had managed to avoid any reference to past controversies and to shift the

[30] Ibid., p. 205 ff.

[31] Ibid. We know that Galileo had written to Father Riccardi in regard to Father Grassi's reply to *The Assayer* (on the question of atomism) thanks to an indirect reference in J. Faber's letter to F. Cesi on September 14, 1624, ibid., p. 207.

[32] *Works*, XIII, p. 210. [33] Ibid. [34] Ibid.

conversation on to Galileo's future activities. Even after having repeatedly met Father Grassi, Guiducci had not been able to pick up any hint regarding his reply to *The Assayer*. Galileo had good reason to be disappointed over the counterproductive results of the precautions that had been taken.

Disappointment was also uppermost in Guiducci's mind at the end of November when, having met Father Grassi at the Collegio Romano, the father removed his mask and "after many compliments and excuses" finally declared openly that he had begun to write a book against *The Assayer*. "He gave me a lot of excuses," Guiducci wrote Galeleo, "that he was forced to write, and that he felt bad about it, and that in any case he did not want to touch on new problems of any kind into which you, Dear Sir, had tried to draw him."[35]

But by now Guiducci had learned at some expense to distrust his interlocutor's ambiguous and dissimulated intentions and was bitter about Father Grassi's perfidious falsity: "having by so many means tried to be my friend, I became convinced that through this new friendship an agreement might be attained, as well as a perpetual silence on past matters. Now having made this mistake, I too have begun to think of vengeance."[36] Feeling that he had been made a fool of by the adept Jesuit, our Guiducci wanted to take his revenge with polemical comments on the invective hurled, as we know, by the Collegio Romano against the new philosophy.

Had Guiducci been informed about the mysterious alarm signals that had reached Galileo the summer before? It is hard to say, because during that period he was not in a position to follow the developments of the inquiries and, later on, having become too familiar with Father Grassi, he had perhaps been kept in the dark by Galileo himself.

Father Grassi's objections to *The Assayer*'s theory of sensible qualities had been spread about many months before. After that, nobody spoke of it. When in April 1625, a year after these rumors, news of *The Assayer*'s denunciation filtered through, Guiducci was probably convinced that the intervention of Count Malvezzi and Marchese Pallavicino had been sufficient to ward off those accusations.

By now, publication of Father Grassi's book against *The Assayer* was expected. But throughout the winter and spring of 1625, perhaps because of the reticence of the Society of Jesus, people would have to wait for that official reply to Galileo.

It was looked forward to with some preoccupation. Would it trot out the initial objections? Would it allude to the mysterious denunciation of the preceding summer, about which Galileo had found it so difficult to obtain in-

[35] M. Guiducci's letter to Galileo, November 8, 1624, ibid., pp. 225-27.

[36] M. Guiducci's letter to Galileo, November 30, 1624, ibid., p. 232 ff.

formation? If Father Grassi had presented a denunciation to the Holy Office at the beginning of the summer of 1624, as the chronology of events we have been following allows us to suspect, then that denunciation was the autograph we have discovered some three hundred and fifty years later. The chronology of events; the suspicions; the futile inquiries; the secret contacts with Father Riccardi—all these jibe with the hypothesis that the author of the mysterious document at the Holy Office was Father Orazio Grassi. In fact, even the handwriting of that "pious soul" who had written the denunciation was very similar to Father Grassi's (Fig. 8).

I knew Father Grassi's writing and style quite well from his letters preserved among the Baliani papers at the National Brera Library, Milan, and from having seen some of his contemporary notes on architectural design preserved at the Superior General Curia of the Society of Jesus. The composition of the manuscript page, the architect's fine, elegant, and personal script—neat and sharp like that of a mathematician such as Cavalieri—made me think from the very start of Father Grassi's handwriting.

Even after a more painstaking analysis, attribution of the anonymous autograph at the Holy Office to Father Grassi seemed to conform to the rationale and proofs of historical interpretation. If identification of the handwriting allowed me to attest to our trail as establishing the attribution of the document's authorship to Father Grassi, a confirmation and dating could be arrived at only by continuing to follow that trail.

Yet, the indirect testimony, clues, and traces that we have thus far brought together lead us to believe that the denunciation of *The Assayer* for Eucharistic heresy was deposited by Father Grassi in the spring or summer of 1624, with a slight delay as regards the promise of answering that book within a few months of the moment when he first saw it in the Sun Bookstore. Besides, in conformity with his previous statements, it was probably at the beginning of the following winter that the father could dedicate himself wholeheartedly to writing his reply to Galileo's book. A year later, in fact, it is learned that the reply is definitively completed and that, probably so as not to compromise the Society of Jesus directly, it would be published in France, rather than in Rome. Meanwhile, in Rome, the first stone of Father Grassi's official work—the Collegio Romano's great church—was being laid.

The news that Father Grassi's book is ready for the press will come from Genoa, in February 1626. It will be discovered, on the occasion of the author's brief sojourn there, when Bartolomeo Imperiali, an acquaintance of both adversaries in the polemic (and a person who must not have been informed about all the background to the dispute), offered himself as an intermediary to compose the quarrel between Galileo and Father Grassi. In any event, Galileo will explain to his Genoan correspondent, in a letter that has not come down to us, the reasons which had induced him "not to accept the reconciliation with Father Orazio Grassi before he had given his book to the printer."[37]

Knowing how things had gone behind the scenes, we can imagine what those reasons might be. For that matter, Bartolomeo Imperiali, in his reply to Galileo, also recognized the seriousness of those motives: he admitted that, in light of the situation, Grassi "deserved to pay the penalty for having been the first to provoke, by opposing the truth."[38] One should note, by the way, that this allusion can hardly be connected with the problem of the comets, since on that question Father Grassi had not initially provoked Galileo.

Still another common acquaintance, the scientist Giovanni Battista Baliani—with whom Bartolomeo Imperiali, as reported by the latter, had conferred concerning the profound reasons for the latest polemic—recognized, "in agreement with the universal view, that Grassi had made a bad blunder." Baliani, moreover, "was disgusted with him for not having brought it up when he was in Genoa, so that he could be pulled aside and convinced of his mistake."[39]

What was the "blunder" Grassi had made? Certainly it was not Tycho Brahe's prestigious comet theory, which Father Grassi defended with plenty of good arguments. But by now it was too late to dissuade Father Grassi from more serious accusations against *The Assayer*, and in any case it would not have been easy, given the fact that, as we know, in the spring of 1624 some influential Roman friends had already tried vainly to persuade him to abandon those accusations. Subsequent events will confirm, with a series of additional clues, the episode we have been reconstructing.

Official Accusations

Grassi's official reply to *The Assayer* was published in Paris at the end of 1626, under the pseudonym of Lotario Sarsi and with a title that declared a dissimulated polemical intention: *Ratio ponderum librae et simbellae* (Fig. 5). In fact, like *The Assayer*, Sarsi's new book was also a work of controversy, weighing one by one each paragraph of the opposing book. But unlike *The Assayer*, which had recourse to the lethal polemical weapons of satire and the new philosophy, the *Ratio* used those no-less-lethal weapons of doctrinal and dialectical retort based on religious and philosophical orthodoxy. Sarsi's book proved that notwithstanding Father Grassi's assurances to Guiducci that he wanted to forget past matters, he had not forgotten either the old or

[37] Bartolomeo Imperiali's letter to Galileo on February 27, 1626, in A. Favaro, "Galileo Galilei e il padre Orazio Grassi," *Memorie del Regio Istituto Veneto di Scienze, Lettere ed Arti* 24 (1891), Nuovi Studi Galileiani, Venice 1891, pp. 203-220, especially p. 215.

[38] Ibid., p. 216. [39] Ibid.

new humiliations suffered on account of Galileo and that, on the contrary, he reaffirmed his points of view.

The book was dedicated to Cardinal Francesco Boncompagni—tied to the Ludovisis' pro-Jesuit entourage—who was the Archbishop of Naples. What leapt to the eye concerning the dedication was the authoritative protection of which the book benefited in the geography of ecclesiastical and political power. But there was another aspect that caught one's attention—one that today, to our modern eyes, might appear irrelevant. Apart from the obligatory privilege of the King of France, the book did not bear any authorization either from the superiors of the Society of Jesus or the ecclesiastical authorities.[40] The absence of the superiors' authorization was partially excusable, given the pseudonym, and was justified by the caution necessary in so delicate a question. The lack of ecclesiastical authorization was much less justifiable. And sure enough, if one looks for the book in Italy, for example in the Library of the Jesuit College in Genoa (today the University Library), there immediately appears, as if by a miracle, the proper authorization. It is signed by two Sorbonne theologians: a certain Antonius Maller and one Aegidius de Amore, a Dominican. At the time, there did in fact exist a prior of the Parisian Dominicans with this name, even if (to be exact) he was called Pierre D'Amour and not Aegidius.[41]

The trick was there for all to see. The authorization must have been inserted afterward because it was dated May 4, 1627—at least six months after publication of the book. It is permissible to suppose that, behind this strange shifting about of sheets bearing the authorization, there must have been an exchange of opinions between Father Grassi and the men of Piazza del Gesú to ensure the book a more authoritative and official distribution in Italy. In fact, in 1627 the book had already appeared in an Italian edition in the city of Cardinal Boncompagni, who had financed it from the start. And this time, as was proper, Sarsi's *Ratio* had a Neapolitan ecclesiastical authorization and, above all, the obligatory license from Father Grassi's superiors.[42] It is

[40] *Ratio ponderum librae et simbellae in qua quid e Lotharii Sarsii libra astronomica, quidque e Galilei Galilei simbellatore de Cometis statuendum sit, collatis utriusque rationum momentis, Philosophorum arbitrio proponitur. Auctore eodem Lothario Sarsio Sigensano*, Cramoysi, Lutetiae 1626. There is no ecclesiastical authorization in the copies I consulted in the Sorbonne and the Richelieu Library (National Library, Paris), or in that copy belonging to Galileo (National Library, Florence). See *Works*, VI, pp. 375-500. In the copy of the Paris edition belonging to the Library of the College of Genoa (now Library of the University of Genoa) is inserted, in substitution for the page of etchings, an unnumbered page carrying the ecclesiastical Paris license for the date May 4, 1627.

[41] See the repertory of Quétif and Echard, *Scriptores ordinis praedicatorum recensiti*, vol. II, p. 496.

[42] L. Sarsi, *Ratio ponderum librae et simbellae*, new ed., Matthaeus Nuccius, Neapoli 1627 (cum licentia superiorum). The imprimatur is signed by the Vicar General of the Archbishopric of Naples, Father Andrea Lanfranco. A warning note from the printer informs us that the entire Paris first edition had sold out rapidly. In *Works*, VI, Antonio Favaro has pointed to (n. 2, p. 18) some of the most significant corrections with respect to the first edition, indicated in pen on the copy in the Collegio Romano's main library.

not difficult to understand that the license of the superiors for the *Ratio*'s Italian edition was something more than a simple formality; it indicated the official solidarity of the Jesuit Order in Italy with its most prestigious scientific exponent. All this was done in a discreet, prudent, and eloquent manner.

Moreover, the second edition was easier to consult thanks to a subject index and marginal notes. Some cuts had eliminated the most traditional and second-hand criticisms, so as to give the book a more original and pungent cast. Of course, the polemical bite of Sarsi's new book did not advance by a single millimeter the scientific discussion on the nature of comets, which had remained at its 1619 starting point. Sarsi did nothing but reconfirm Tycho Brahe's theory with respect to the exiguous parallax of that phenomenon. For the rest, the book countered the ideas of *The Assayer* and Mario Guiducci's *Discourse* with often quite convincing dialectical arguments.

During the last two years, Father Grassi had been chiefly occupied with the project of the Church of St. Ignatius and certainly had not had an opportunity to re-examine the observation data on the 1618 comets. The book should not be read as a work of scientific polemic: by now, the polemic was purely personal and philosophical and, from this point of view, this was undoubtedly a successful and original work. The true purpose of the *Ratio*, more than that of restoring Tycho Brahe's astronomical authority and the authority of the Collegio Romano, was to discredit the official support that had gathered around *The Assayer*, publicly unmasking the heretical nature of that book's ideas.

From its exordium, the leitmotif of Sarsi's new book was the will to denounce *The Assayer*'s heresy—using a crescendo of tones, from the sibylline allusion to the reticent connivance from which Galileo had benefited, all the way to an out-and-out appeal to the Holy Office. In fact, it began with an ambiguous evocation of Calvin's name,[43] then moved on to a re-evocation of Galileo's quotations of Copernicus's condemned theory. Sarsi's voice becomes more shrill when he treats of natural philosophy: namely, the problem of the physical nature of comets, in relation to the paragraph in which *The Assayer* took up the defense of Cardano's and Telesio's philosophy.[44]

Galileo had energetically defended the reputation of these two "venerable Fathers of natural philosophy," branding Sarsi's theological challenge to their ideas as "demented" as well as an example of dissimulation and immorality. Thus, Galileo had referred to materialist, atheist authors, officially condemned by the Church, as "venerable"; and Sarsi now replied "much more freely than the first time" to this person who had the gall to impart lessons of Christian morality:

[43] See *Works*, VI, p. 393. [44] See *Examen*, IX, ibid., p. 397 ff.

> What sort of piety is it, Galileo, to defend those whose philosophy—condemned from so many sides as hardly consonant with what is Catholic—should instead be buried in eternal oblivion? Such human and demonic doctrines on the spirit of worldly knowledge, to repeat Tertullian, are meant for the avid ears of men that God, calling this madness, has chosen as worldly follies to confound that same philosophy: that is, arguments of a worldly nature rashly used as interpreters of the divine nature and disposition. He who wishes to praise Cardano and Telesio, let him do so. I turn to a more religious kind of praise.[45]

To cite Tertullian's *De praescriptionibus ad haereses* in a scientific work, as Sarsi does here, was tantamount to trying to intimidate by means of a code of heresiological procedure. A little further on, in confirmation of these intentions, the *Ratio* branded Galileo's atomism as the doctrine of an author whom Galileo had made sure not to defend (Epicurus, "who completely denies God, or at least his providence") and so on in the same vein.

Up to this point, however, Sarsi was only repeating in a harsher, more detailed form accusations that were already old to someone who had read the *Libra*. But the reader who was not aware of the events that had taken place during those preceding three years had the pleasure of knowing an original denunciation. In fact, in correspondence with *The Assayer*'s most interesting and newest part—that is, the famous paragraph 48 on the theory of sensations and the atomic structure of matter—Sarsi this time presented a much more precise accusation (Fig. 6).

In the *Ratio*'s "Examination XLVIII," *The Assayer* was denounced for Eucharistic heresy. For the convenience of the modern reader, I translate that section of Sarsi's book (see Documents). This will facilitate a comparison with the anonymous denunciation deposited with the Holy Office.

Anyone familiar with that denunciation understood the significance of the accusation which the *Ratio* made a matter of public domain in 1626. This is so true that some of the more attentive of Galileo's historians recognize it in the official version given in that book. But since these historians could not have known the background of the denunciation, to them—as to less informed contemporary readers—that accusation, at least *expressis verbis*, must appear to be rather vague and malicious, a superficial and routine dialectical expedient. Thus it has been that those among Galileo's historians who have

[45] Ibid., p. 397 ff. Sarsi considered Galileo a Telesian philosopher because *The Assayer* subscribed to the thesis of the optical nature of comets and to that of the corpuscular nature of heat. At this point, in an annotation at the foot of the page in his copy of the *Ratio*, Galileo questioned whether such propositions were truly "so impious and damned as to deny divine providence or God himself." Galileo exclaimed: "Well, you ought to be ashamed of yourself. And beneath what mask are you trying to hide your rabid malignity?" Galileo seems, therefore, not to recognize the legitimacy of putting Telsio's *De rerum natura iuxta propria principia*, which goes back to 1596, on the Index.

reported it—Dijkterhuis, De Santillana, and Shea—have not dwelt on it and have reported it without devoting to it more than three lines of commentary.

Sarsi began by recalling his theory of the production of heat by friction, by way of explaining the fire of the comet. First of all, he interpreted Galileo in light of Aristotle's *Meteors*, attributing to him Anaxagoras's theory of celestial fires. Then Sarsi immediately faced head-on the atomist and dynamist version of heat presented by *The Assayer*, submitting himself directly to the judgment of those who "watch over the safety of the faith in its integrity"; that is, the cardinals of the Congregation of the Holy Office.

Now this allusion to a repressive measure had no justification whatsoever in itself. *The Assayer*'s imprimatur was impeccable from all points of view. Without an explicit denunciation, much more explicit than this allusion, the Holy Office simply could not proceed to a documentary inquiry into Galileo's book. However, the allusion becomes coherent and legitimate if one is aware that there had been deposited, on just this point, a denunciation against *The Assayer*, shelved but still pending like a Sword of Damocles. To facilitate its fall on Galileo's head, Sarsi had publicly circulated certain doctrinal scruples that jibe exactly with the denunciation in question.

When the formula of Eucharistic consecration pronounced by the priest acts so that the substance of the bread and wine is transubstantiated into that of the body and blood of Christ, the external appearances—color, taste and heat—are still preserved by divine intervention. Now, for Galileo, these are pure names. "Would perhaps a perpetual miracle be necessary to preserve some pure names?"

This is the exact version of Sarsi's denunciation, which sums up in essential, more meditated, and manifest form the content of the denunciation at the Holy Office. But here, too, one cannot understand fully the sense of the accusation, unless one takes into account the denunciation previously put forward.

The Assayer, in fact, is accused of "expressly" affirming the subjective theory of sensible qualities, according to Sarsi, in the case of Eucharistic accidents, as well. Now that was manifestly false. As everyone could see, *The Assayer* had abstained from pronouncing itself "expressly" on the delicate question of the nature of Eucharistic accidents. Even if it was quite clear to a contemporary reader that Galileo could not elude the question, it still had to be explained that the Eucharistic paradox could violate Galilean corpuscular physics.

Sarsi omits the explanation. But he could proceed directly to an identification of the problem, for it had already been raised and debated among those whose duty it was to do so. Thus, it was enough to refer officially to the serious accusation. And this is so true that, immediately thereafter, implicit mention is also made of the subterfuge resorted to in Roman Galilean circles to steer *The Assayer* clear of its responsibilities. In fact, the *Ratio* presents Father Guevara's defense (of course, without citing it), which, one may rightly argue, had also posited the possibility of interpreting the concept of a Eucharistic "species" in different terms from those of the dominant theology.

Only by keeping all of this in mind can we appreciate the important specification that Sarsi added to this point as a warning and reminder to the Holy Office. Indeed, Sarsi objects to those obscure justificatory subterfuges, asserting that if it has not been granted that one may freely interpret the earth's motion—basically a secondary problem in regard to faith—then, all the more, must one not grant any possibility of a heterodox interpretation of the dogma of the Eucharist, since this is incommensurably more important than the question of whether the earth moves. Father Grassi was perfectly right to insist on this point: the Eucharistic dogma is the fundament, the essential postulate of Catholic faith.

Nobody better than someone who had known Bellarmino's ideas could evaluate this difference. One could perhaps be Catholic and Copernican, but one could not be Catholic without respecting the Tridentine postulate of the Eucharist. After this, the *Ratio* adds a series of considerations, of a logical and physical order, on the falsity of the atomist conception of heat and on Galileo's claim to offer a new exegesis for the Aristotelian proposition that "motion is the cause of heat."

Sarsi exposes the attempt to say that *The Assayer* had spoken of atomism only in regard to heat. *The Assayer*'s ambiguous and cautious terminology must not deceive us, Sarsi says, since the Democritan atomism professed by Galileo holds for all sensible phenomena, not just for heat. Let us assume that the sensible qualities are produced by the shape of the atoms of the various natural elements; then, comments Sarsi, it becomes unthinkable that a substance, even after long trituration, preserves unaltered the geometric forms of its atoms, maintaining its characteristics of taste, smell, and color. On the contrary, we perceive that a very fine grinding of sugar, for example, does not alter its properties of color or taste.

As for light—composed according to Galileo of indivisible atoms, more mathematical than physical—in this case, logical contradictions arise. Such indivisible atoms must be finite or infinite. If they are finite, mathematical difficulties would arise. If they are infinite, one runs into all the paradoxes of the separation to infinity which had already caused Aristotle to discard the atomist theory: between the extremities of a line composed of infinite atom-points, there cannot logically exist an intermediate point. But if the light cannot be separated, neither can it penetrate into substances by separating. Besides, it is a function of its very high velocity of propagation that light must be the most penetrating phenomenon. Yet, one observes that heat, owing to what today we call its conduction, traverses substances impenetrable to light.

Unlike this last criticism, the criticism of infinity was certainly not up to the level of Father Grassi's intelligence and originality, since it exhumed a discounted argument, copied word-for-word from almost any scholastic philosophy textbook. We have, for example, Father Aversa's recent volume, which used exactly the same words.

The Jesuit mathematician Father Guldin, great opponent of the geometry of indivisibles, and an excellent Roman friend of Father Grassi, must have dissuaded him from repeating such obvious objections. Thus the second edition of the *Ratio*, the Neapolitan edition of 1627, omitted as superfluous the whole section on indivisibles.[46] On the other hand, the second edition kept intact all the previous arguments and accusations.

Instinctive Defenses

What had remained until then a confidential piece of information, to be kept as secret as possible owing to its gravity—that is, the denunciation of that Eucharistic heresy descried in *The Assayer* by Father Grassi—was now in the public domain. There was no immediate danger, since the almost fleeting allusions of that book were not enough to set in motion a judicial procedure blocked by *nolle prosequi*. Nevertheless, one must take into account that this denunciation was expressed not by some ordinary monk or solitary Aristotelian university man, but rather by the outstanding personality of the Collegio Romano's scientific culture.

At the start, however, the circulation of Sarsi's official reply was very discreet. The book reached Galileo and Guiducci, perhaps at Florence, only around the spring or summer of 1627. But this is only a hypothesis, based on the fact that the first reference to the reply harks back to Galileo's letter to Castelli in August 1627.[47] Yet, this hypothesis is supported by the sequence of events. Galileo received a copy of the first edition, the Parisian one, at the end of 1626, and thus was induced to make a serious error of underestimation. That copy of the *Ratio*, preserved in the Galilean section of the National Library in Florence, carries the annotations by Guiducci and Galileo which were to be used for a possible reply. These are commingled marginal notes; also, those in Guiducci's handwriting must have been agreed upon, if not dictated by Galileo himself.

Each time that Sarsi raised a suspicion of heresy, Galileo appeared so wor-

[46] The omission, in the Neapolitan edition of 1627, pertains to p. 490, ll. 12-35 (*Works*, VI), which are replaced with the phrase, "that should be quickly said on a not very clear question."

[47] The *Ratio* circulated in Rome until the winter of 1626; see N. Aggiunti's letter to Galileo, containing C. Accarigi's revelations, December 23, 1626, *Works*, XIII, p. 210 ff.

ried as to jot down long defensive arguments. And when Galileo had—before his eyes, in print—the most serious denunciation, the one relating to the Eucharistic dogma, instead of being scandalized by his adversary's "malign anger" as on previous occasions, this time he was frightened, as we can see from the note relating to this point in the book. His instinctive reaction was to consider the protection that he could count on in Rome and to reassure himself with that patronage, in lack of other arguments. In fact, we can read in Guiducci's handwritten note:

> This scruple we leave to you because *The Assayer* is printed in Rome, with the permission of the superiors, and dedicated to the supreme head of the Church; it has been reviewed by those who "watch over the safety of the faith in its integrity," who, having approved it, will also have thought of the way in which one can remove that scruple. Nor will they be dismayed by your saying that "sly and adept minds will excogitate some way of avoiding this": a manner of speaking typical of those who publish, as you have done, without the license of superiors, and who display ill feeling toward the Roman publication. The impudence for which you yourself feel no shame makes me ashamed, as the gentleman said.[48]

Galileo was right to count on the immunity that had already been fully granted him in Rome. But he was mistaken in the naive persuasion that an imprimatur printed on the verso of the book's frontispiece would inviolably sanction that immunity. A few years later, as we all know, that naive persuasion will bring about his downfall.

Above all, Galileo was mistaken in evaluating his adversary's authority on the basis of that almost clandestine edition. Galileo could not know it, but the following year the *Ratio*'s denunciation will reappear with all the official sanctions. In fact, when the Italian edition of Sarsi's reply was issued, Galileo's initial underestimation was replaced with fear and caution.

Galileo had sent a copy of the book to his direct Roman correspondent at the time, Father Castelli. Yet, at the beginning of 1628 (the letter has not come down to us, but is known indirectly), confronted by the threat of that newly and authoritatively comminated denunciation, Galileo urged Castelli to take a close look at the accusations in terms of their dangerousness and to explore the matter with Father Riccardi, the trusted consultant in the Holy Office.

Father Castelli answered:

> Since you command me I will go back and read it, and I will be with

[48] *Works*, VI, p. 486.

> Father Monster [Father Riccardi's nickname], who *at other times has told me that those things did not bother him at all*, and that he had enough courage always to defend your side.[49]

By the following month, however, Father Castelli had not yet been able to obtain the requested assurances.[50] Galileo had to insist; and finally, on February 26, 1628, Castelli wrote:

> In the presence of Signor Ascanio Piccolomini, I spoke to Father Monster [N. Riccardi] asking that he tell me his opinion concerning Sarsi's objections. [Riccardi] said that your opinions were not otherwise against the Faith, being simply philosophical, and that he would serve you in all that you commanded—but that he would not wish to come out openly whenever you are given trouble by the Holy Office's tribunal, where he is a qualificator, because if he were to declare himself, he would not be able to speak later. He also mentioned that his friars had given him some trouble, and *in fine* he concluded that he was on your side and that if you were to send him in detail the doubts on which you are in need of a reply, he would resolve them.[51]

Father Riccardi, let us not forget, was compromised just as much as Galileo. His enthusiastic theological approval of *The Assayer*, which perhaps had facilitated his career with the Holy Office at that marvelous moment, exposed him now to the criticisms of his order over the thorny question raised by Father Grassi's official rebuttal. It was completely understandable that he did not want to jeopardize himself in public once again.

Should the matter be dropped, or the consequences faced openly? The decision was a difficult one. Galileo therefore asked the opinion of the Academy of Lynceans; that opinion, communicated to him on September 9, 1628, was not to pick up the challenge.

In fact, Prince Cesi, having heard Monsignor Ciampoli and the other Roman "paladins and literati," told Galileo that he "was not obliged to descend into the arena, or enter the palisade,"[52] and that he should profit from the continuance of extremely favorable conditions—not wasting time on this question, but devoting himself to "other major writings."

At so great a distance of time, it is permissible to think that this agreed-on decision was probably ill-advised. Even more favorable conditions would have allowed the problem to be clarified once and for all, with the greatest probability of a success that would have removed the suspicions of heresy then thickening around *The Assayer*.

But perception of the problem's importance was conditioned by the fact

[49] B. Castelli's letter to Galileo, January 22, 1628, *Works*, XIII, p. 388 ff.

[50] Ibid., p. 393. [51] Ibid. [52] Ibid., p. 448.

that "the impertinences" of Father Grassi remained a dead letter with respect to the already shelved proceedings. The only cause for embarrassment was the publicity given that incrimination. But even this publicity was offset in a very reassuring manner, as we know, by the constant growth of consensus around Galileo's ideas and the Academy of Lynceans.

The Collegio Romano's campaign of pressure and Father Grassi's intimidations heightened, rather than averted, the Roman battle of which *The Assayer* was the manifesto. The Galilean party exerted increasing pressure and gained enough ground not merely to pound against, but actually to penetrate, the walls of the opponent's fortress: the walls of the Collegio Romano.

Academic Defenses

What a great festival for the Roman innovators when, at the Collegio Romano on September 3, 1625, Marchese Pallavicino, favorite of the Cardinal of Savoy's academy, defended his memorial doctoral thesis in philosophy! Here was "a new Pico della Mirandola," capable even of making one forget the erudition of Virginio Cesarini. The Marchese Sforza Pallavicino—departing from the custom of the Collegio's wealthier students, who had the professor write their theses—had, through a three-day session, answered all the questions on logic, metaphysics, and natural philosophy, as well as all the arguments of the Collegio's professors.

As was his custom, the diploma candidate's godfather, the Cardinal-Prince Maurizio of Savoy, had spared no expense. The Collegio's Great Hall had been transformed into an even more sumptuous theater than usual, with beautiful singing and music and tapestries of unprecedented richness; the Great Hall was "decorated from top to bottom." At this session of an exceptional diploma candidate, twenty cardinals participated, as well as all the prelates and intellectuals from the academies and even the pope. Only Cardinal Francesco Barberini and those friends who made up his diplomatic entourage could not participate, as they were involved in the legation to France.

In the autumn of 1628 (we do not know exactly when) the festival was repeated, this time for Marchese Pallavicino's doctoral thesis in theology. This was another inebriating string of defenses, disputes, questions, and confutations—protracted for five days—on all forms of theology: dogmatic, Scriptural, and controversial. It was the same spectacle, the same audience.[53] But this time Cardinal Maurizio of Savoy was absent from the festival, for reasons beyond his control (he had had to leave Rome precipitously the year

[53] See Anonymous, *Origine del Collegio romano e suoi progressi* (ms. 143, Archive of the Pontifical Gregorian University), p. 49 ff. On the grand diploma-bestowing celebration for Sforza Pallavicino (September 3, 1625), see Aranea, *De universa philosophia . . . publice asserta*.

before because of debts). The great protector of the prestigious diploma candidate was still, however, the pope, to whom theses were dedicated.

Strong in his protectors, his knowledge, and the Roman intellectual climate in which he was a recognized protagonist, the twenty-year-old marchese spoke *ex cathedra*, as was suitable at such academic rituals. He even went so far as to touch on subjects of current interest, holding forth with youthful presumption on the untimely theological polemics stirred up by some Collegio Romano professor more competent in mathematics and architecture than in dogmatic theology. This was understood when the candidate reached that point of the doctoral marathon where comment was expected on cases connected with fundamental articles of Session 13 of the Council of Trent, concerning the Eucharist. The audience of admirers did not miss a single remark.[54]

Marchese Pallavicino had had good thesis advisers: Father Vincenzo Aranea and Father De Lugo. The arguments of the candidate were most up-to-date with respect to theology. No longer valid were the traditional Thomist ideas on transubstantiation by means of a passage (*transitus*) from one substance to another or by absorption. Transubstantiation was now presented in the style of Scotus and nominalist theology, modernly re-evaluated by the Jesuits: *accessio*, the abductive or progressive theory of substantial conversion.

But, having reached the second canon (that on the fateful Eucharistic accidents), the diploma candidate, already famous for his philosophy and culturally up-to-date, did not prove to be a slavish follower of the dominant theology, which interpreted those accidents as absolute, without subject. In fact, he admitted that there might be two theories, depending on the physics one adopted: the theory of minimal components of substances or that of substantial qualities. The first is less probable than the second, because the latter is clothed in the authority of tradition; but this is not to say that anyone who believes in minims must be heretical with respect to the dogma in question:

> those who assume minims in natural species must affirm the same thing concerning the material of the Consecration, since the Consecration cannot take place except in determined species. However, those who, with a more probable opinion, deny these minims in natural things, must equally affirm it concerning the material of the Consecration.[55]

[54] Pallavicino's academic defenses are praised by the doctoral candidate's friend Monsignor Ciampoli, in an ode which is very interesting for the study of the Roman intellectual environment of those years: G. Ciampoli, *Scelta di poesie italiane*, Venice 1686, p. 116. See, also, P. I. Affò, *Memorie della vita e degli studi del cardinale Sforza Pallavicino*, Parma 1794.

[55] P. Sforza Pallavicino, *De universa theologia*, Romae 1628, bk. VIII, p. 139 and 149 (on Canon 2 of the Eucharistic dogma).

Nothing more could be said. On the substance-quantity question, too, the candidate pronounced a brilliant, moderate, and conciliatory thesis—adopting the definition of local impenetrability, which made it possible not to convert the Eucharistic dogma into a tribunal of philosophical theory.

The term "species," used in the formulation of that dogma, served Marchese Pallavicino to admit as a unique and certain truth the fact that all qualities of the Eucharist, whether sensible or hidden, were maintained miraculously in order "to produce a sensible effect." And that was all: the dogma did not impose an interpretation of this "sensible effect."

These brilliant academic defenses—sustained in the very bosom of the Collegio Romano, with all the official support that accompanied them—must add their weight to the scales, allowing Galileo to know on September 9 that it was not deemed necessary to defend *The Assayer* openly and continue an endless polemic. Actually, then, Father Grassi's accusations seemed perfectly unfounded in the eyes of Galileo's Roman friends. But perhaps, like Galileo, they were too accustomed to underestimate the authority of tradition. And in the dispute over the Eucharist, that authority was of great concern.

SEVEN. THE DISPUTE OVER THE EUCHARIST

> Comme l'astre brillant qui porte la lumière
> au déclin d'un beau jour, en la saison d'été,
> luit sur son char penchant avec tant de clarté
> qu'elle éblouit encore la plus forte paupière.*

Light and Heat

The marble inlay on the ornamental altarslab is worthy of the refined artwork in semiprecious stones of fifteenth-century Florence. It depicts an indefatigable interlacing of knots: a symmetrical labyrinth of loops and crisscrosses which is never interrupted and which returns on itself *ad infinitum.*

At the center of the altar, planted with impressive solidity, stands a chased-gold monstrance applied with cast decorations. From its triangular base rises, in the midst of these decorative motifs, a stem with a round knop surmounted by a cross. And, above the cross, there is a gilded circular frame with small, engraved winged heads. At the center of the frame, supported by an invisible wedge, the Consecrated Host bears imprinted on it the customary image of the crucified Christ.

More than a particle, that Host is a parcel of light. It does not reveal its own contours. It is perceptible only because it stands out sharply like a halo of transparent and immaterial light against the white light of the sky behind the altar, which forms an empty background for the monstrance.

The observer's gaze cannot avoid concentrating on this transparent Host. That quiet "still life" seems to emit vibrations of its ineffable inner light. On it, in fact, converge from right and left the gazes of theologians, doctors of the Church, saints, prelates, popes, and the faithful—all convoked here to testify for us that the Eucharistic mystery is the focal point of the faith.

From that focal point, along both sides of the altar, one's line of vision tends to follow a sinuous arc of personages of the teaching Church: St. Gregory the Great, St. Jerome, St. Ambrose, St. Bonaventure, and St. Thomas. Dante and Savonarola are also recognizable.

Almost everyone is holding a book in his hands. Some of them recite from it aloud. Others, with St. Bonaventure, scrutinize it with absorbed, exeget-

* "Like the radiant star which brings light, / on the decline of a limpid summer day / glitters in its chariot at sunset with such splendor / as to dazzle even the strongest eyes."—Antoine Godeau, *Institution de l'Eucharistie, Poésies chrétiennes et morales*, Paris 1663, tome III, 1

ical reflection. Other books and treatises are scattered at the feet of that exceptional assemblage, and those books are also concentrated near the altar at the center.

The animation of the great medieval theological disputes between Dominicans and Franciscans on the mode of the real presence can be read in the animation of the personages, in their gazes and their hands. Gazes and hands are directed at that point of light, that symbol of the mystery.

Having thus returned to concentrate on the altar, one's gaze, following an arm's imperious gesture, is now propelled on high along the vertical axis of the monstrance. And it is proceeding in that direction that the circular presence of the Host's luminous halo gradually appears to be amplified, with a series of concentric circles that develop *ad infinitum*, like those which are caused by a stone skimmed on the flat surface of a lake.

Indeed, from the very first, immediately above the first circle of the monstrance's frame, the Host's circularity is amplified in that of a gilded *tondo* which surrounds the dove of the Holy Ghost. Then, the circular motif expands to the dimensions of a large nimbus that surrounds the entire figure of the resurrected Christ, repeating the decorative motif of the monstrance's frame.

Finally, this superimposition of concentric circles which rises from the Host to the sky ends with a crown of vapors and light, of which only a part is visible: the sky of the glory of angels, containing in its golden light the Eternal Father in the act of blessing. This ultimate celestial circle, infinite, is marked off by rays, as in the spokes of a wheel, or (better put) by a dome covering the lower part.

And, as in a church, the altar's Sacrament is set at the center of the orthogonal axes of the structure of this theological Catholic universe. Above this extraordinary armillary sphere of the faith, as though to dominate it in the world of ideas, appears an allegory of theology, personifying revealed truth. Alongside it there is another allegory, that of fire, as though to signify that light had an essential symbolic function in a theology centered vertically on its Eucharistic nucleus.[1]

The axial disposition of the Trinity, vis-à-vis the Host, thus culminates in the fire's light. It reveals to us that this theological universe is inspired by doctrines associating the theological vision of that Eucharistic system with a vaster cosmological sentiment. Here fire, the primary source of light, has a sphere which is also perfectly integrated, intensely associated, with that system which revolves around a single point of enormous attraction: the miraculous Consecrated Host.

What we have been trying to read up to this point is the decoration of the

[1] We are speaking here of the *Allegory of Fire*, a panel (not by Raphael) in the ornamented ceiling of the Stanza della Segnatura, Vatican Palace.

eastern walls and ceiling of the Stanza della Segnatura in the Vatican Palace, frescoed by Raphael in 1509 (Fig. 9). That very well known room was the hall for the Tribunal of the Segnatura Gratiae, a privileged place of power of a Church intent on realizing a new universal order. That room's complex decoration was guided by a constant effort at intellectual symmetries, symbols, reminders, and relationships skillfully orchestrated by the great humanistic and scientific culture of the author, Raphael, and of the person who commissioned him, Julius II.

For the historian, the room is a doctrinal speculum of uncommon intellectual value. On the history of Eucharistic theology, on its inseparable connections with the history of philosophy and science, one might gather together here an entire library of pre-1509 works. Instead, this fresco by Raphael[2] itself sums up, in its luminous image, the Eucharistic mystery in the cultural vision of European man. In that room it becomes clear that in 1509, when Raphael frescoed it, there did not exist an insoluble antinomy between classical naturalism and Christian transcendentalism, between pagan philosophical materialism and Catholic sacrality. Directly opposite this triumphal exaltation of the Church, Raphael has in fact painted the Ficinian temple of profane science, *The School of Athens*: the classical world reconstructed as a temple inhabited by Plato, Aristotle, Socrates, Epicurus, Averroës, Heraclitus, Euclid, and Ptolemy. On one side is the Bible, on the other the *Timaeus*.

The two images stand face-to-face—without contact, but also without prevarication. It was without shame that these ancient mathematicians and naturalists (among whom is perhaps hidden Democritus, as well) could look at the triumph of the Church. Raphael's Neo-Platonic religiosity coincided with the liberalization of a Roman court sensitive to the fascination of a Renaissance culture that precluded nothing when confronted by the wealth of historical and intellectual problems.

But only forty years later, if he had lived until then, Raphael would no longer have been able to paint a fresco in that room with the same feeling of equilibrium between faith and philosophy; because only forty years later, in 1551, the thirteenth session of the Council of Trent gave a definitive and combative version of the truth of the Eucharistic mystery, one which inevitably changed that equilibrium.

And after Trent, the cardinals, prelates, and popes who frequented that room of the Tribunal of the Segnatura also looked with different eyes at Raphael's fresco. Originally it was probably entitled *The Triumph of the Church*, or *The Triumph of the Eucharist*, since it positively celebrated rational

[2] Raphael, *The Dispute concerning the Holy Sacrament* (1509-1510), fresco on the lower part of the Stanza della Segnatura, ivi (see Fig. 9). See the interpretation of this fresco in J. H. Beck, *Raphael*, New York 1982.

inquiry and debate in theology. Now, however, in an age of inviolable dogmas, of controversies and intellectual oppression, their view of that particle vibrant with light had also changed; nor could it be otherwise. In the seventeenth century, the original title of the fresco was changed to *The Dispute concerning the Holy Sacrament*, which is still its title today.

Instrument of Faith

In Galileo's time, the dispute over the Eucharistic sacrament raged as never before. But it was no longer the one depicted in the Stanza della Segnatura: its frontiers had shifted. Raphael had illustrated that intellectual debate within the Church as a spiritual ferment. Now, after the formulation of the Tridentine dogma, the dispute could no longer take place within the Church, but rather along its besieged frontiers—against those who stood outside the Church, where there is no salvation. The serenity of that cycle of frescoed mirror-images was the swan song of a Platonized Christianity at its sunset, illuminated by the light of faith in that Host which was like a star in the serenity of a midsummer day's sunset.

In that large fresco concerning the Eucharist, something presaged an ineluctable problem. A symbol: that solitary and imposing monstrance that Raphael had carefully meditated upon and elaborated to give it a liturgical significance in keeping with the times.

When Raphael made a pen-and-ink sketch of his first idea of that symbol, as we can see in the study of the *Dispute* preserved today in the British Museum,[3] he had set forth the altar of the Eucharistic supper and also a supporting mount. He had been satisfied, however, with representing the Sacrament in its most mystical and ancient form, as he had done when painting Faith among the theological virtues: a chalice, surmounted by the Host (Fig. 10).[4]

The monstrance depicted in the Vatican fresco, on the contrary, was conceived in accordance with the most modern design of Catholic liturgical implements. How much more modern this was compared to the pinnacled monstrances dominant until the sixteenth century, which were reliquaries adapted for the elevation of the Sacrament. The monstrance chosen by Raphael was new, shaped like the sun; of the original chalice, only the base remained, while the cup had atrophied and was transformed into a sun.

History of implements, history of ideas: that kind of monstrance was the symbol of a new form of the Eucharistic cult completely dissociated from the ritual of the Mass; it was the symbol of the supper as Protestants understood

[3] See the drawing for *The Dispute concerning the Holy Sacrament*, London, British Museum, vol. 63, 1900 (8/24/108), published in G. Castelfranco, *Raffaello. Disegni*, Florence 1977, tav. 42.

[4] Raphael, *The Faith*, panel of the predella of the Baglioni Tavola (1507), Vatican Pinacoteca, Room VIII.

it. It served the worship of the Sacrament under a single species—complete, of absolute value in itself, the center of a special and independent cult. In the seventeenth century, that cult will attain a frenzied ostentation, and that sun-shaped monstrance will attain monumental dimensions beneath huge Baroque domes.[5]

In the seventeenth century, in this regard, the Catholic faith no longer needed, as it did in medieval times, to understand itself. It needed to impose its great Tridentine dogma—the keystone of the entire faith of the Catholic Reformation, as unshakable as the enormous block of white marble painted by Raphael in the background of his fresco, perhaps to summon up the image of St. Peter's great edifice.

Words and Problems

The Eucharist is the most important sacrament of the Christian religion. Among all the signs that express man's participation in the divine life, the Eucharist is in fact the only one to render Christ not only really present amid men, but also integrally present.

The celebration of this mystery, in the form of the sacrifice of the Mass and Communion, had been the central practice of early Christianity. Its significance had been modulated by lively patristic discussions on the Christological theme. It was a venerated practice without the imposition of an explanatory doctrine: "a mystery that cannot be reasonably investigated,"[6] as St. Augustine will say. But, from St. Augustine to Galileo, there unfolded the long history of a word, "transubstantiation."

The term had appeared only quite late in the eleventh century and will be imposed as a "canonical" doctrine even later, during the succeeding two centuries, when Scholastic theology assumed a juridical style in its argumentation. In fact, canonical collections used for the pastoral instruction of bishops and preachers made the fortune of this word, with its effective didactic and polemical characteristics.

The neologism entered sermons, liturgical works, and polemical treatises. Thus, well before being sanctioned by the Lateran Council of 1215 and centuries before being affirmed as dogma, the doctrine of transubstantiation had become an unofficial dogma by linguistic virtue of a term that eluded

[5] See the colossal monstrance in the Cathedral of Enna (Sicily) by Paolo Grilli (sixteenth century). On the history of the monstrance as a Catholic liturgical implement, see M. Righetti, *Manuale di storia liturgica*, 2nd ed., Milan 1950, I, p. 474 ff.; H. Bremond, *Histoire littéraire du sentiment religieux*, IX, Paris 1932. The art of liturgical implements carries to its peak the identification of the Word incarnate, the light, and the sun (as suggested by John and Paul) which depicts the convergence of intellectual, religious, philosophical, and architectonic tensions in the Baroque age. See C. Costantini, *Dio nascosto. Splendori di fede e d'arte nella S. Eucaristia*, Rome 1944; and E. Mâle, *L'art religieux après le Concile de Trente*, Paris 1936.

[6] St. Augustine, *Liber sententiarum* . . . , VIII, bk. IV, dist. XII, 2. See G. Lecordier, *La doctrine de l'Eucharistie chez saint Augustin*, Paris 1930.

any sort of intuitive and rational grasp and that was therefore ideal for designating a mystery. But this virtue had its other side, one that accompanied it for a long time, tracing an often dramatic history of the irreconcilability between doctrinal theology and a vision of reason for which the value of sensible experience in physics was decisive.

It was a long and difficult history, that of the conflict between reason and faith. In comparison, the opposition to the heliocentric astronomical truth seems like a short-lived, marginal episode.

Why this dogma and not others, such as the conceptual mystery of the Trinity? The answer is easy: transubstantiation was the only dogma to render flagrant the antinomy between the testimony of the senses and the affirmation of doctrinal faith. In fact, the dogma imposes sensible phenomena (color, taste, and smell) and mechanical and chemical properties identical to those of everyday experience, but postulates alongside this identity of experience a radical change in the substance of the consecrated bread and wine. It was not difficult to foresee that this would result in casualties among philosophers and scientists.

A dogma is for a religion what a postulate is for a hypothetical deductive theory: nobody claims to demonstrate it; but it is legitimate to ask whether its formulation is intelligible, or whether it is self-contradictory. Hence, we see an effort to interpret this mystery rationally, to understand its significance.

The doctrine of transubstantiation posed many challenges to human comprehension. Among them are two great orders of questions. How is this transformation of substance produced—by annihilation, or by conversion of the original substance of the bread and wine into the body and blood of Christ? The second question, on the other hand, concerned the permanence of the original sensible data. This was a miraculous permanence, to be sure, but was subjected to the common experience of the senses. How can these phenomena and their perception in physics be explained?

Questions of the first type were at the heart of the dueling between the Dominican followers of St. Thomas and the Franciscans, followers of Duns Scotus for several centuries, in a debate of exquisitely theological character. The second order of problems directly influenced theories of knowledge and philosophies of nature in an inextricable commingling of arguments, wrongs, and rights.

Their commingling was inevitable; for "faith if it is not thought is nothing,"[7] as St. Augustine had said when announcing the adventurous search for that intelligence of faith which will characterize culture and men in medieval and modern Europe.

[7] St. Augustine: "Fides si non cogitatur nulla est," *De praedestinatione sanctorum*, n. 5, in J.-P. Migne, *Patrologiae . . . Series II: Ecclesiae Latinae*, Paris 1844-1855 (hereafter, Migne PL), 44, sec. 63.

A great philosophic destiny could easily be assigned to transubstantiation. Indeed, it is clear that if the idea of substance had been separated from that of sensible phenomena, it would have been easier to rationalize the idea of the permanence of sensible phenomena whose substance was changed. On the contrary, as is recognized by modern apologists, any philosophy introducing into the notion of substance quantitative elements in the form of extension, number, and mechanical properties would have rendered more difficult—if not contradictory—the condition of substance existing in the Sacrament. Thus, it is understandable why no philosophy is more in consonance with the exigencies of faith than the Aristotelian-Thomist metaphysics, the so-called hylomorphism.

When this great advantage of St. Thomas's philosophy became obvious, all the preconceptions and reluctances initially felt against adopting not only the logic but also the philosophy of nature of a pagan philosopher such as Aristotle disappeared. In any event, the choice was almost obligatory, since the other philosophies and cosmologies all proved to be incompatible with the formulation of that great principle of faith. The sole alternative solution would have been the radical choice of stipulating the autonomy of science and faith—which, obviously, will be hard to imagine until a time very close to our own. This presented a difficulty on both sides, given the absolute importance of the cultural patrimony of Catholic dogmas, and therefore of the theological background of the theories of knowledge and the world, at least until the seventeenth century.

So there was a long conflict within Christian culture—not against it—in the minds of men who were all equally Christian, but who were divided by their ideas and their responsibilities toward the intellectual problem of the Eucharist. On one side were the interdictions of a consecrated doctrine, dense with philosophical consequences; on the other, epistemological hypotheses that rejected the value of abstract concepts, such as that of substance, and instead upheld sensible knowledge, or the atomist explanations—the hypotheses of dialecticians, nominalists, and atomists. All the latter, in different ways, heralded a problem that leads us right to Galileo.

They all collided, as will also happen to Galileo, with the problem of the Eucharist. Many of them were to regret it; but what counts is that through their discussions, even if they later recanted, they established a movement in the direction from which modern science will be born.

From Chartres to Trent

The history of the intellectual dispute over the Eucharist is, so far as we are concerned, a history that repeats itself again and again in the course of several centuries, through recurring circumstances: philosophy, science, and politics. That profound tension begins at a great intellectual and spiritual pole: at Chartres, in the first quarter of the eleventh century.

In that famous school there was an innovative theologian named Berengarius, who showed his intolerance toward the principle of authority, protesting that the senses and reason—the reflection of divine wisdom—could delve even into the depths of that mystery. He was successful, becoming archdeacon at Angers and a brilliant *scholasticus* at Tours, in the School of Saint-Martin, so lively as to provide competition for the authority of the famous school of the Norman Abbey of Bec.

Berengarius decided to take advantage of a favorable situation. His protector, Eusèbe Brunon, had become Bishop of Angers in 1047; and the Count of Angers, the antipapist Geoffroy Martel, supported him unconditionally. So Berengarius said publicly what he thought: the senses perceive the Eucharistic appearances—colors, odors, and tastes. But since, in good dialectics, these are inseparable from their substance, the bread and wine continue to exist after the Consecration. Berengarius was already a nominalist: the colors and tastes were colored and tasteful substances, and all this was in the name of "the eminent role of reason."[8]

The powerful abbot of the authoritative Abbey of Bec denounced him, reproaching him for wanting to reduce "the entire Eucharist to a pure symbol,"[9] that is, to only the name of a sacrament. The history had begun.

Seven councils condemned Berengarius. Repeatedly Berengarius anathematized his heresy, and punctually he retracted his retractions. Finally, at the Lateran Council of 1079 he acknowledged once and for all that he believed in the canonical dogma of transubstantiation.[10]

The seed of the dispute, however, continued to germinate. The school at Chartres, medieval cradle of many other heterodoxies, such as philosophical atomism, lined up with Berengarius. Many sects traditionally hostile to transubstantiation (the Waldensians, Cathars, and Patarins) came forward. The Benedictine Order, guardian of the official theological tradition, had to mobilize its best cultural centers (such as the Breton Abbey of Mont-Saint-Michel) alongside the school at Bec and form a bulwark against the new ideas.

But they penetrated the formidable walls of the abbeys and cloisters and began to filter into the towns and countryside. So, as a warning against

[8] Berengarius of Tours, *De sacra coena*, ed. A. F. and F. T. Wisher, Berlin 1834, p. 53. See T. Heiz, *Essai historique sur les rapports entre la philosophie et la foi de Bérenger à saint Thomas d'Aquin*, Paris 1909.

[9] Lanfranco of Bec (or of Pavia), *De corpore et sanguine Domini adversus Berengarium*, nn. 6 and 7, Migne PL, 150.

[10] That is to say: "that the bread and wine . . . are transformed into the same flesh, true and vivifying, and into the blood of Jesus Christ . . . according to their true substance" (H. Denzinger, *Enchiridion symbolorum, definitionum et declarationum de rebus fidei et morum*, 32nd ed., Freiburg 1963 [hereafter Denzinger], n. 355). See R. Redmond, *Berenger and the Development of Eucharistic Doctrine*, Tynemouth 1934.

Berengarius the heretic, the elevation of the Host began to be introduced into the liturgy; the great Eucharistic procession that still takes place at Angers was another consequence. Much more important, however, were the immediate intellectual consequences.

"Illam summam controversiam,"[11] as Abelard will still call it in 1120, nevertheless had the effect of establishing the theological terminology in an incontrovertible sense, consecrating for apologetic reasons the contested word "transubstantiation." That word, destined to become one of the most often pronounced and written in European history, appears at this moment, perhaps for the first time, precisely under the pen of St. Peter Damian, great critic of the rational examination of theological arguments. And since words are weapons in the history of ideas, that neologism proved so effective as to spread and endure in time.

But the adversary's weapons also help the cause: the "reason" invoked by Berengarius the heretic is used as an orthodox instrument. The dialectical minutiae serve to prove transubstantiation through the grammatical analysis of the pronoun "hoc" in the Eucharistic formula and sift an abundance of quotations from the patristic texts to confirm the new terminological clarification. The *summae*, the *sermones*, the manuals for the use of inquisitors, and the apologetic treatises did the rest. In 1215, the Lateran Council sanctioned a *fait accompli*.

The antirationalist controversy had the effect of uniting theology with canon law and of clearing the way for St. Thomas. For polemical, pedagogic reasons or for reasons of catechism, the theory of transubstantiation marks the passage of St. Augustine to Scholasticism. The contemporary Augustinian rationalism is replaced by a speculative spirituality, propelled in an apologetic and polemical spiral toward an intransigent metaphysics. Metaphysical realism and defense of the principle of authority were from this moment on indissolubly joined.

Already in Berengarius's abjuration of 1079, the notion of substance had made its appearance. Guitmund of Antwerp, a pupil of Berengarius's denouncer, will perfect the doctrine according to which "taste, color, and the other sensible accidents subsist."[12] With the support of the council of 1215, the road inevitably led to a philosphy of matter according to which bodies have a fundamental constitutive reality and a reality that appears to the senses. The first part is called substance; the second is called quality, species, *proprietates*.[13] But are Eucharistic species objective reality or subjective impressions? This was the problem.

[11] On whether the Host is the shape or substance of the body of Christ, see Abelard, *Theologia*, Migne PL, 178, col. 1120.

[12] Guitmund of Antwerp, *De corporis et sanguinis Christi veritate in Eucharistia*, Migne PL, 149, col. 1481.

[13] See Denzinger, n. 430.

Another innovator, another dialectician full of faith in experience, suggested for this philosophical problem related to theology a mental experiment that we might call the experiment of the apple. A pupil of Roscellinus, his name was Abelard; and he proposed that since these phenomena could not exist in the void, they must subsist in the encircling air—just as the smell and taste (color?) of an apple persists even if it is no longer there (for example, when it has been eaten). His theory was condemned. But the problem had been raised.[14]

It was in great measure owing to the success with which it allowed one to confront the philosophical and logical problems of the Eucharist that the Aristotelian doctrine of matter adopted by St. Thomas—hylomorphism, a neglected word today—had a more lasting success in official philosophical instruction than almost any other theory of the constitution of natural bodies. Indeed, it is still taught in Catholic universities at the present day.

A body, according to this theory, is composed of two metaphysical principles: matter, which gives the body its extension; and form, a qualitative principle that confers on it activity and specific properties. Substance is the product of these two principles. Now, in the Eucharist, there was a substance different from its extension, since the substance of the body of Christ did not evidently coincide either with the extension of the Host or with its sensible properties.

The great advantage of the hylomorphic theory was of an economic character: it permitted the mystery of the Eucharist to be reduced to a single miracle; that is, the single miraculous separation of a body from its extension. With matter thus separated from its extension (that is, from quantity), transubstantiation became a perfectly rational doctrine, enjoying all the theoretical advantages that St. Thomas illustrates in the very famous questions 73-83 concerning the Eucharist, in his *Summa theologica*. Much later the fathers at the Council of Trent will base themselves on that work in formulating the fateful dogma.

"All the substance of the bread is transmuted into the body of Christ. . . . therefore, this is not a formal conversion but a substantial one. Nor does it belong to the species of natural mutations; but, with its own definition, it is called transubstantiation."[15] All the substance: that is, the matter and the form. The fact that even today Catholics know that to take Communion the species of bread alone is sufficient, is another result of St. Thomas's philosophical and theological perspicacity.

[14] Gathered from the accusations of Guillaume de Saint-Thierry, *Disputatio adversus Abelardum*, c. IX, Migne PL, 180, col. 280; and from the letter of St. Bernard, *Epistola* (190), *De Erroribus Abelardi*, Migne PL, 182, col. 1062. See *Capitula errorum Petri Abelardi*, Cousin, Paris 1849-1859, vol. II, p. 768.

[15] St. Thomas, *Summa theologica*, III, q. 75, a4. Ample bibliography can be found in J. R. Armogathe, *Theologia Cartesiana*, La Haye 1977.

The consequence that interests us, however, has to do with the problem of physical phenomena according to the new doctrine. And here the consequence was metaphysical realism.

St. Thomas had the intellectual audacity to affirm that which his predecessors—from Albert the Great to St. John of Damascus and Algero—had only timidly suggested: that Eucharistic phenomena are sensible phenomena separated from substance, accidents without a subject. Hence, the quantity (extension) of the Consecrated Host is not sustained either by the material of the bread or by the surrounding air. It persists, miraculously, without substance. So, too, with all the other accidents that adhere to extension: the notorious "color, odor, and taste." These persist and act "as if" they depend on a substance, but in reality they persist and act without substance.

The metaphysical theory of transubstantiation made possible the solution, in a logically unexceptionable fashion, of some traditional theological puzzles that people had debated for centuries: What happens when the priest breaks the Host? And when the Consecrated Wine begins to turn sour? What if a Host has been nibbled by mice, does such a profanation also injure the sacred body of Christ? It is on problems like these, through centuries of subtle argumentation, that the zest and rigor of logical discussion had been forged. The Thomist solution cast a very rational metaphysical light on these contradictions and questions.

With the Thomist theory of "accidents without subject,"[16] the doctrine of transubstantiation had been given a rational shelter against the risk that a philosopher in the mood for dialectical discussion would appeal to the problems surrounding the sensible experience of heat, taste, and smell. It also offered protection against the risk that skepticism might trouble the consciences of the faithful if a mouse—probably a not infrequent occurrence—should nibble a Consecrated Host.

Aware of this inestimable apologetic value, the Church officially appropriated the idea of "accidents without subject." Obviously, however, this doctrine of faith also went hand-in-hand with a metaphysics of matter. Substance was being in itself, without any need of subjects in order to exist. Accidents were an independent reality tied to the substance—but virtually, not in fact—so as to permit the substitution of this state of things with the enunciation of the words of the Consecration. On the "consecration" of Thomist metaphysics, however, not all Scholastics were in agreement.

From the halls of the Sorbonne and later from the cloisters of Oxford, the Franciscan John Duns Scotus expressed his critical reservations concerning both a rational theology of that type and the intellectual monopoly that St. Thomas's order, the Dominicans, thus claimed to exercise—not only on the-

[16] *Summa theologica*, q. 77.

ological but also on philosophical and cosmological questions. We all know what influence the school formed within the Franciscan Order by Duns Scotus had on pre-Galilean philosophical and scientific tradition;[17] but not everyone knows that the touchstone and stumbling block of that great Franciscan intellectual tradition was always the keystone of Catholic faith, the Eucharistic mystery.

History repeats itself. The costumes and places may change, but the drama remains the same. Now, as in the time of Berengarius, there is again a religious intellectual order acting as watchdog over doctrinal orthodoxy. But in the thirteenth and fourteenth centuries, it is the black-and-white habit of the Dominicans that plays the role enacted in the past by the Benedictines.

On the other hand, there is still a teacher and a group of innovators who follow him. Their challenge is always the same in the culture of Catholic Europe: the relationship between faith and reason, between theology and cosmology. The endeavor now is to liberate the study of the laws of nature from subjection to the totalitarian system of Thomist Aristotelianism and the intellectual monopoly of Dominican culture.

From this issued a very brilliant methodological and scientific chapter, that of fourteenth-century Scholasticism, through the ideas on dynamics of the Parisian physicists and the mathematics of the *calculatores* of Merton College, Oxford. Their conceptual innovations in logic and physics enjoyed a contested success in European university instruction and would still be praised in Italian universities in the late sixteenth century, as in the case of Francesco Bonamico's courses at the University of Pisa, when Galileo was still a student.

Ockham—a Franciscan like Grosseteste, Roger Bacon, and Duns Scotus—also enjoyed a great popularity within that tradition which had made a common cause of the privilege of experience and a spiritual ideal of aversion to the political supremacy of the papacy. Ockham's thought had been a great war machine against the Aristotelian-Thomist physics and metaphysics. Its objective had been to detach cosmology from the theological problem, giving free rein to a series of probable hypotheses on the nature and infinity of the world, the observation of phenomena, and logic—a series of premonitory questions destined to reappear, like a stream after a long subterranean course, some three centuries later in Galilean scientific and philosophical thought.

The most radical attack on Ockham's logic had struck at the theory of matter, space, and motion. And since, as we have seen, the Aristotelian-Thomist system was based on the idea of matter and form united in substance, Ockham tried to invert it, declaring that one can know only individual aspects of

[17] See A. C. Crombie, *Robert Grosseteste and the Origins of the Experimental Science, 1180-1700*, Oxford 1953; id., *Augustine to Galileo: The History of Science,* A.D. *400-1650*, London 1952, 2nd ed. 1957.

a substance, not its universal metaphysical principles. In the case of fire, for instance, the only knowledge is the sensation of heat, which does not allow one to demonstrate the existence of the substance of fire.

As a good Franciscan who had studied at Oxford, Ockham inherited anti-Aristotelianism as his natural vocation, as well as a notable dose of intellectual intolerance for authority—so much so as not to renounce the privilege of intelligence even if his illustrious predecessors Grosseteste and Roger Bacon had, for various reasons, been excommunicated or condemned. Those fellow members of the order had, among other things, interpreted heat as the result of a dispersion of particles and had shared the corpuscular ideas on light of pseudo-Dionysius the Areopagite.

Banned by Scholasticism, Democritan atomism now owed to Ockham its philosophical revival in grand style. For Ockham—as for Democritus and, three centuries later, Galileo—the appearances of qualitative differences and of natural mutations were much better explained by thinking of an extended material whose parts, by uniting, caused all phenomena and could replace those mysterious metaphysical entities (that is, qualities which are neither imaginable nor clearly conceivable).

Ockham, in fact, could not conceive of matter without distance between its parts, without extension. Substance, as Ockham stipulated in his *Summulae in libros Physicorum Aristotelis*, is therefore knowable only by its extensional attributes:[18] it can grow or diminish, down to the dimensions of a mathematical point. In short, substance is not imaginable intuitively except as *res quanta*; it is the same as quantity.[19] When Galileo, a student, heard of Ockham, he must have been durably persuaded by this intuition.

Ockham, however, knew very well that to undo the knot that kept Aristotelian-Thomist cosmology and theology indissolubly united, it must be cut at the point of orthodox Eucharistic doctrine. It was for this reason that Ockham later developed his theory of knowledge in a treatise entitled *Tractatus . . . de sacramento altaris*.

The Eucharistic question had already appeared in previous works, the *Sentence Commentary* and *Quodlibet*, and was one of those which caused a preliminary hearing to be held by the censors of the papal court of Avignon concerning some fifty suspect Ockham propositions. This was in 1326 and Ockham was also at Avignon, in the eye of the imminent political and religious hurricane that would soon lead Ludwig of Bavaria to declare the deposition of Pope John XXII for heresies of which the Friars Minor of St. Francis accused him.

The preliminary inquiry against Ockham was similar to that against

[18] William of Ockham, *Summulae in libros Physicorum Aristotelis* (Venice 1506), published Romae 1537, reprinted London 1965, I, 14.

[19] "Substantia est quanta." Ibid., chap. XVI.

Copernicus and Galileo many years later, in 1616. The theological commission of inquiry condemned his theses (among them the idea that substance and quantity are identical) as reckless and dangerous; but strangely, perhaps for reasons of political opportunism, owing to a delicate moment in relations between the papacy and the Franciscans, Ockham was not subjected to an official condemnation. He will be condemned officially only in 1349—but for his political doctrines, after the failure of his impossible attempt at schism and at having a pope condemned by his successor.

In the *Tractatus . . . de sacramento altaris*, Ockham did not deny his philosophical convictions. The Eucharistic accidents—color, taste, and weight—like all sensible phenomena, are also "quanta, because endowed with length, breadth, and depth."[20] Accidents, including those of the Eucharist, are a quantum reality, not qualities. Nonetheless, Ockham pointed out that he believed completely in transubstantiation, submitting himself in each case to the authority of the Roman Church in the matter, and added that his rejection of the permanence of real qualities without subject "belonged to logic, more than to theology."[21]

Avignon's theological experts, however, were not fooled by the Oxford doctor's formidable dialectic and had understood that if quantity was identifiable with substance, then in the permanence of the quantity of Consecrated Bread the substance also remained. Transubstantiation, consequently, became a dangerous "consubstantiation."

The ecclesiastical authorities, however, more concerned at that moment with canon law than with philosophical heresies, had let Ockham's ideas pass. But when a heresy was not persecuted from its first manifestations, it spread in an uncontrolled manner, creating centers of infection that later proved impossible to stamp out, like the Black Death. The University of Paris became one of these centers of infection of the nominalist-empiricist heresy, the citadel of the "modern way" of the "nominales." By the time the authorities realized this and took the necessary steps, it was too late.

While Ockham was staying at Avignon plotting against the pope together with his general and Ludwig of Bavaria, Nicole d'Autrecourt, a brilliant student at the Sorbonne, obtained his degree in the arts and his baccalaureate in theology. He too was an innovator, a skeptic galvanized by Ockham's corrosive criticisms. In a series of letters to Bernardo d'Arezzo, he set forth his

[20] *Tractatus venerabilis inceptoris Guglielmi Ocham de sacramento altaris* (Strasbourg 1491), critical ed. T. B. Birch, Burlington (Iowa) 1930, chaps. 23 and 26.

[21] Ibid., chap. 30. See A. Pelzer, "Les 51 articles de Guillaume de Occam," *Revue d'Histoire Ecclesiastique* 18 (1922), pp. 240-70, especially p. 261. The proposition "substantia et quantitas sunt eadem res" was censured by theological experts as "contra communem sententiam sanctorum doctorum et philosophorum." See P. Duhem, *Études sur Léonard de Vinci*, Paris 1909, vol. III, p. 341; L. Rougier, *La scolastique et le thomisme*, Paris 1925, p. 632. Note that the 1564 edition of the *Index librorum prohibitorum* does not mention the *De sacramento altaris*.

antimetaphysical convictions—on the impossibility of deducing the existence of knowable entities, the value of intuitive knowledge, and the reduction of so-called substantial mutations to the local motion of "corpora athomalia";[22] that is, atoms like those of which pupils at Chartres, such as Adelard of Bath and William of Conches, had spoken before.

Nicole d'Autrecourt extended his criticism of Thomism-Aristotelianism into natural philosophy and demonstrated the greater probability that, rather than generations and corruptions, there were aggregations of atoms endowed with the same virtues by which "the magnet attracts iron."[23] In addition, D'Autrecourt advanced the possibility of a corpuscular theory of light: "light is nothing else but bodies"[24] (since perhaps the instantaneous velocity of light was also only apparent). Finally, D'Autrecourt transferred atomism from the physical to the spiritual, in order to explain mechanistically the perception of sensible qualities. His criticism of real qualities brought him finally to hypothesize that sensible qualities revealed nothing other than dispositions and motions of atoms, as Lucretius had said about the nature of color.

Still, in natural terms, Nicole d'Autrecourt brought out of oblivion a famous Lucretian slogan: "*ex nihilo nihil fit* [nothing is born of nothing]" (*On the Nature of Things* I, 265). All this was stated as probable, in a plainly dialectical style and in contradiction to the teachers of official culture.

Nicole was the sort of person who, in the seventeenth century, would be called a follower of the "new philosophy." In the treatise that presented his scientific point of view, *Exigit ordo executionis*, he in a sense anticipated *The Assayer* by three centuries—claiming, *contra* the Schoolmen, the right to innovation, to knowledge proceeding from known things and not from deference to authority. He rejected the dogmatism of those who compiled *quaternos* full of Aristotelian quotations. He pointed out that, unlike Aristotle's modern followers, dogmatic and intransigent when it came to the famous Ockhamist thesis of the identity of quantity and substance, Aristotle himself (in book III of the *Metaphysics*) had shown a greater critical sense, defining the quantity-substance problem as one of the most difficult to examine.

In the end, Nicole d'Autrecourt protested his fidelity to the Church, invited the men of his time to turn their thought to natural things, and announced a future in which, free "from logical sermons and from Aristotle's

[22] See J. Lappe, "Nicolas van Autrecourt sein Leben, seine Philosophie, seine Schriften," in *Beitrage zur Geschichte der Philosophie des Mittelalters*, VI, fasc. 2 (1908), p. 38 ff. In it were also published the two remaining letters to Bernardo d'Arezzo. The ms. treatise "Exigit ordo executionis" (Bodleian Library, Oxford) is published in J. R. O'Donnel, "Nicholas of Autrecourt," *Medieval Studies* (1939), pp. 179-280. It was analyzed previously in P. Vignaux, "Nicholas d'Autrecourt," in A. Vacant et al., *Dictionnaire de théologie catholique*, vol. XI, cols. 561-87, with extracts from the ms.

[23] Ibid., col. 572.

[24] Ibid.

obscure propositions, they will show to the people the intelligence of divine law."[25]

Challenged on the principle of authority, the ecclesiastical power reacted, but this time too with scant determination. In 1340, Nicole d'Autrecourt was cited with six other Parisian students to appear before Benedict XIII at Avignon to answer for the spread of erroneous doctrines. But six years were to pass before the trial ended with a clear condemnation of some sixty propositions, among which were included the identity of quantity and substance, the corpuscular theory of light, and atomism. This time, however, the condemnation was official: on November 25, 1347, D'Autrecourt had to abjure publicly his philosophical ideas and burn his writings.[26] He was divested of the title of *maître ès arts* (that is, of philosophy) and of his diploma in theology.

The condemnation of nominalist atomism and the Eucharistic heresy of the association of quantity and substance were thus ratified with a sentence of historic significance, which later had repercussions in the persecution of Parisian "nominales." And yet, Ockhamism had not really been the object of a total and unequivocal condemnation. The procedure was fraught with delays, hesitations, and judicial hairsplitting. The ecclesiastical power had intervened only against certain precise, circumscribed theses, because they compromised the Eucharistic dogma. Thus, that power proved it did not have an accurate perception of the actual danger of the "modern way."

The real danger was not the individual theses of some dozen propositions. The true danger was the absolute distinction between faith and dialectic—what later would be called reason. In other words, the danger was the Ockhamist vindication of the right of reason to go to the extreme consequences of disputation in the philosophic and naturalistic field, while denying, however, that this dialectic involved objects of faith and stating that it concerned only the theory of nature or logic. This was the irreducible question.

By not intervening radically right away, the Church condemned itself to prosecuting repeatedly every single philosophical and scientific thesis that would inevitably conflict with doctrinal dogma. Even today, when modern Thomists lament the initial laxity that permitted the empiricist heresy to proliferate, only slightly hindered by those specific hairsplitting objections, they do not seem from their point of view to be entirely mistaken. The Reformation will come and make apparent the greatness of this danger. And the Council of Trent will come to thwart it.

[25] Ibid., col. 581. See P. Duhem, *Le système du monde. Histoire des doctrines cosmologiques de Platon à Copernic*, 10 vols., Paris 1913-1954, vol. VI (Paris 1954), pp. 665-70, and VII, pp. 20-23.

[26] See the condemned propositions of D'Autrecourt in Lappe, "Nicolas van Autrecourt"; and in Denzinger, nn. 1028-1049. The proceedings of the trial are in Denifle-Chatelain, *Chartularium Universitatis Parisiensis*, II, Paris 1891, pp. 576-87.

Since Ockham had not been formally condemned for Eucharistic heresy, some heterodox theologians and nominalists, such as Robert Holkot at Oxford and Cardinal Pierre d'Ailly,[27] felt authorized to believe that the question could be limited to a purely philosophical conflict between two antagonistic concepts of quantity. So they advanced, though not as a hypothesis, a criticism of the metaphysical realism of "color, order, and taste."

Here there appears, perhaps for the first time, the possibility that those accidents were, instead of qualities, permanent psychological impressions, deriving from an imaginary substance to which the reality of Christ's body had been added by the Consecration. It was only a hypothesis, but it guaranteed a minimum of discussion. Besides, the official Thomist doctrine was not yet sanctioned by a dogma. At the end of the fourteenth century, however, the situation suddenly changed and became one of total rigidity.

Among the many admirers of Ockham and of the nominalist ideas there was in fact a subversive, an irreducible and extremist adversary not only of the hierarchy and ecclesiastical power, but also of Thomism. This subversive Ockhamite, or, to put it as the Scholastics did, this execrable and diabolical heresiarch was Wyclif.

Wyclif had begun teaching at Oxford that "heat, odor, taste, and the other accidents existed in the subject." Then, in reply to those who asked him for clarifications, he said that it was a matter of a "corporeity of a mathematical character" (*corpus mathematicum*);[28] and everyone understood that it was a *res quanta*, amenable to mathematics, but also quite corporeal. In other words, Eucharistic accidents depended on the permanence of quantum matter.

It was the heretical doctrine of "Christ made bread" in the Eucharist, welling up with extraordinary polemical violence from its distant Ockhamite sources, and dragging with it an avalanche of accusations of idolatry addressed to the supporters of accidents without subject. Caught off guard, the tutors of orthodoxy reacted to the "made into bread" heresy with a first censure of the chancellor of the University of Oxford, William Brenton. But the affair was very serious and passed from Oxford to London and from here to the highest ecclesiastical jurisdiction: the Council of Constance in 1415.

Wyclif was already dead. He lived only long enough to know a condemnation on the part of the London Synod in 1382. But the doctrine did not die with its author and, indeed, had created an endemic hotbed of Eucharistic heresy in Bohemia, where it took root extraordinarily well. From that

[27] R. Holkot, O.P., *Super quatuor libros Sententiarum quaestiones*, Lugduni 1497-1518; P. d'Ailly, *In IV Sententiarum libros*, ivi, 1500. The latter accepted the Ockhamist theory of consubstantiation (IV, 6).

[28] See W. of Woodford, "De censis condempnacionis . . . J. Wyclif, Contra trialogum Wiclefi," in O. Gratius, *Fasciculi rerum expetendarum et fugiendarum*, Coloniae 1535 (ed. E. Brown, Londini 1690), I, pp. 190-265, especially, p. 191.

moment on, Bohemia will be the great thorn in the side of the Church of Rome. Only the Jesuits, in the seventeenth century, following Father Pazmany and the soldiers of Ferdinand II, will succeed in smothering the conflagration that Wyclif's ideas had lit on the banks of the Elbe and Moldau rivers two hundred years before.

The Council of Constance, then, had to take vigorous measures. Wyclif was dead, but one could try his follower, the nationalistic Bohemian leader John Huss, author of a treatise *De corpore Christi* in which he gave proof of following Wyclif's heresies.

To ascertain the grade of transubstantiation admitted by these innovators was humanly difficult. Their positions, as one saw at Constance, were rather obscure. There was not yet a precise dogma, and Huss defended himself against the accusation of not believing in transubstantiation. In short, the secret convictions of Wyclif and Huss were not the principal point at issue. The decision to proceed was determined by an apologetic necessity of a general character.

Between 1409 and 1412, flames had burned the works of Wyclif in the courtyard of the archiepiscopal palace in Prague.[29] At the University of Oxford and in London, some of Wyclif's political followers were also condemned to the stake. On May 4, 1415, sentence was pronounced on the memory of this heretic who had died thirty years before. It was a historic sentence.[30]

Huss, however, was still alive. On July 6, during Session 15 of the Council of Constance, the condemnation of Huss's Eucharistic heresy was pronounced. The text of the condemnation officially adopted as its own the doctrine of accidents without subject.[31] Perhaps unwittingly, Bishop Berthold von Windungen, who that day read the condemnation of Huss, in the same moment authenticated the hylomorphic theory of the Eucharist's sensible phenomena. So, in the eyes of many contemporaries, the Council of Constance had put Aristotle on Catholic altars.

From this moment on, it was no longer possible to continue around the mystery of the altar a choral theological debate such as the one later depicted by Raphael in the Stanza della Segnatura. Now, around the sacrament of the altar, the Church became the "body of the lacerated Christ," and for consciences the laceration was deep.

[29] Wyclif's errors in Denzinger, nn. 1151-1195 (581-625), go back to the treatise *De Eucharistia* (1379) and to the Twelve Articles taught in 1382. See *The Latin Works*, Wyclif Society, London 1883-1922 (reprinted New York 1966). On the Eucharistic problems in Wyclif's heresy, see G. Leff, *Heresy in the Later Middle Ages*, 2 vols., Manchester 1967; F.-X., Jansen, "Eucharistiques (accidents)," in *Dictionnaire de théologie catholique*, vol. V., cols. 1360-1452, especially col. 1399.

[30] Denzinger, n. 1151 (581).

[31] See J. D. Mansi, *Sacrorum Conciliorum nova et amplissima collectio*, tome 27, Venetia 1784, p. 747, concerning the Council of Constance, sess. XV, July 6, 1415, the day on which rejection of the Scholastic theory of accidents without subject was officially declared by the Church to be "blasphemous dementia."

"My conscience," wrote Luther, a pupil at the Ockhamite University of Erfurt, "confirms in me the opinion that in the Eucharist, along with the true flesh and true blood of Christ, there is true bread and wine."[32] The dispute was no longer philosophical-theological, but was for or against the truth of faith. The nominalist (or nominalist-origin) idea of consubstantiation, which admitted the copresence of the original substances with that of the body and blood of Christ, automatically became synonymous with the Lutheran heresy, *ipso facto*.

Luther did not intend to stipulate a canonical theory on the disappearance of the substance of the bread and wine. His was the claim of returning to the Augustinian conception of the mystery, eliminating as "sophistical subtlety"[33] the controversy on transubstantiation in order to assume a simply sacramental vision of the real presence concomitant with perceptible phenomena. These proposals were a radical inversion of the tendency that, in 1537, Luther and Melanchton wished to submit to the new ecumenical council. The reply came with the anathemas hurled at the conclusion of the thirteenth session of the Council of Trent, which reaffirmed the idea of transubstantiation, repeating almost word-for-word St. Thomas's *Summa*.[34]

Nonetheless, instead of the by now prevailing expression in orthodox theology, "accidents without subject," the council fathers gave evidence of caution by subscribing in their formulation of the dogma to the permanence of "species," a more generic and also less binding concept. Yet, the caution had been rather ambiguous and could not prevent others from seizing on that formulation to affirm that the theory of real accidents without a subject was a conclusion of faith, as would indeed happen.

After Trent

In the sixteenth and seventeenth centuries, the role played first by the Benedictines and later by the Dominicans had passed (after the great reawakening of Dominican Scholasticism at Salamanca) to the theologians of the Society of Jesus, for whom Eucharistic apologetics were an essential

[32] See M. Luther, *De captivitate Babilonica Ecclesiae Praeludium* (1520): "One takes iron and fire, two substances, as they are mixed in the heated iron, in such a way that it becomes iron and fire together, so why then could not the glorious body of Christ be in both parts of the bread's substance?" (text proposed for the examination of the Council of Trent). See *Luthers Gesammelte Werke*, Weimar ed., vol. VI, 1888, pp. 497-573, especially p. 507, for the rejection of Aristotle's ideas.

[33] Ibid., VI, p. 510.

[34] See *Conciliorum Oecomenicorum decreta*, Bologna 1973, pp. 693-95 (October 11, 1551, Sess. XIII). See H. Jedin, *Il Concilio di Trento*, Brescia 1973, vol. III, p. 379 ff. Above all, the imperial bishops spoke out against the provocative dogmatic definition of the term "transubstantiation." To the objection of the Bishop of Vienna, the Bishop of Bitonto replied that it was necessary to preserve the controversial word for the good reason that the Protestants rejected it (ibid., p. 390). One should remember that this great doctrinal decision in the history of Catholic liturgy is embodied in the cult of the Feast of Corpus Domini.

preoccupation—the great inspirational motive for their Counter-Reformation struggle, their banner. The Benedictines and Dominicans had substantially failed; they were unsuccessful in preventing the spread of the most dangerous heretical opinions on the Eucharistic mystery. The Society of Jesus, for its part, had the immense advantage over its predecessors of being able to range its forces behind the inviolable bastion of a dogmatic formulation. It would have sufficed to remove from that dogma its final traces of lexical generality, making it unequivocal, to prevent any possible deviation. Vazquez, Suárez, Toleto, Bellarmino, Lessius, Tanner, De Lugo: the finest names in Eucharistic theology between the sixteenth and seventeenth centuries were names from the Society of Jesus.

It is a fact that the Council of Trent had put back into circulation a great Scholasticism, which had languished for more than a century because of Ockham. With Suárez and Fonseca, metaphysics regained its voice after a long silence, and Aristotelian metaphysics and theology were again fostered. Science, philosophy, and theology were again inseparable.

The apologetic vocation and supreme pragmatic wisdom that animated the Society of Jesus tried to force the hand of the fathers at the Council of Trent, with the demand of defining as a matter of faith the traditional thesis of Scholastic theologians on accidents without a subject. More than any other, it was the word of Father Suárez that made "the voice of all Scholastics" resound, as Bossuet will say.

To engage in polemics with Protestants, however, a special intellectual sensitivity was demanded of the Jesuits. They welcomed St. Thomas as a guide, but reserved the right to an ideal and a method of concreteness distant from the rarefied abstractions of Scholastic speculation. Suárez's philosophy, like other aspects of Jesuit culture, tried to maintain contact with concrete reality in its espistemology, too. In Suárez's philosophy, consequently, there prevailed in respect of pure Thomism an Ockhamist opinion on the epistemological value of the singular and of the individual experience.

Suárez, the great commentator on St. Thomas at the Collegio Romano from 1560 to 1585, was the most authoritative and most widely read philosopher during the first half of the seventeenth century, even more than Aristotle and St. Thomas. He had been educated at Salamanca under the guidance of an Augustinian and nominalist philosopher, the Augustinian Father Juan de Guevara. Suárez was a Scholastic, but a post-Trent Scholastic. He aspired to be the restorer of Scholasticism in metaphysics and epistemology. But for this to be done—to defeat atomism—the new times demanded that one become an Ockhamist in order to defeat sensist Ockhamism; to become a negator of the spiritual world, of faith and rational theology, of hylomorphism.

Suárez's philosophy tended to an ideal, eclectic synthesis of the whole tradition of Catholic thought preceding the Council of Trent—St. Thomas,

Duns Scotus, and Ockham—with the declared apologetic purpose of demonstrating that all Catholics were in agreement among themselves. The price of this synthesis was the abandonment of St. Thomas's theory of free will in order to admit direct knowledge of the individual—experience, senses, and intuition. In epistemology Suárez rejects raw material: he is an Ockhamist who proclaims "Omnis substantia singularis"; every entity is individual, and therefore every accident, every substance, is individual.[35]

In Suárez, father-founder of the Society of Jesus' "theological philosophy," the theologian prevailed over the philosopher. Even his most original conceptual formulations were illuminated and directed by controversial exigencies of the Tridentine faith. The theory of substantial modes originally elaborated in Suárez's philosophy served to revitalize metaphysics and Scholastic epistemology, above all in defense of the Eucharistic dogma against interpretations of a nominalist and atomist stamp.

The problem, as we know, was twofold. In the first place, the permanent Eucharistic appearances were to be saved rationally and protected from atomism. In the second place, there was the question of the relationship between substance and quantity. The theory of "substantial modes," already proposed by Egidio Romano to justify rationally the union between the Word and the man in Christ, was applied by Suárez to the problem of speculative theology concerning Eucharistic accidents without substance.

Between substance and accidents, Father Suárez's philosophy suggested the existence of metaphysical entities—the "modes"—by virtue of which the accidents adhere to the substance (accidental modes) and the forms to the respective materials (substantial modes). In the case of the Eucharist, whose accidents exist without substance, it suffices to think of a miracle capable of separating the substantial mode of the accidents of the bread and wine from the matter of the bread and wine, without that matter being destroyed. The accidents, thanks to the suppression of the substantial mode, maintain their reality (real species), but only in appearance is there of bread and wine, since those accidents are no longer united to their substance. The value of this solution was that it united Scholastic metaphysics with a principle of great economy as regards miracles, in homage to nominalist thought.

On the second question, relating to quantity, Father Suárez was so comprehensive and so sensitive to nominalism as to recognize that the Ockhamist identification of substance and quantity was very reasonable—but, the rights of reason had a limit, and that limit was, obviously, the Eucharistic dogma. God would never have separated substance from quantity to deceive men (that is, if *ex natura rei* the substance were not separable from quantity). Such separation

> although it is not rationally demonstrable [Suárez's dispute XL af-

[35] Suárez, *Metaphysicarum disputationum* (1597), Paris ed. 1619, 5, 3, 8.

> firmed, and this is quoted in the complaint against *The Assayer*], must however be absolutely preserved. We become convinced nonetheless of its truth on the basis of the principles of theology, above all in relation to the Eucharist.[36]

In other words, just as will be pointed out to Galileo, the substance-quantity problem is not a philosophical question, but a question of faith and orthodoxy. The nominalists say no, that there persists only the quantity of color, odor, and taste and not the quantity of the substance to which these accidents are attached:

> this solution [Suárez retorts] first of all is repugnant to the common opinion of theologians, who hold that after the Consecration there remains the quantity of the bread's substance and that it is the subject of the other remaining accidents.[37]

As to the theological authority there can be no objection. However, Father Suárez also proposed to the "nominales" an experimental ascertainment of the orthodox theory: "it can also be proved," he added, "with the effects that we experience in the Consecrated Species that otherwise would be impossible to save without many and continuous miracles. The first is that the Consecrated Host is a quantity and is extended in the place occupied by it, and cannot be in that same place together with another Host, nor be by another compenetrated. . . ."[38]

This conception of the Eucharist, then, joined speculative theology and physics in an inextricable and indissoluble manner. Father Grassi, as we know, will draw his words and arguments directly from this in order to incriminate Galileo. But even later, throughout the entire first half of the seventeenth century, this theme of Suárez's became a leitmotif whenever atomism and corpuscularism came up.

Father Suárez's theological philosophy blocked the new physics, but also the mathematics; for in support of his Eucharistic apology, that philosophy turned to the traditional Aristotelian argument against geometric indivisibles. The impossibility of lines formed by points without extension, the absurdity of the physical existence of an indivisible point of contact between a plane and a sphere, and all the paradoxes of geometric infinity were part of that apology.

Father Suárez's philosophy was the official philosophy of the Society of Jesus, the one followed by the most orthodox and representative fathers. It was especially cultivated, applied, and promulgated by the Collegio Romano. There the ingenious speculative intuitions of Suárez's metaphysics

[36] Ibid., pp. 365-406. [37] Ibid., p. 369. [38] Ibid.

had been taught contemporaneously with the methods of Scriptural and theological Eucharistic controversy of Father Bellarmino.

Bellarmino's teaching was not speculative, but applicative, operative. It was based chiefly on rigorous linguistic analysis of the sacramental formula as an enunciation that, by its very linguistic significance, was able to change a state of things, if the necessary parameters were respected: matter, the distance from it, the intention, and the sacerdotal nature of the speaker.

In his controversies course at the Collegio Romano, from 1576 to 1588, Bellarmino had developed the polemic against heretics by elaborating not only his "linguistic proof" of transubstantiation, but also the famous theory of the destruction of the Host in the Mass as sacrifice.[39] But from the philosophical point of view, which is what interests us more than any other, the father's ideas and Cardinal Bellarmino's coincided with the presuppositions of Suárez's theological philosophy: "God can do everything, except that which implies contradiction."[40] The Eucharistic miracle therefore requires a philosophy equal to the faith. The thesis of real accidents without subject is necessary and suited to the purpose. For Bellarmino, but also for Father De Lugo, who like Bellarmino had a career as teacher at the Collegio Romano and then became a cardinal, the Scholastic theory of Eucharistic accidents as real qualities was *de fide*.[41] But if it was *de fide* that those accidents were real qualities, and if it was for centuries that one said it was *de fide*, how was it possible to imagine that non-Eucharistic color, taste, and odor—exactly identical to the Eucharistic ones—were not real qualities, without being automatically against the faith?

With Bellarmino and Father De Lugo we are now at the end of a historical dolly shot that has brought us all the way to the battle between the new philosophers and the Jesuits, all the way to the incrimination of Galileo by Father Grassi and the Collegio Romano. After this journey through the past, following a history which is not all that obscure, we know that the sensation of "color, odor, and taste" and that of the quantum substance were for centuries compromising words and issues precisely because they were compromised in Eucharistic theology in an essential manner.

[39] See R. Bellarmino, *Disputationes de controversiis Christianae fidei adversus huius temporis haereticos*, 17 vols. Ingolstadi 1589-1593, vol. III (1590), p. 1365b; id., *Philosophia Eucharistica de potentia et voluntate Dei*, Amburgae 1604. On Bellarmino's great work of positive and controversial Eucharistic theology, see Father J. de La Servière, S.J., *La théologie de Bellarmin*, Paris 1908, p. 390 ff.

[40] "The accident cannot in any fashion subsist by itself; in fact, in the Sacrament there can be accidents without subject. In any case, they do not subsist by themselves per se, as substances, but are sustained by God in a supernatural fashion" (Bellarmino, *Disputationes*, vol. II, bk. IV, and vol. III, bk. III, chap. 24). This is one of the innumerable Bellarminian expressions that form the foundation of his Eucharistic theology in the categories of Scholastic philosophy.

[41] See G. de Lugo, *De sacramento Eucharistiae*, Lugduni 1644, dispute X, sec. I, p. 363. See, also, L. Lessius, *De sacramentis et censuris*, Lugduni 1636, 2nd ed. 1645; and *De perfectionibus divinis*, Antverpiae 1620.

Galileo did not speak of the Eucharist in *The Assayer*, and at so great a distance of time we may be surprised at a doctrinal objection apparently so far from physics. But in the seventeenth century, as we have seen, "color, odor, and taste" were cultural terms that designated before all else the daily experience of the Eucharistic miracle. They were words of the theological language and of everyday religious life. In short, behind those words stood a century-old erudition, a set of intellectual presuppositions, a whole mentality.

With *The Assayer*, however, Galileo had wanted to attack that erudition, that principle of dogmatic authority, that consensus subjected to tradition—above all, that inextricable connection, full of equivocations, between reason and faith. Neither he nor his Roman editors, experts in theology, could ignore the dangerousness of the words "color, odor, and taste," or the delicacy of the relationship between substance and quantity. But their contemplative faith also compelled them to abandon the commingling of Scholasticism and physics on that delicate, most important point. Ockham had dared. Three centuries later, Galileo also dared; but certainly under more difficult conditions, because now there was an official dogma, the Tridentine dogma *par excellence*.

Moreover, interpretations of the dogma in a rigidly Scholastic sense must appear to Roman Galileans of the new pontificate as one of the numerous authoritarian prevarications of Jesuit arrogance. In fact, the text of the dogma also allowed hope for a liberating distinction between natural philosophy and faith in regard to "color, odor, and taste."

The possibilities existed and were tied not only to the theological debate, but also to reasons of power, necessity, and renovating will. At the moment of *The Assayer*, these variables had been favorable. Would they be definitive?

Christian Ockhamism had decreed the ruin of medieval Scholasticism. Was the philosophy of *The Assayer* announcing the ruin of the new Scholasticism?

EIGHT. THEATER OF SHADOWS

When I behold this goodly Frame, this World,
Of Heav'n and Earth consisting, and compute
Thir magnitudes, this Earth a spot, a grain,
An atom, with the Firmament compar'd
And all the numbered Starrs, that seem to roule
Spaces incomprehensible (for such
Thir distance argues and thir swift return
Diurnal) merely to officiate light
Round this opacous Earth, this punctual spot*

Catastrophe

We must weep, Blessed Father, for frightful destruction and an immense ruin. The edifice that with its hands Divine Wisdom had erected, that eternal temple of peace between God and men is demolished by impious pillagers, destroyed, razed to the ground.

How truly atrocious it is to witness the scene of imminent ruin. Those instruments, those levers, those machines, the workers—all is prepared and ready for the frightful destructive work. . . . The custodian of the temple, new Levites, sleep a profound sleep. . . . But terror shakes them now from their profound sleep. The mob of furious pillagers advances. . . . Already the veil of the temple, at the separating of the soul from the body of Christ, is torn; already the entire structure lists to one side, and with such a deathlike clangor that even if sleeping they are now driven to awake. Sacred implements are trampled underfoot, the altars overturned, the temple in ruins: where will we find shelter, where, I say?[1]

It was the afternoon of April 18, 1631. The final words of the solemn sermon of Good Friday fell as heavy as the stones split away from that temple in ruins. They resounded grimly under the ceiling of the Sistine Chapel, immersed in the purple light of Easter mourning. To underscore their dramatic echo there then came, as always, the dolorous notes of Allegri's *Miserere*, which on that holy Friday must have seemed more transfiguring than ever.

With striking realism, that lamentation inspired by Psalm 73 had evoked

*John Milton, *Paradise Lost*, 1667, bk. VIII, ll. 15-23

[1] O. Grassi, "Divini templi excisio" (1631), in *Orationes quinquaginta de Christi Domini morte habitae in die sancto parasceve a patribus S.I.*, Romae 1641, p. 596 ff.

a threatful image, worthy of a prophecy by Campanella and a catastrophic painting by Monsú Desiderio. It had evoked with great efficacy the creaking that heralded the collapse of a Church mortally undermined by heretics. Not by chance was that scene of ruin summoned up by a professional architect, who had addressed directly to the pope that disquieting description of catastrophe in the form of a severe warning and an arraingment of all tolerance for enemies of the faith.

As it did every year, it fell to the Jesuit fathers of the Collegio Romano to pronounce the solemn, prestigious oration in the course of the papal celebration of the great liturgy of Good Friday. That year, the oration had been given by the consultant to the Collegio Romano, Father Orazio Grassi, the architect of St. Ignatius. Ordinarily, the Jesuit fathers exhibited themselves in inimitable, virtuoso performances of ascetic oratory. This time, though, the sermon, with its metaphysical tones, had the character of a threatening political prophecy, as its author will explain when dedicating it to the pro-Spanish Cardinal Ludovisi.[2] To Urban VIII's ear, Father Grassi's spinechilling sermon must have sounded like one of the many recriminations that the intransigent party had never failed to direct at him in recent months. However, this last sermon must have seemed particularly sinister, given the circumstances and the fact that the horizon was truly grim.

At the gates of Rome, a dense smoke rose into the sky from fires lit to fumigate household goods and the mail. The deeds of witchcraft and diabolical phantasms, the piles of unburied corpses of the year before at Milan were recent stories in the gazettes. But now the great anguish of the plague had arrived at Rome, isolated from the horror of contagion by measures of quarantine and sanitary precaution. Moreover, the plague held even the most lucid minds in the grip of ancient superstitions and new disbelief. And there was a war on. Italy—because of the unhappy outcome of the Mantuan succession, which had pushed Rome's politics into a blind alley—had now become the theater of operations of the Thirty Years' War.

When some months later, at the end of the year, after a hundred and thirty years of absolute calm, the summit of Vesuvius suddenly burst open with an immense explosion that spread night over Naples, it truly seemed that the presages of misfortune of all those comets which had appeared in 1618 were now going to come true in Italy, too. Irresistible as a lava flow from Vesuvius, Gustavus Adolphus's agile Swedish army inundated Europe, disrupting with one blow the military and diplomatic European balances. The sweeping arrival of the great Swedish strategist and his unbeatable Protestant army on the theater of the Thirty Years' War definitively pulled the rug out from under the feet of the fragile, ambiguous, and contradictory papal

[2] See the first edition of Father Grassi's *oratio*, Romae 1641, in 4°, of which I have examined the copy owned by Father Paolo Casati, Grassi's friend, in the Palatina Library at Parma.

desire for a balancing action among the Catholic powers, an action that in practice meant a pro-French policy.

Since the month of January, Cardinal Richelieu had allied himself with that new master of European warfare. But Richelieu could afford to have no scruples about allying himself with an anti-imperial Protestant force; the pope could not. The road to a pro-French alliance was no longer viable. The dream of arbitrating European diplomacy by staying above the fray had ended, and so had the ambitious hope of preserving Roman moral and political independence. To save the Counter-Reformation faith it was imperative to form a common front and line up with the Hapsburgs, renouncing in Italy every ambition of independence from Spain.

Father Grassi's tirade on the day of Christ's Passion seemed the very voice of the Society of Jesus, the group most exposed to the danger of defeat on German territory. It reproached the pope for a culpable negligence in the custody and vigilance of the Tridentine Church's fundamental values.

In any event, since the end of the previous year, every Thursday meeting at the Holy Office had been the setting for clashes between Pope Urban VIII and Cardinal Borgia, the Spanish ambassador. Every occasion offered the pro-Spanish party a chance to accuse Urban VIII of protecting heresy at Rome by his excessive tolerance. They demanded energetic action. They tried to make the pope understand that the days of carefree intellectual liberality that marked the beginning of his pontificate could not continue, and that it was the moment to take up, without reticence or shilly-shallying, the leadership of a Catholic crusade against heresy and the subversive new ideas. Urban VIII did not have to wait long for the realization of the warning announced to him on that holy Friday by Father Grassi's sermon.

The contradictory tangles of the Church's political management under the Barberini pontificate came to fore less than a year later, when Urban VIII's papacy experienced its most serious political crisis. Until that moment the protests of the intransigent wing of ecclesiastical power had been contained in semiofficial form. But on March 8, 1632, in the new Hall of the Consistory, while a secret consistory was being opened (that is, a council of state of the Church), the pope was confronted by the open denunciation of Cardinal Borgia, protector of Spain, backed by all the cardinals of his party: Ludovisi, Colonna, Spinola, Doria, Sandoval, Ubaldini, and Albornoz.[3]

The denunciation was made in person by Cardinal Borgia, who took the floor to announce the reading of a statement "of the greatest interest for religion and the faith." He began to read: it denounced the heretical alliance with the Swedish king, declared that Madrid wanted the pope to make his apostolic voice ring out like a trumpet of redemption. Urban VIII cut him off, ordering him to remain silent and threatening to depose him. But Bor-

[3] See Pastor, *Storia dei papi*, pp. 1025-1028.

gia wished to speak not only as a prince of the Church, but mainly as the representative of the Catholic king. Antonio Barberini, the Cardinal of Sant'Onofrio and brother of the pope, was the only one to react to the grievous offense, lunging in Borgia's direction as though to attack him. But he only seized his arm because Cardinal Colonna (representative of the empire) and all the Spanish and pro-Spanish cardinals had clustered around Borgia to allow him to read the statement.

The tumult increased. Attendants intervened. The consistory was interrupted, but Cardinal Borgia had enough time to distribute copies of the statement among the cardinals while the pope said the last word: "The care of the Catholic religion, over which we have stood guard and continue to stand guard, is our task."[4] It was a formal commitment for the future.

The news of the serious scandal in the consistory, in which the pope had been accused of tolerance toward heresy, went the rounds of all the chancelleries in Europe. One well-informed observer, the Florentine ambassador, Francesco Niccolini, reported to Florence that the protest hinged on the accusation against the pope of an easygoing attitude toward the enemies of religion, and that from now on—still according to this authoritative diplomatic source—intransigence toward heretics and innovators and surveillance of orthodoxy in Rome itself would be the Spanish party's instrument of political and ideological pressure on the Curia.[5] What other means of pressure, if not ideological, might be brought to bear on the pope in order to extort from him money and anathemas to further the political interests at stake in the religious war?

The pope could still order Cardinal Borgia to be silent; but he could no longer silence the Spanish protest, much less depose Borgia. Cardinal Borgia had dominated the consistory and isolated the pope, who knew that he could now count only on his relatives.

The day after the scandal, Rome sent a note of dignified protest to Madrid. But on March 11 there was already another clash between the pope and Cardinal Borgia at the Holy Office. From Naples and Madrid arrived threats of direct intervention on the side of the cardinal–Spanish ambassador. The extremists, like Cardinal Ludovisi, threatened to depose Urban VIII, protector of heresy.

These are days of dramatic political crisis, during which the pope tries to strengthen his authority and reduce the irresistible shift in the balance of forces. Unable to obtain satisfaction in regard to Cardinal Borgia, the untouchable new master of the situation, on March 18 Urban VIII strikes at

[4] See Cardinal Francesco Barberini's account to the nuncios Ceva and Grimaldi (ms. Barb. 8376, p. 85, Vatican Library), published in Pastor, *Storia dei papi*, p. 1028.

[5] See the diplomatic report transmitted by the Florentine Ambassador Niccolini (Medicean Series 3351, State Archive of Florence), in Favaro, "Amici e corrispondenti di Galileo. Giovanni Ciampoli," p. 121 ff.

Borgia's Italian accomplice, Cardinal Ludovico Ludovisi, expelling him from Rome.[6]

Cardinal Ludovisi leaves his stupendous chancellery palace and, before departing, goes to the Church of Gesù to bid farewell to his protégés. He will not live long enough to see either the Church of St. Ignatius or the new political and cultural climate of whose advent he had been a victim.

At the end of March, there arrives in Rome the great cardinal and former Jesuit Peter Pazmany, the indomitable Catholic restorer and special representative of the Hapsburgs, in order to ask of the pope the same requests made to Madrid and to blackmail him with exorbitant requests for money to finance the war against Gustavus Adolphus of Sweden.

The situation is indeed urgent. On April 7, Gustavus Adolphus reaches Bavaria, the heart of German Catholicism. The Jesuit colleges are plundered, and the fathers are expelled from the cities occupied by the Swedes. Caught between the contradictory opposition of the requests of Philip IV and Ferdinand II, who support Borgia, and Richelieu's urgings to break with Spain, Urban VIII is unable to choose. But time decides, and quickly. For by May 1632 Gustavus Adolphus has reached the Grisons and is ready to cross the Alps and descend on Rome. Will he be content, as rumor has it, with capturing the Palatino Library and taking it back to Heidelberg?[7]

By then, the choices had already been made. When the political crisis is at its peak, under the threatening pressure of Hapsburg intimidation, the pope is forced to yield and furnish ample guarantees and satisfaction to the Spanish party. The official measure of the political and ideological turn which took place in the Curia are the diplomatic instructions sent to the Nuncios by the cardinal-nephew. On May 1, the pope bitterly complained of

> very false suspicions and conjectures without foundation, . . . a thousand judgments of distrust, calumnies, and erroneous judgments—many of which with time and truth have been clarified. And the same can and must be believed and said about the others; that is, that they will always be found to be mendacious and without any substance.[8]

Against those "contrary or malign or ignorant persuasions,"[9] the pope was obliged to promise from now on a greater rigor and clarity about his real intentions ("the counsels and sentiments of hearts") regarding the protection of orthodoxy. The adventurous pro-French experience was over, as was

[6] Besides Cardinal Ludovisi, the pope took his revenge on another pro-Spanish extremist leader: Cardinal Roberto Ubaldini. On the political crisis of 1631-1633 and the *volte-face* of the Barberini pontificate, see A. Leman, *Urbain VIII et la Maison d'Autrice de 1631 à 1635*, Lille and Paris 1919.

[7] See Pastor, *Storia dei papi*, p. 463.

[8] See Cardinal Francesco Barberini's instructions, May 1, 1632 (Vatican Secret Archive, a.3, tome 47; and ms. Barb. 2629, p. 135, Vatican Library), published in Pastor, *Storia dei papi*, pp. 1029-1033.

[9] Ibid.

the inebriating atmosphere of free and unprejudiced patronage, of that mystical optimism in reason which had made the Rome of the new pontificate the capital of the innovators. In any event, had not Cardinal Maurizio of Savoy, the pope's great elector and leader of the pro-French party, also changed parties; had he not, after the Treaty of Cherasco the year before, also abandoned France to go over, with an agile political pirouette, to Spain?

Cardinal Ludovisi's defenestration had been Urban VIII's personal revenge—a way of saving papal dignity offended by Borgia—before surrendering and accepting Spanish and imperial dictates. With the outwardly imperceptible, but actually substantial, change in the Roman political picture, the situations of the innovators and virtuosi in Rome also changed. The season of indulgent patronage, of Florentine and pro-French intellectual worldiness and sophistication had ended. The Jesuits took over, returning in grand style to the Curia's sphere of cultural activity. Bernini is at work on the great baldaquin in St. Peter's. The pope's poetic work, as if to sanction the changed intellectual climate, is published by the Collegio Romano in a sumptuous edition illustrated by Bernini. In Rome the Baroque begins.

Poussin, Cassiano dal Pozzo, the community of innovative literati and artists must step aside; indeed, many of them leave Rome. Tommaso Campanella can no longer benefit from the impunity that had released him from prison, and he prepares his flight to France. The "marvelous conjuncture" is over: we shall see what the consequences will be for the main figures of that miraculous season.

A "Pythagorean and Democritan" Book

In February or March 1632, during the same period in which power and politics had secretly imposed an atmosphere of intolerance and suspicion at Rome, the *Dialogue* was published in Florence. No moment could have been more inopportune, but after so many delays it seemed necessary to make up for lost time.

It had taken six years of negotiations, agreements, and corrections for the *Dialogue* to see the light of day in such a form as to offer no surprises. The *Dialogue* had been prepared with infinite precautions: basically, the content of this great pedagogic work was the same as that outlined since 1624, in the fateful reply to Monsignor Francesco Ingoli, with the addition of tides as a further argument in favor of the earth's motion. As early as 1616, however, one knew that geophysical argument would become a mainstay of Galilean cosmology. In fact, Galileo had suggested the title *Of the Flux and Reflux of the Sea*, a title he had then agreed to change completely, in obeisance to the ecclesiastical authorities' prudence, who preferred the more anodyne title that was in fact adopted. The preface had been read, reread, and corrected by Father Riccardi, Master of the Sacred Palace; the text had been examined in Rome by Father Visconti[10] for the Roman imprimatur; and then, when it was printed in Florence, another Dominican, Father Giacinto Stefani, had also examined the new work. They were all obliging censors—Galileo's friends—but they were also authoritative theological experts of the Dominican Order.

The printing, whether by design or because of the plague which hindered communication with Rome, had been fraught with difficulties. It was adorned with the Roman imprimatur without having really complied with existing rules governing the deposit of printed copies. But, after all the agreements on a personal level between the author, the pope, and the Master of the Sacred Palace, that breach was purely formal. In any case, the assurances of May 1630, when Galileo had personally submitted his manuscript to Father Riccardi, confused the situation. What counted, though, were people's attitudes, and these had been clear until the end.[11]

During that last Roman visit, Galileo had seen the loyalty of the pope's and Cardinal Francesco Barberini's support. For even when an attempt was made to involve Galileo, together with Campanella, in the incident of a funerary horoscope announcing Urban VIII's imminent death, the pope—as the cardinal-nephew had amply assured—had not allowed himself to be manipulated by Galileo's enemies and still treated him with the same friendliness as before.

Galileo, therefore, did not have to fear the threats of "rivals" and their slanders.[12] Galileo's "rivals," however, anxiously awaited the *Dialogue*, though there would be nothing new to discover in that work of cosmology. It would honor the promises and agreements and would present Copernican astronomy with the understanding good will of scientific reason accompanied by the vigilant protection of theological prudence. The only *suspense* [English in original—Trans.] was the new book's belated arrival in Rome, owing to the sanitary precautions then in effect.

The *Dialogue*'s first complimentary copies reached Rome at the end of May (1632), and Cardinal Francesco Barberini was the first to receive the book.[13] In August, other copies arrived and were distributed by Filippo Magalotti.[14] There was not enough time, however, to distribute it to the bookstores.[15]

Just the same, it was in a Roman bookstore—we do not know exactly which one—that the state of mind of one who had anxiously been awaiting

[10] See Father Raffaello Visconti's letter to Galileo, June 16, 1630, *Works*, XIV, p. 120.

[11] See Michelangelo Buonarroti's letter to Galileo, June 3, 1630, ibid., p. 111.

[12] See Father Riccardi's letter to F. Niccolini, April 25, 1631, ibid., p. 254.

[13] See B. Castelli's letter to Galileo, May 29, 1632, ibid., p. 357.

[14] See Father Campanella's letter to Galileo, August 5, 1632, ibid., p. 366 ff., especially p. 367.

[15] See Filippo Magalotti's letter to M. Guiducci, August 7, 1632, ibid., pp. 368-71.

the book was publicly displayed. The scene seemed to repeat the one that had taken place nine years earlier at the Sun Bookstore. This time, too, the protagonist was a scientist connected with the Collegio Romano, Father Christopher Scheiner, who, hearing the *Dialogue* praised by a Galilean—the Olivetan Father Vincenzo Renieri—"got all excited, his face changing color"[16] and said that he was willing to pay as much as ten scudi for a copy of the *Dialogue* in order to answer it immediately.

From all that we have learned, we are now inclined to take seriously Jesuit threats of denunciation voiced in a bookstore. But what could there be to rebut in a book so minutely planned, examined, and corrected as the *Dialogue*? Copernicanism? Obviously, but let us not forget that Galileo's initiative had not been spontaneous. He had received an unofficial authorization to go ahead, and he had striven to get his book officially accepted and approved. He had been given permission and could talk about Copernicanism provided he did so in a manner that was "hypothetical and without reference to Scripture."[17] In the *Dialogue*, "Scripture" had certainly been avoided; as for the first warning, however, it was more difficult to evaluate the work's degree of sincerity. Nevertheless, in order to avoid misapprehensions, there had actually been inserted, as in Copernicus's *De revolutionibus*, an *ad hoc* introduction and a final summation with the intent of preserving appearances from any reasonable suspicion. But we know there were unresolved suspicions, and perhaps Father Scheiner could have found pretexts for slander in Galileo's work.

It should be said that, in April of the previous year (1631), Father Scheiner had at last been able to see his great book *Rosa ursina* (which had been ready for the press since 1626) published by the Orsini house. This book contained many things but, above all, made a bitter claim to priority in the discovery of sunspots. A notable part of the book, though, was devoted to the necessity of reforming the ancient theory of the incorruptibility of the heavens, in connection with the Jesuits' new astronomical discoveries.

Rosa ursina made public the results of the scientific debate, held in 1612 at the Collegio Romano, on the cosmological consequences of that discovery with respect to the Aristotelian teaching which stipulated the inalterable nature of the celestial substance.[18] The Jesuit fathers, in order to maintain the validity of Aristotle's world and at the same time take into account the phenomena of change that the sky revealed to the telescope, arrived at an agreement on the honorable compromise of a new theory; though celestial substance was incorruptible, some accidental changes could nonetheless take place. For example, accidental regroupings of the stars could be formed, to the point of giving the impression of sunspots. The substance remained un-

[16] Ibid., p. 359 ff. (June 19, 1632). [17] *Works*, XIX, p. 327.

[18] See Scheiner, *Rosa ursina*, p. 656 ff.: "De coeli et idem naturali corruptibilitate."

altered; only the accidents changed. This was the opposite of what happened with the Eucharist, but in this case, too, there came into play the delicate philosophical problem of the relationship between substance and accidents.

Galileo knew of this compromise solution, which made all the new astronomical discoveries acceptable while preserving unchanged the Aristotelian cosmology and epistemology (that is, the vision of a distinction between the celestial and terrestrial worlds and the doctrine of substances and accidents). Now, for Galileo, it was essential to defend the homogeneity of the heavens and earth, both of which were formed by a common nature, a nature entirely and everywhere knowable through an identical rational observation of phenomena, or "accidents." Also in the *Dialogue* came the need, first of all, to challenge Aristotelian philosophy, attacking the separation between substance and accidents.

Galileo now knew very well that in so doing he could not easily avoid the problem of "Eucharistic proof," which had hung over him like a menacing Sword of Damocles ever since Father Grassi, in his book, had denounced the heretical consequences of the new philosophy. But on this delicate point, too, Galileo had taken precautions—requesting, as we have seen, authorization from Father Riccardi on the matter and receiving from him something like a reassuring promise of protection. It was an oral promise, nothing more; so far as we know, Father Riccardi had not put anything in writing.

Galileo dared. He dared, but he respected the norms of prudence by staying within "natural terminology" and not once uttering the word "Eucharist." On this point, it will be recalled, Father Riccardi had been definite. Thus it was that in the *Dialogue* Galileo discussed atomism in natural philosophy, and only there, even taking care to emphasize that this was not a matter of Eucharistic theology.

At the very start, at the beginning of the *Dialogue*, during the impassioned First Day in which the homogeneity of the celestial and the terrestrial world was defended, Galileo attacked Aristotelian ideas on the generation of substance. To criticize Aristotle's *Of Generation and Corruption* was no trifling matter. It meant denying the whole world of Aristotelian culture, as Galileo insisted on saying, putting into Simplicio's mouth the decisive opinion: "This way of philosophizing tends to subvert all natural philosophy, and to disorder and set in confusion heaven and earth and the whole universe. But I believe the fundamental principles of the Peripatetics to be such that there is no danger of new sciences being erected upon their ruins."[19] Indeed, we know that those fundamentals were solid, planted in the truth of certainties more unshakable than those of nature.

But Galileo would not accept this. It was like reading *The Assayer* all over

[19] See *Works*, VII, p. 62.

again: "demonstrations" in philosophy are learned from books of mathematical demonstrations, not from books of Aristotelian logic. Aristotle was "a great logician, but not very expert in knowing how to use logic."[20] And from his critique of the Aristotelian idea of celestial incorruptibility, Galileo went on to criticize the *sancta sanctorum* of Aristotelianism: the theory of life on earth, through generation and corruption produced by contrary elements. "I have never been able to understand fully this transmutation of substance," said Salviati, Galileo's mouthpiece, and immediately he went on to specify (still remaining within natural terms) "whereby matter has been so transformed that one must needs say it has been totally destroyed and nothing of its previous being remains in it and that another body, very different from it, is produced."[21]

Next, having contested the Aristotelian physics of substantial corruption as destruction and annihilation, Galileo (who evidently believed that "nothing is born from nothing") has Salviati also say that it was not "impossible" for natural changes to take place "by a simple transposition of parts, without corruption or anything new being generated, because we see such metamorphoses every day."[22]

The earth, in reality, was the realm of similar corruptions and generations: the earth showed changes from hot to cold, from one color to another, from one substance to another, a new plant, a new animal, iron that changed into rust, and so on. Scholastic philosophy looked upon these changes as qualitative. Galileo followed Ockham's path: he rejected the possibility of a natural corruption by destruction and proposed the possibility of a local movement of parts of matter. Like Ockham, Galileo clearly identified corporeal substance with its qualities by means of quantity—a quantity of shaped matter. Like Ockham, Galileo rejected Scholasticism and relied on Democritan atomism.

Exactly as in *The Assayer*, the *Dialogue* too opened with the philosophical and mystical idea of the contemplation of "the book of nature . . . creation of the Omnipotent Artificer."[23] Then Galileo discredited the logical certainty of Aristotelian discourse, saying that Aristotle "is planning to switch cards on us and fit the architecture to the building, instead of constructing the building in conformity with the precepts of architecture."[24] Galileo

[20] Ibid., p. 60. Aristotle's paralogism was related to the theory of natural motions, keystone of the scientific truth of geocentrism. See A. Koyré, "Galilée et la loi d'inertie," in *Études galiléennes*, Paris 1939, 2nd ed. 1966, new ed. 1981.

[21] *Works*, VII, p. 64 ff. Somewhat before, there recurred the well-known Galilean example of the generation of flies in wine.

[22] Ibid., p. 65. [23] Ibid., p. 27.

[24] Ibid., p. 43. On the style and pedagogic and confutative methodology of the *Dialogue* with respect to Aristotelian cosmology, see Koyré, *Études galiléennes*, 202 ff.; Geymonat, *Galileo Galilei*, p. 157 ff.; E. Garin, "Galileo 'filosofo,'" in *Scienza e vita civile nel Rinascimento*, Bari 1965, pp. 147-70. Shea, in

thought he could remake the world of Aristotle by "establishing the prime fundamentals . . . with better-considered precepts of architecture.[25] And at the end, in the book's third dialogue, one also learned that "to put this universe in order it would be necessary to get rid of many axioms commonly accepted by all philosophers, as nature does not multiply things needlessly"—Ockham's principle of economy, so as to show the validity of the Copernican system.[26]

Either out of caution or at the suggestion of his theological advisers, Galileo did not speak of atoms in the *Dialogue*. Nevertheless, he did keep open the option of denouncing Aristotelian physics in the name of the local motion of matter, which explained transformations in nature. The prudential specification that his discourse dealt only with "natural terms," and not with the Eucharist, revealed Galileo's awareness of the risk of explaining in material terms "substantial transmutations"; that is, transubstantiations, or something very similar.

But even with all his dialectics, Galileo was hiding behind a finger. In fact, Eucharistic accidents also become corrupted, adapt and change in accordance with "natural terms." The Scholastic distinction between substance and accidents (or substantial modes) actually served and was necessary to justify rationally the conciliation of transubstantiation and natural philosophy. If, however, one claimed that natural transformations were caused by the local motions of matter, that conciliation would be impossible. Yet, Galileo had to run the risk; for, in order to validate a new alliance of a contemplative character between reason and faith, he had per force to go that way and cut the contorted knot which tied Peripatetic philosophy to theology.

It is certain that when Father Scheiner got hold of the *Dialogue* he must have looked immediately at the pages which concerned him most; that is to say, those on the problem of the incorruptibility of the heavens. Thus, he immediately found before his eyes that material theory of "substantial transmutation" and, biased as he was, must have thought immediately of Father Grassi.

There does not exist any other historically documented proof that Father Scheiner, Father Grassi, a Jesuit, or anyone else put into circulation, as regards the *Dialogue*, Father Grassi's public and secret denunciations against *The Assayer*. And no proof exists simply because the denunciations brought against the *Dialogue* (if, as seems logical and obvious, there were any) are unknown; and perhaps we shall never know them. Nevertheless, the historian is not authorized to underestimate the possibility of reactions owing to

---

Galileo's Intellectual Revolution, emphasizes the problem of the demonstrative character of the theory of tides. See, for the Galilean methodology developed in the *Dialogue*, M. Clavelin, *La philosophie naturelle de Galilée*, Paris 1968.

[25] *Works*, VII, p. 43. [26] Ibid., pp. 149 and 423.

the atomist effect of the *Dialogue* among contemporaries, even if today, for us, this book is above all a great pedagogical work on the Copernican system.

Indeed, that is the point. One of the rare testimonies on the brief and unhappy "fortune" of the *Dialogue* in Rome reveals in fact that the work was received, whether for good or ill, as a resumption of *The Assayer*'s earlier materialist formulations in physics.

The testimony comes from Tommaso Campanella. He was an idealist and perhaps did not quite understand the new political and ideological situation that was created in Rome and that by now threatened the impunity from which he benefited exceptionally during those golden years when the innovators had enjoyed the Curia's protective indulgence. Not understanding this, Campanella welcomed the *Dialogue* with enthusiasm, followed its fortunes closely, and even thought he could intervene authoritatively on Galileo's behalf when things took a turn for the worse. Fortunately, his zeal fell on deaf ears. With his solid reputation as a heretic, Campanella's unsolicited help could only have worsened the situation, terribly compromising Galileo, his friends, and his old protectors.

Nonetheless, Campanella, having taken a personal interest, was at least well informed on the Roman echoes provoked by Galileo's new book. And since he considered himself the official prophet of the new philosphy, he was also perhaps a little jealous of anyone who wanted to go him one better in innovation. Thus it was that with one of his usual enthusiastic, disastrously sincere, and perfectly imprudent declarations, Tommaso Campanella told Galileo, in a letter on August 5, 1632, about the nature of the triumphal success the book justly deserved: "novelty of ancient truths, of new worlds, new stars, new systems, new nations, beginning of the new century." Up to here there was nothing new in the exultant Campanellian hosanna; but, carried away by his fervor, Campanella continued, saying something that perhaps he would have done better not to put in writing at a moment when, under the excuse of the plague, letters and packets were being opened at the city's gates. In fact, the letter described in detail how, during all those preceding years, many literati and innovators in the academies had made the ideas of Galilean philosophy their own—almost as though they had invented them—and that now the *Dialogue* restored them quite deservedly to their Galilean paternity. And those ideas of the *Dialogue* already so much in fashion in Rome "were of the ancient Pythagoreans and Democritics [Democritans]."[27] This Campanellian declaration is eloquent proof of the sensibility

[27] Letter (see note 14, above), *Works*, XIV, p. 367; T. Campanella, *Lettere*, ed. V. Spampanato, Bari 1927, p. 241. Some time before, in 1614, Campanella had suggested to Galileo that he leave the atoms in the dark (*Works*, XII, p. 32): "fortify the style of perfect mathematics, and leave the atoms for later";

that saw in the *Dialogue* a renewed proposal of those atomist opinions which had already earned *The Assayer* its intellectual success.

Campanella's impression should be taken, even if with reservations, as a rare trace of the reactions, polemics, and denunciations which accompanied the publication of the *Dialogue* and which are no longer extant. At any rate, such polemical reactions and denunciations caused the rapid deterioration of a situation that, for other reasons, was already compromised and difficult. Confronted by these criticisms, for us mysterious, the multiple authorizations that adorned the *Dialogue* seemed to count for nothing, and the same goes for the unofficial agreements.

A theater of shadows moves behind this episode between late July and August 1632—efforts to bring about Galileo's incrimination by seizing on the occasion of his new book and stirring up yet another of the many scandals rampant in Rome and so force the pope's hand. Father Riccardi, Galileo's friend and accomplice in his works, is obviously involved. His chief preoccupation, now that things are taking a bad turn, will be to dissociate himself or defend himself. Still, it is Father Riccardi who sounds the first alarm, revealing that the shadows acting against Galileo are projected by Jesuit fathers, who have returned aggessively to the center of the Roman stage in Cardinal Borgia's footsteps.

In Chapter Two, I cited other later testimonies that will eventually confirm the first suspicions revealed by Father Riccardi. Father Riccardi, however, is quite careful not to say which Jesuits and on the basis of which accusations the persecutions against Galileo are taking place. But, eventually, the accusations must have been serious enough to justify immediate measures to stop the distribution of the book, almost as if desperate attempts were being made to correct it before a scandal could spread.

In any case, by July 25 the *Dialogue* has been unofficially banned—for Father Riccardi, on that date, sends a rather sibylline letter to the Inquisitor of Florence, asking him to intercept at customs all copies of Galileo's book before they leave for Rome. As to the reasons for this confiscation, Father Riccardi only hints that mysterious and burning questions are at stake and that the initiative comes from on high where, as usual when an attempt is made to quash a scandal, no one wants to speak up:

> Signor Galileo's book has arrived in these parts, and there are many things which are not liked and which the masters want in every way to correct. Meanwhile, it is our lord's order (though no name but mine should be mentioned) that the book should be stopped, and not let through by you until he sends what has to be corrected; nor should it

but later, as we know, the situation changed considerably. On the atomist image of Galileo in Campanella, see Garin, "Galileo e la cultura del suo tempo," *Scienza e vita civile*, pp. 109-146, especially p. 135.

> be sent elsewhere. Your most revered self will please discuss this with the most illustrious Monsignor Nuncio and, acting with caution, see to it that everything turns out as desired.[28]

Act with "caution"; that is, avoid a scandal.

After all that has been said, we may safely assume that the pope at that moment had other things on his mind besides the astronomical systems of Ptolemy and Copernicus. So if it is true, as everything leads us to believe, that it was the Jesuits who started an action of protest against the pope's old friend and his new book in the summer of 1632, then we must ask ourselves what were the concerns that moved the Jesuits to level an accusation at their adversary.

As has been said, to this day we do not know which of the Jesuits' denunciations set off the preliminary inquiry against Galileo. Galileo's historians have surrendered when faced with the big gaps in the evidence that afflict the development of Galileo's incrimination in Rome. They have mentioned only motives of personal hostility to Galileo, which must have "fanned the fire," so that the *Dialogue* would be incriminated on the basis of Copernicus's doctrine.

I believe, however, that the decision to denounce Galileo, taken by the Jesuits at the end of July 1632, had more important and profound reasons than purely personal hostility. Indeed, if we do not know the nature of the precise accusations brought at that moment against Galileo, we do have a very precise clue concerning the fundamental concerns of the Society of Jesus at that moment—serious concerns and prohibitive, severe dispositions which, however, had nothing to do with Copernicus's doctrine.

The Preoccupation of the Moment

Just at that moment, when mysterious accusations are brought against the *Dialogue*, on August 1, 1632, the Society of Jesus severely prohibits the doctrine of atoms.[29] The condemnation would thereafter be repeatedly stressed, but until that moment it had not had official precedents. It was issued by general censors of the Society, and it was for "internal use"; that is, it was intended to be a strong warning to professors of philosphy in the Society's colleges and schools.

We already know very well that Father Orazio Grassi, official spokesman of intellectual orthodoxy at the Collegio Romano, had authoritatively condemned atomism and the theory of indivisibles. But why was the need now

[28] See Father Riccardi's letter to Father Clemente Egidi, *Works*, XX, *Supplemento al Carteggio*, p. 571 ff.

[29] See Historical Archive of the Superior General Curia of the Society of Jesus (Rome), *Fondo gesuitico* 657, p. 183, cited in C. Costantini, *Baliani e i gesuiti*, Florence 1969, p. 59.

felt in Rome to circulate the official proscription of atomist ideas to all Jesuit centers of official Catholic instruction?

Actually, a reason existed that allows us to understand how urgent it was for the Roman general staff of the Society of Jesus to eliminate officially all tolerance toward philosophic and scientific atomism. It was not just a question of Galileo. To understand this reason, we must briefly leave Rome and travel to the imperial capital, Prague.

"To go to Prague and hear Arriaga" had become a set phrase, because at the Jesuit University in Prague the philosophical and theological teaching of Father Rodrigo de Arriaga enjoyed European fame. Father Arriaga was among the most notable representatives of the Jesuits' modern and open-minded Scholastic philosophy. He was a direct pupil of Father Suárez at Valladolid and Salamanca, and he had learned from his great teacher to look ahead, beyond Aristotle; to wed the metaphysical exigencies of the old Scholastic with the methodological instances of modern nominalist tradition.

In philosophy Father Arriaga followed Suárez and Ockham, but in natural philosophy he prized the modern scientists—above all, the great physician and scientist Francisco Vallès, very famous in Spain for having been Philip II's physician. In 1625, Vallès's *Controversiarum medicarum et philosophicarum libri decem* (1582), which presented the newest and most stimulating substantialist and corpuscular ideas on heat, appeared in Lyon.[30]

Beginning in 1623, Father Arriaga taught philosophy at the Collegio in Prague. In his *cursus*, a demonstration of the probability of the Ockhamist thesis in cosmology (that is, that the essence of matter consists in extension) was usually accompanied by illustrations from physics on corpuscles of matter. Father Arriaga published these opinions as only probable (but certainly heterodox with respect to those of the Jesuits' theological philosophy) in his important *Cursus philosophicus*, dedicated to the Emperor Ferdinand II and actually published in Prague in 1632.

Father Arriaga's book was a work naturally animated by the most sincere desire for Scholastic restoration and by an effort to go beyond the stage of purely polemical and controversial opposition to the moderns by freely appropriating their new ideas on nature. In fact, Father Arriaga subscribed to the idea that heat was constituted by fiery corpuscles; that rarefaction was caused by the intromission of corpuscular substances of vapor and earth mixed with atoms which also penetrate the air and water. Father Arriaga even went so far as to declare that "light is something real." Adding that "I do not agree it is an accident," he therefore proposed "with other authors"[31] that light was a substance. Immediately after this, Father Arriaga confirmed

[30] See Francisco Vallès, *Controversiarum medicarum et philosophicarum libri decem*, Lugduni 1625, pp. 272-74.

[31] R. Arriaga, *Cursus philosophicus*, Antverpiae 1632, p. 508 ff.

his interest in atomism by proposing that even substantial generation should no longer be viewed as creation from nothing, the production of an entity *ex nihilo*, but could perhaps be explained as the production of some other subject. Identification of substance and extended quantity and atomism: the unprejudiced will for a renewal of Scholasticism on the part of the University of Prague's chancellor came very close to the Galilean idea of the new philosophy.

The danger was very serious, because the young Jesuit philosophers and scientists looked upon Father Arriaga as those of the preceding generation had looked upon Father Suárez: he was a modern teacher, a creator of new intellectual perspectives. Thus, there was the risk that behind Arriaga a wave of research favorable to atomist ideas would arise within, of all places, the Society of Jesus. And this the Collegio Romano and Piazza del Gesù could not permit, at a moment when in Rome Galilean atomist heresy was being fought with the weapons of the Eucharistic controversy.

It was for these reasons that the Society of Jesus, on August 1, 1632, sent out the proscription of the doctrine of atoms in physics. In 1633, Father Arriaga ceased teaching philosophy at the college in Prague; and from then on he will write no longer about philosophy, but only theology. However, his submission to the official Aristotelian positions championed by Rome will not save him, later on, from other severely critical reprimands because of those heterodox proposals of 1632.[32]

So we know that, for the Roman Jesuits in the summer of 1632, atomism was the preoccupation of the moment—so much so that when the *Dialogue* reached Rome, their concern was so grave as to persuade them to censor within their ranks any shadow of a corpuscular propensity in the teaching of natural philosophy. How could other peoples' atomism be condemned if it were to be discovered that atomists were nurtured in one's own house?

The clue to the Society of Jesus' state of mind in the summer of 1632 is very important, and the internal measure of condemnation that I have mentioned is the only move that can give us any inkling of the mysterious intentions which animated the Jesuits' censorious zeal. However, since the initiative had been confidential, and Galileo's historians have until now been unaware of it, it was difficult until today to relate these preoccupations to the Jesuits' secret plots against the *Dialogue*.

In any case, it was only a probable motive. It was a clue that revealed nothing of any certainty about the nature of the anti-Galilean charges contained in those denunciations which, during the Roman summer of 1632, gave rise

[32] On Father Arriaga's natural philosophy, see L. Thorndike, "The *Cursus philosophicus* before Descartes," in *Archives Internationales d'Histoire des Sciences* 4 (1951), pp. 19-24. On the censure that struck at Arriaga's corpuscular positions, see G. M. Pachtler, *Ratio studiorum et institutiones scholasticae Societatis Iesu per Germaniam olim vigentes*, III, Berlin 1890, p. 76.

to the case against the *Dialogue*. The fact is that not one of those denunciations ever appears, and so historians are led to underestimate them or even to doubt that they exist, preferring the hypothesis of a machination that popped ready-made and all of one piece from the pope's ire. We will probably never know which charges prompted the preliminary investigation into the *Dialogue*, because if it is reasonable and necessary to suppose that they were brought to bear, it is certain that they met with a very strange fate.

The Church had set up the appropriate tribunal—that of the Holy Office, with a prestigious Congregation of Cardinals, among whom stood out the rigorous Cardinal Borgia—in order to deal with questions of this kind, matters of orthodoxy and faith. Indeed, this tribunal was the perfect instrument for the purpose and was presided over personally by the pope, who thus had every guarantee that cases, even the most difficult, were prepared according to the rules and with due rigor. And, in fact, if a work was to be prohibited, a theory condemned, an author declared a heretic, the most convenient thing to do was, as we have seen, to lodge a denunciation with the Holy Office.

Under Urban VIII's pontificate, however, we have seen that Galileo, the pope's official scientist, was not like every other man on this earth subject to the universal jurisdiction of that sacred tribunal. This time, too, instead of being referred to the Holy Office and instead of seeing his trial follow the predictable course, Galileo had the right to exceptional treatment.

From the very first, the trial was removed from the jurisdictional competence of the Holy Office and was conducted by two of Galileo's most authoritative friends in the Curia: the pope, who did not want to appear, and Cardinal Francesco Barberini. Indeed, at the beginning, rather than arranging for a clear-cut investigation into the case, these highly placed protectors had moved to sidetrack the affair before it was born, in the vain hope of gaining the time to requisition all the copies of Galileo's new book, and then perhaps see to a more thoughtful correction without causing a scandal.

But it is too late. On August 1, when the functionaries of the Florentine Inquisition showed up at the Landini print shop with the sequestration order, all copies of the book had already been sent out. On that same day, that fateful August 1, a desperate attempt is also made in Rome to get rid of the cause of scandal. Father Riccardi met with the bearer of complimentary copies of the *Dialogue* at Rome, Filippo Magalotti, with the intention of getting the books from him. But here again it is too late: all these copies are in circulation, and what is more in the hands of important persons, even Jesuits. It is truly too late; it is no longer possible to pretend that the *Dialogue* had never been published.

Filippo Magalotti, however, puts to Father Riccardi our own question. Of what has Galileo been accused? The matter must be serious, because this time Father Riccardi either could not say anything—secret of the Holy Office—or has been kept in the dark owing to his involvement with Galileo's ideas.

To justify himself, Father Riccardi comes up with two painful excuses, one unbelievable and the other false: the typographic symbol of three dolphins, with a satiric allusion to Barberinian nepotism, and the absence of arguments suggested by the pope to temper the truth of the Copernican system. But Magalotti knew very well how much Galileo had always benefited from the protection of the Barberini family in power. As for the pope's argument. Galileo had put it in the *Dialogue* as requested, albeit in the mouth of the Aristotelian Simplicio. Magalotti, on the other hand, asks if there is any truth in the "rumors" that are said to have greeted the book's appearance in Rome, and who these "malicious people" might be. Father Riccardi confirms the "rumors," but claims that he does not know their content. He says, however, that he does know their authors and that, in view of the present situation, this is the most serious truth. They are, as Magalotti will write to Galileo, Jesuits working behind the scenes: "The Jesuit fathers must be working very valiantly for the book to be banned, and this he told me himself [Father Riccardi] with these words: 'The Jesuits will persecute him bitterly.' "[33]

Since by now the cause of the scandal can no longer be got rid of, everything must be done to prevent it from providing new arguments to those who have already brought ideological accusations against the pontificate. But there is one thing that absolutely cannot be done: namely, bring the case before the Congregation of the Holy Office, where every meeting offers an opportunity for the arrogant Cardinal Borgia to insinuate his intimidatory criticisms of the pope's permissiveness, of his imprudence and the absence during all those years of a firm protection of the Counter-Reformation.

If Galileo, the scientist so beloved by the pope, protected to the point of averting with scandalously obliging procedures the gravest denunciations of heresy—as had happened only a few years before—was now under serious suspicion of heresy, then taking his case to the Holy Office would have been political suicide. A serious denunciation of heresy against Galileo, if revealed within the Congregation of the Holy Office, would have meant offering on a silver plate to Cardinal Borgia—with an enormous doctrinal scandal—proof of the pontificate's scant religious vigilance, its ambiguous tolerance for innovators. An imperative arose: recall the case.

The Special Commission

There is, then, a simple device that can perhaps explain why in the summer of 1632 the "rumors" against Galileo were removed from their natural ju-

[33] See F. Magalotti's letter to M. Guiducci, August 7, 1632, *Works*, XIV, p. 368 ff.

diciary seat and submitted to a special commission under the direct control of the pope (through Cardinal Francesco Barberini, who was called to preside over the commission's work, which took place under the cover of the greatest confidentiality). Such an extraordinary commission could be justified only in cases of exceptional gravity, but above all in cases of a difficult theological nature. To find another example of an extraordinary commission comparable to this one, we must wait until the 1650s, when before following up on the Jesuit-invoked condemnation of Jansen, Pope Innocent X presented the incriminating propositions to a theological commission, which included Jesuits.

Our commission, too, was motivated by very serious concerns of a theological nature, for when the pope reveals its existence to the Florentine Ambassador Niccolini on September 4, 1632, he will justify his decision by saying that "it was a matter of the most perverse material that one could ever have in one's hands . . . doctrine perverse to an extreme degree."[34] It was not permissible for the pope to say more.

The extraordinary theological commission met in the middle of August. The official version given by the pope and by Cardinal Barberini claimed that its purpose was to study the possibility of not bringing the *Dialogue* before the Holy Office. Actually, given the final outcome, Galileo's historians tend to consider this justification unconvincing. But perhaps, at bottom, the pope was not lying, not even in regard to what happened afterward.

However, news of the theological commission had already reached Florence in a less official manner on August 21, when Father Campanella sounded a serious alarm with a letter to Galileo written in his customary style of dramatic prophecy. This was a case, he wrote, of "irate theologians"; and in that commission "there is not a single person who knows mathematics"[35]—which, if the commission had dealt with Copernicanism, was in fact rather scandalous.

Campanella did not know mathematics either, but he advised Galileo to bring diplomatic pressure to bear from Florence so that he, too, together with Father Castelli (who was an authoritative mathematician), might enter and form part of that gathering of experts. But the next day, the Florentine ambassador's dispatch from Rome presented a much different picture of the situation. Cardinal Barberini had reassured the ambassador, guaranteeing him, first of all, the greatest reserve on the subjects dealt with and specifying that there were nonetheless signs of good will toward Signor Galileo.[36] Cardinal Barberini had already given proof of his loyalty to Galileo. Other assurances of loyalty will be given later by the pope as well as by Father Riccardi, who will insist that diplomatic pressure must be brought to bear and

[34] See F. Niccolini's letter to A. Cioli, September 5, 1632, ibid., pp. 383-85, especially p. 384.

[35] Ibid., p. 373.

[36] Ibid.

that the work of the special commission should not be hindered if one really wanted to help Galileo.

At the beginning of September, Magalotti, who continued to probe the state of Roman feelings, also calmed things down, assuring Galileo that even if the pope meant to condemn the Copernican theory by the verdict of that commission, the Holy Office, "where questions concerning dogmas are principally handled,"[37] would not have declared so controversial a question heretical. Of course, but what if it was a matter of "questions concerning dogmas"?

What did the extraordinary commission deliberate upon? On what charges and on what *corpus delicti*? Only those theologians, Cardinal Barberini, and perhaps the Master of the Sacred Palace (Father Riccardi, who must have received those denunciations) knew.

The commission met five times,[38] and in the middle of September, after one month, it had finished its task, having evaluated, weighed, and discussed the reasons for Galileo's incrimination. The most absolute secrecy surrounds the proceedings of those meetings. For the historian, that commission is like a "black box." We know what came out of it in the middle of September, but what had gone into it?

The commission presided over by Cardinal Barberini furnished the Tribunal of the Holy Office with a preliminary inquiry that was perfectly fashioned for a quick trial against Galileo on the basis of a carefully circumscribed accusation: the violation in the *Dialogue* of the injunction not to defend the Copernican theory condemned by the Holy Office, which had been imposed on Galileo by Cardinal Bellarmino in 1616.

The dossier of the preliminary inquiry which issued from the commission's work was complete: the *corpus delicti* was the *Dialogue*; the major official charge was disobedience, high treason by dint of an infraction of Cardinal Bellarmino's injunction.[39]

There was documentary evidence, even if it had no juridical value. This was a transcript supposedly dug out from among the dossiers of the Holy Office, from which it appeared that Cardinal Bellarmino, in the presence of witnesses and the father commissary of that time, had actually enjoined Galileo from defending or in any manner dealing with the recently banned ideas of Copernicus. The document was a more than sufficient basis for accusation; it was perfect. But it was not signed. However, if it had no legal value, it was nevertheless very useful to set the trial going on the path of a well-defined conviction, not too serious for either the defendant or his protec-

[37] See F. Magalotti's letter to Galileo, September 4, 1632, ibid., p. 382 ff., especially p. 382.

[38] See Francesco Barberini's letter to Monsignor Giorgio Bolognetti, nuncio at Florence, September 25, 1632, Works, XIV, p. 397 ff. (see note 42, below).

[39] See *Works*, XIX, doc. XXIVb, pp. 324-27, especially p. 327. The accusation of disobedience is the last; most serious imputation.

tors: inquisitional heresy, as juridical specialists in the field say, was not doctrinal heresy. In short, there was an infraction of a decree, not a true and proper heresy concerning "perverse material . . . to an extreme degree."

Several minor crimes were indicated. These ranged from the typographical symbol used in the preamble to the disrespectful treatment reserved for consecrated authors, from the improper demonstration of the earth's rotation with the example of tides (ridiculing Ptolemaic arguments) to the abuse of the Roman imprimatur and the illicit comparison between human mathematical reasoning and divine intelligence.[40]

All were pertinent crimes that could easily be imputed to the accused, but also—let it be said—very venial crimes. If it had not been for that juridical pretext, retrieved who knows how (that is, the transcript of Cardinal Bellarmino's injunction in 1616), all the charges would have been infractions capable of correction. Almost any consultant to the Holy Office would have been able to identify them easily. The accusation of Copernicanism did not demand a great effort of the inquisitorial imagination from any theological expert witness in 1623. Had that whole mountain of mystery and those suspicions of "perverse doctrine to an extreme degree" been constructed only to give birth to a report like this one, based on a series of infractions and irregularities? A great secret theological jury is not assembled for crimes so clear, nor for material so incontrovertible as the respect due Cardinal Bellarmino's official injunction or the norms of printing. It is hard to believe that those "irate theologians" had gone through a month of examination and five meetings simply to pronounce themselves on the *Dialogue*'s all-too-obvious Copernican propensity.

But let us not judge the work of a commission too hastily, solely on the basis of its results and without knowing what was discussed within it. If one looks closely, in fact, the report of the preliminary investigation did contain the minimum necessary for a trial according to the rules and not very compromising to anyone; and that was sufficient. The report exonerated the judges from any further preliminary inquiry. It furnished them with a perfect juridical pretext for accusing Galileo on the basis of his infraction of an ecclesiastical precept. Nor was it a matter of putting Copernican theory on trial again, which would have permitted Galileo to use his fascinating skill at dialectics and perhaps even to perform experiments with buckets of water at the Holy Office, converting his very judges to Copernicanism. It was not a question of confronting thorny theological problems, since this was not the moment to repeat another Bruno trial: Galileo was only being asked to face the consequences of having broken, more or less by dissimulation, Bellarmino's prescriptions. Following these trial lines, the pope would have given proof of dissociating himself officially from Galileo, of no longer offering

[40] *Works*, XIV, p. 373.

complaisant permissiveness to novelties—without, however, giving satisfaction to those who suspected that his scandalous protection involved more serious heresies of Galilean philosophy.

So let us not judge too hastily this "black box" from which we see the *Dialogue* emerge, without knowing what had entered it at the start. Because this is the doubtful point. What was submitted to the commission as the *corpus delicti*? The *Dialogue*, historians have spontaneously thought until now. But perhaps this is also only one of the many retrospective illusions of the Galileo affair. Perhaps the *Dialogue* was not the only book of Galileo's to be denounced.

Father Campanella, who did not know the charges, had the right to protest the absence of scientists on that extraordinary commission. But he also told Galileo, in the letter quoted above, that that commission composed only of theologians had been formed "against *your books*";[41] that is, to examine at least two, perhaps all, of Galileo's books.

Cardinal Francesco Barberini, who knows everything but says nothing, writes a letter to the nuncio at Florence, Monsignor Giorgio Bolognetti, on September 25, 1632, after the commission has completed its work incriminating the *Dialogue* for Copernicanism. The letter is covered by diplomatic secrecy, yet the cardinal-nephew does not reveal any details about the secret discussions that took place within the commission. However, he does reveal at least one important detail as to what had been the object of that initiative: "Some suspect things having been discovered, our lord in deference to the Grand Duke has charged a special congregation to examine them and see whether one could avoid bringing them before the Sacred Congregation of the Holy Office."[42]

From these words in the dispatch to the Florentine nuncio it would also seem that the *Dialogue* was not the only one of Galileo's works incriminated in the summer of 1632. It is also true, however, that Cardinal Barberini's letter is known to us through a draft: the plural could be an error in transcription. But—apart from its repetition and the following passage, coherently relating to a single "work" then actually charged by the commission of inquiry—we cannot forget that this allusion by the Commission president to several incriminated "works" of Galileo initially coincides with Campanella's analogous reference.

The fact is that if several works by Galileo had been incriminated, everything changes. If, for example, *The Assayer* was also among those works, it would be easier to understand why a special secret commission had been mobilized. For in such a case, besides the Galilean sympathy for Copernicus,

[41] Ibid.

[42] Ibid., p. 397, letter cited in note 38, above (Cod. Barb. Lat. 7310, cc. 34-35, Vatican Library). This is a draft written by Pietro Benessi, the cardinal-nephew's private secretary.

"one will run into"—as Father Riccardi will say in this connection to the Florentine ambassador—"many dangers to the Faith, not dealing here with mathematical matters but with Holy Scripture, religion, and faith."[43]

Copernicanism did involve Holy Scripture, but if one spoke of faith, then it was something else: it was really a matter for the Holy Office and, as Filippo Magalotti had feared, it was a dangerous matter, a matter of dogma. In that case, the problem was really serious in those days of restoration, and the scandal was really dangerous for everyone, not just for the person who set it in motion. Perhaps those who knew how things had really gone in the preliminary inquiry had their own good reasons for saying that that theological commission was useful for eliminating all the initial denunciations, whether old or new, instead of attaching them to the final report on the crimes of the *Dialogue*.

Three Faces, One Secret

The books that have been written on the trial of Galileo would fill several shelves in a library, but all the historians seem to have shared Campanella's disdainful intolerance of that extraordinary commission charged with the secret inquiry. In fact, devout and apologetic historians also experience a certain embarrassment at imagining Galileo's grand cosmological synthesis in the hands of theologians whose culture and intellectual activity did not confer on them sufficient scientific title to judge the *Dialogue*'s astronomical and physical discussions, not even when Galileo was mistaken.

Since we cannot know anything about what was said during those meetings, let me at least introduce, on the basis of their competence and their positions, the experts called to decide on those "works" by Galileo. The concordant testimonies of Campanella and Father Riccardi reveal that the commission was composed of three theologians.[44] The most important was certainly Monsignor Agostino Oreggi, probably appointed by the pope himself, since he was his personal theologian. Consultant to the Holy Office and to Rites, pupil at the Collegio Romano (where he had been taught by Bellarmino, Father Eudaemon-Johannes, and Father Vitelleschi), Oreggi had been close to the pope since the days of the Bologna legation. He was therefore intimately linked to the Barberini pontificate, which had showered him with honors and sinecures in Rome. He was a prestigious theologian, a specialist on the subject of the Eucharist, and he had written a large work on *Theologia*, which will be published a few years later.[45] After the conclusion

[43] See F. Niccolini's letter to A. Cioli, September 11, 1632, *Works*, XIV, p. 388 ff.

[44] Ibid., p. 373. See Father Riccardi's revelations in Niccolini's letter of September 11, ibid., p. 389.

[45] See A. Oldoini, S.J., and A. Ciaconio, O.P., *Vitae et res gestae pontificum Romanorum et S.R.E. Cardinalium*, Romae 1677, vol. IV, cols. 593-96.

of the Galileo affair, as a reward for his work as consultant, Urban VIII will make him a cardinal. But what was the precious assistance he contributed in this particular case?

Oreggi never took, either before or after that inquiry, any polemical initiative against Galileo or the Galileans, something at any rate that would have been impossible in view of the general agreement of intellectual and religious ideas which had developed in the Curia at the start of the pontificate. Just the opposite. In his book *Theologia*, Oreggi showed how far away he was from the constrictive theological philosophy of the Jesuits, with whom he had public theological debates. Monsignor Oreggi was not at all an "irate theologian," nor was he an obscurantist and a Scholastic given to controversy. He did not at all believe in the ambiguous confusion of natural philosophy and theology practiced by the Jesuits, whom he had supplanted as the pope's confessor. His most recent published work, *De opere sex dierum* (1632) was a book of Mosaic physics. Oreggi stated that one could affirm with certainty only that which was written in the Bible; but it was in regard to Aristotle that he had made that statement. The fundamental premise of his new and more important *Theologia* was the distinction between "what belongs to physics, what concerns mathematics, and metaphysics."[46] Having posited this general premise, Monsignor Oreggi's book lengthily developed the theory of "quantity, shape, color, and other accidents," obviously only in reference to the great Eucharistic theme.

And in this case, too, the reader of Oreggi's *Theologia* will be surprised to encounter a mental and religious atmosphere very distant from that of the Jesuit theologians: no apologetic and controversialist animosity here. The problem of the Eucharistic nature is entrusted to "the intellect illuminated by faith."[47] And although Monsignor Oreggi subscribed to the theology of St. Thomas, Bellarmino, and Suárez on the problem of the generation and corruption to which the Eucharistic accidents are susceptible, he nonetheless adopts a sufficiently revelatory terminology. Transubstantiation, "this reproduction of matter, must seemingly be defined as creation."

In sum, we are in a sphere of thought very far from that of natural philosophy and Aristotelian logic. We are in the spiritual perspective of creation, rather than in the optics of a rational interpretation of transubstantiation and of accidents without a subject. If Monsignor Oreggi had been questioned in connection with Galilean atomism, he would not have fallen victim to a restrictive Scholastic vision of the Eucharistic problem.

The second theological expert on the commission of inquiry was also one

[46] A. Oreggi, *Theologia, Pars prima. Physica*, Romae 1637, p. 509. On Oreggi's Mosaic cosmology, see L. Thorndike, *A History of Magic and Experimental Science up to the Seventeenth Century*, 8 vols., New York 1929-1958, V, p. 58 ff.

[47] Oreggi, *Theologia, Pars tertia*, p. 206.

of the pope's men: Father Zaccaria Pasqualigo, a thirty-year-old Theatine of Sant'Andrea della Valle—Barberini's church—in whose college he was a young and brilliant professor of theology. He was not yet a theologian of great reputation like Monsignor Oreggi, but he was a prominent figure in the pope's cultural-theological entourage, pro-French and Augustinian.[48]

In order to understand why, it will be useful to examine quickly the relations between the Theatines, protected by the new pontificate, and the Jesuits. To say that the Theatines and the Jesuits of 1632 got along like cats and dogs would be putting it mildly. In 1632, there raged a white-hot furious controversy between these two profoundly dissimilar orders. The Theatines, an order with a vocation for charitable works of Augustinian inspiration, and in the good graces of Richelieu's France, were heirs to the great spirituality and cultural tradition of the Oratorio della Vallicella—quite different, it would appear, from the Jesuit Order.

At that moment, between Sant'Andrea della Valle and the Piazza del Gesù, an already-old debate concerning St. Ignatius had flared up. The Theatines had revealed that the fiery Knight of Loyola had in his time been politely rejected by the Theatine Order, to which he had first aspired, and had been advised by them to found a congregation of his own, "more active" and more in keeping with his character. The Jesuits denied this, and the debate spread with reciprocal accusations, leading to an open competition between Theatine and Jesuit missions in the Middle East. In the end, the Holy Office will be forced to intervene authoritatively between the two parties in 1644, silencing the Jesuit protests. But since, to the detriment of the image of the Catholic apostolate, the dispute gave no sign of abating, the Holy Office will again intervene—more drastically—imposing silence on both belligerents.[49]

To this state of affairs was added the aggravating circumstance that Father Pasqualigo was engaged in an almost personal debate with Jesuit theologians. Today the matter might seem completely irrelevant and of no interest for our time, whereas, looked at more carefully, it certainly retains some aspects of contemporary import. At stake was the following problem of extremely controversial moral theology. Can a man dispose freely of his own body? Can he decide to keep a childlike voice in order to make certain of a profession and provide great liturgical edification with his singing? In other words, it was the moral controversy over the *castrati*. A century later, not even the great moralist St. Alfonso de' Liguori was able to say yes or no to this delicate problem of the admissibility of child castration.

[48] See A. F. Vezzosi, *I scrittori de' chierici regolari*, pt. II, Rome 1780, p. 156 ff.

[49] The polemic was started by Father Giovanni B. Cataldo, *De B. Cajetani Thienaei cum B. Ignatio Loyola consuetudine . . . epistola*, Vicentiae 1618. On the Jesuit side, the controversy was fought chiefly by Fathers Negrone and Rho.

Even more than his predecessors, Urban VIII was very proud of the *castrati* who sang in the choir of the Sistine Chapel. The Jesuits did not agree: they appealed to the authority of theological tradition to affirm that man was not given ownership, but only the use of, his body. The Jesuits criticized the great number of *castrati* in the Curia, calling Rome a new Constantinople. Yet the problem was controversial and very delicate because the musical requirements of Catholic liturgy were certainly important. So the ecclesiastical power, parading equally valid theological arguments, approved the custom of the castration of young singers.

Pasqualigo held that man was free to use his own body, provided that it was for a useful and worthy purpose. He will officially take up the defense of the papal custom against its Jesuit detractors. In any event, the latter denounced Father Pasqualigo's moral theology as too lax and will later succeed, though only after Urban VIII's death, in having it condemned by the Index.[50]

Apart from this burning controversy, Father Pasqualigo was a theologian who specialized in the subject of the Eucharist. He was a theologian for the new age of the marvelous conjuncture at the beginning of the pontificate—a mystical extoller of reason. In the history of Italian Eucharistic theology. Pasqualigo's work has left a significant imprint. He had adopted, in polemic with the Jesuit theological school,[51] an Augustinian theory of the Eucharist which closely recalled the mystical concepts of Eucharistic sacrifice elaborated at the beginning of the century in the mysticism of Cardinal De Bérulle and the French theological school.[52] "I do not see why physics and theology must be confused in a single science," Father Pasqualigo wrote at that moment.[53] Father Pasqualigo was also, like Monsignor Oreggi, the right person to relieve the pope of the embarrassment of a denunciation of Galileo on the subject of the Eucharist.

The third member of the commission of inquiry was a Jesuit. The Jesuits, who had promoted the case in the summer of 1632, and who enjoyed re-

[50] See Anonymous (Father T. Raynaud, S.J.), *Eunuchi, nati, facti, mystici*, Lutetiae Parisiorum 1655, containing the denunciation of Father Pasqualigo's positions. Pasqualigo's book of moral theology, *Decisiones morales juxta principia theologica ad sanas atque civiles leges*, Verona 1641, will be put on the Index in 1684. On the *castrati* it must be recorded that it was actually Father Inchofer who first intervened in the controversy against Father Pasqualigo: M. Inchofer, "De eunuchismo dissertatio ad Leonem Allatium," in L. Allaci, *Opuscula Graeca et Latina*, Coloniae 1653, in *Supplementum Historiae Byzantinae*, Venetiis 1733, pp. 73-80.

[51] See Z. Pasqualigo, *De sacrificio novae legis. Quaestiones theologicae, morales, juridicae*, Lugduni 1552, p. 328, in which Pasqualigo combats Father Pereira's natural philosophy.

[52] Ibid.

[53] Pasqualigo was then going to press with his *Disputationes metaphysicae, theologia*, Romae 1634-1637, p. 48. "The object of physics must not be subordinated to metaphysics" (ibid., p. 182). Father Pasqualigo adds that when there are several compounded problems, we must rely on our degree of ignorance to guess how they are articulated.

newed power to put pressure on the Curia, must evidently be included in the commission called together by the pope, since he was intent on stifling any suspicion of ambiguous tolerance toward the new ideas. So, in order to silence suspicions and slanders, there was also a representative of the Society of Jesus.

But the pope had intelligently reserved the right to choose which Jesuit would join the special commission and had also charged Father Riccardi with finding the most suitable person for this third role in the secret affair. And Father Riccardi had fortunately found in Rome the required Jesuit: his personal friend, Father Melchior Inchofer.

The top people at Piazza del Gesù could find no fault with this choice. The commission was of a theological nature, and Father Inchofer had the official qualifications of a theologian, had mathematical and astronomical knowledge—certainly superior to that of the other two members of the commission—and was indeed known as being ferociously anti-Copernican.[54]

But in truth I cannot believe that the Jesuits were absolutely enthralled by Father Riccardi's clever choice. In fact, there is no doubt that the Jesuits could have done better when it came to theologians and mathematicians.

Their representative actually cut a poor figure in that commission when compared with the embattled theological authoritativeness of an Oreggi or a Pasqualigo. His limited scientific prestige was worth very little when it came to dealing with delicate theological problems, in regard to which Father Inchofer unfortunately was not on the same level as his interlocutors. Father Riccardi had chosen him not only because he considered him a reliable acquaintance—as he also assured the Florentine ambassador—but also because Father Inchofer was at that moment the least offensive Jesuit among all those present in Rome.

The judge was actually in a position analogous to that of the accused: Father Inchofer was in Rome because he was being investigated, cited by the Congregation of the Index. This was a rather uncomfortable position, certainly not the best one for speaking out in a loud voice. In reality, Father Inchofer thundered mightily against the Copernicanism of the *Dialogue*, when during the course of the trial the commissioners were asked to write a report on the amount of Copernicanism in the *Dialogue*; but by then the bets were already down.

Father Melchior Inchofer was a professor of theology at Messina, a Spanish city and the seat of a university run by the Society of Jesus. At Messina,

[54] See the titles of Father Inchofer's astronomy manuscripts, cited in Sommervogel's collection, vol. IV, pp. 562-66. The Casanatense Library (Rome) has a ms. entitled *Vindiciae sedis Apostolicae SS. Tribunalium auctoritate adversus neo-pytagoros motores*, by Father Inchofer, analyzed in D. Berti, *Il processo originale di Galileo*, Rome 1878, p. 76 ff.

Inchofer had published his first book, in 1629, with the title *Epistolae B. Virginis Mariae ad Messanenses veritas vindicata*.[55] In it Father Inchofer, with many erudite arguments, claimed to demonstrate the historical truth of a local Marian legend: a short letter allegedly written by Mary to the people of Messina in which all the principal Catholic dogmas were listed. The letter bore the date 62 A.D. Father Inchofer had definitely not invented that letter, with its most controversial attribution: even today arriving in Messina from the mainland, one can read at the entrance of its port the letter's closing words.

Now, for the Jesuits, Marian apologetics and hagiography were a strong point in the struggle against heretics and Jansenists. Unfortunately, though, in the case of the latter, the distinguished historian Cardinal Baronio had declared that the authenticity of that pious document should be defended and considered only as a hypothesis, not as historical truth, just like other Marian apocrypha. And since Cardinal Baronio had, in the eyes of the Congregation of the Index, an authority in the historical field comparable to that of Cardinal Bellarmino in the theological field, Father Inchofer was cited in Rome. With respect to that letter, his book was guilty of the same mistake Galileo had made vis-à-vis Copernicus in the *Dialogue*: an excess of zeal. The book had been validated by the Society, which was forced to reverse itself and, one year later, in 1633, publish a completely revised edition, such that Mary proves to be only the probable author of the letter.[56]

It is also probable that Father Inchofer brought to the special commission all the aversion of the Roman Jesuits for Galileo. And it is further probable that the other members of the commission, who did not share that state of mind, allowed him to blow off steam in the direction of Copernicanism. This will later become apparent from the trial documents and from a subsequent initiative of Inchofer's. During the trial, the commissioners were asked for an expert opinion on the amount of Copernicanism in the *Dialogue*.[57] While the opinions of Oreggi and Pasqualigo were quasi-bureaucratic certificates, Father Inchofer used to the fullest his talent as a fine polemicist as well as his astronomic knowledge in an exhibition of great inquisitorial zeal. After the trial, Father Inchofer will publish a *Tractatus syl-*

[55] On the gnarled affair of the letter of the Madonna of Messina, see Father E. Aguilera, S.J., *Provinciae Siculae S.J. Ortus et res gestae ab anno 1546 . . . ad annum 1672*, Panormi 1740, tome II, p. 247 ff. The original edition of Inchofer's book was put on the Index on March 19, 1633. Father Melchior Inchofer was a specialist in controversies and polemics, both outside and inside the society of Jesus. Later on, in fact, his shadow and his talent for polemics will appear behind the denunciations and the intimidations suffered within the Society by Father Sforza Pallavicino. On Inchofer's personality, see L. Allacci's profile, *Apes Urbanae*, Romae 1633, ad nominem.

[56] See M. Inchofer, *De epistola B. Virginis Mariae ad Messanenses conjectatio plurimis rationibus et verosimilitudinis*, Romae 1631.

[57] See *Works*, XIX, p. 348 (Oreggi); pp. 356-60 (Pasqualigo); pp. 349-56 (Inchofer).

lepticus against Copernicanism, a work of chiefly exegetical character in line with his theological-literary rather than astronomical background. Nonetheless, the book is revelatory of an attitude: it insists that certain principles inspired by pagan or heretical philosophers be banned from the Catholic religion. Father Inchofer called for the punishment of all those philosophical assertions, even those of Aristotle, which conflicted with Catholic faith.

The slogan of *Tractatus syllepticus*, repeated several times, is Pope Leo X's injunction at the Lateran Council: "Teachers of philosophy must castigate with the manifest truths of religion the principles or conclusions of Philosophers who deviate from the true faith."[58] Copernicanism did not "deviate" from the faith.

All this emphasis may surprise the reader of Father Inchofer's work in reply to the *Dialogue*: after all, Copernican ideas in conflict with Scripture had been duly punished at the trial of Galileo. But, evidently, Father Inchofer had not been completely satisfied; perhaps not all the denunciations had been accepted by the preliminary commission whose majority was hostile to him. More than this he could not say.

Having become closely acquainted with the three faces with the tightly sealed lips who participated in the commission's meetings, all that remains to ask is why it took them a good five meetings to come to an agreement on their final report. The only thing that can reasonably be thought is that they were not unanimously in agreement. For the rest, we can only regret the complete confidentiality surrounding the commission's proceedings. Its results reflected the adeptly measured balances within it, so that the scandal of other eventual "materials suspect in the highest degree" would be eliminated at birth.

All that a historian could do was wait patiently—perhaps until the death of Urban VIII, who had decided that the commission should be constituted in that fashion and enveloped in that secrecy. Sooner or later, the truth always wins out.

Extrajudicial Measures

The official results of the commission of preliminary inquiry became the object of diplomatic revelation at the end of the month of September 1632. The identified crimes left room for a correction of the *Dialogue*, but the pope could not further violate normal procedure. Finally the report was sent to the Holy Office, as required, so that the judicial procedure that we all know about could be initiated. Obviously, in order to avoid scandal, the "suspect materials in Galileo's books" that had initially been denounced, as well as the

[58] See M. Inchofer, *Tractatus syllepticus, solisque motus vel statione*, Romae 1633, p. 57 ff.

denunciations themselves, were not submitted to the tribunal. But the pope, in secret audiences with the Florentine Ambassador Niccolini, was less reserved.

On September 4, the ambassador requested an audience to deliver to the pope his government's note of protest concerning the make-up of the commission, erroneously regarded in Florence as a machination against the *Dialogue*. Urban VIII exploded in anger, saying that "Galileo had dared enter where he should not, into the most grave and dangerous subjects that one could possibly raise at this moment."[59] The pope also added that those theologians he had assembled "are examining every item word by word" because it was a question of "the most perverse material." He pointed out that he had done Galileo the great favor of not having submitted this material to the tribunal, but "to a special congregation, newly created, and that is something."[60]

Everything adds up: we know what "this moment" so difficult for the pope was, we can suspect what subjects were being examined "word by word" by the theologians, and we also know that the sort of anxieties and subjects that Galileo had no right to delve into could not be Copernicus's theory, which he had unofficially been authorized to broach. A week later, as we have already mentioned, Father Riccardi confirmed with further assurances the pope's substantially favorable attitude, inviting Florentine diplomacy to desist from counterproductive intrusion in the work of the commission, which there was no reason to consider hostile.[61]

Here, too, things add up. In fact, on the morning of September 18, Ambassador Francesco Niccolini is granted an audience with the pope. The purpose is make use of the good diplomatic relations between the Holy See and Florence to block the commencement of Galileo's trial before the Holy Office.

The ambassador, however, went to that audience with more worries than hopes. The fact is that the pope had made a point of communicating with great confidentiality, through his private secretary Pietro Benessi, that it was impossible to spare Galileo an official trial hinging on the *Dialogue*. To the protests of Ambassador Niccolini, who brought up the ecclesiastical permission obtained by the book in Rome, Secretary Benessi had mentioned obscure reasons of absolute necessity and gravity. In fact, one could not risk

> that religion might suffer harm . . . and put Christianity in danger with certain sinister opinions, and that His Holiness has told him that in dealing with dangerous dogmas His Highness the Grand Duke of Tuscany, having set aside all respect and affection for his mathematician, would be glad to concur in sheltering Catholicism from all danger.[62]

[59] *Works*, XIV, p. 383. [60] Ibid., p. 384. [61] Ibid., p. 388 ff.

[62] F. Niccolini's letter to A. Cioli, September 18, 1633, ibid., p. 391 ff.

Dangerous dogmas? Danger to the faith? But had not the *Dialogue* been explicitly presented as a scientific exposition of Copernican theories that was more than respectful of ecclesiastical decisions in theological and Scriptural matters?

These were the perplexities which must have agitated Ambassador Niccolini's mind that morning. And almost as a confirmation of his fears, he will learn from the pope's own mouth, under the strictest bond of secrecy, not only that the trial could not be put off, but that the most serious imputations—much more serious than those one could imagine—were at work behind the episode.

Pope Barberini had not forgotten his old friendship with Galileo and "that Signor Galileo was still his friend." Yet, as if to justify himself before a witness quite aware of the Roman political situation at the moment, that trial was imposed by necessity "of the interests of faith and religion. These are subjects," the pope said, "which are troublesome and dangerous . . . and the subject is more serious than His Highness believes."[63] The ambassador tried to keep to the subject of "opinions." Again the pope imposed the obligation of secrecy on him, "by express order and under penalty of censure." In the College of Judges, the pope "hinted, through gritted teeth," that Borgia and the other pro-Spanish cardinals watched his every move. So one must, he added,

> try to be a bit alert, and I should immediately explain to His Highness that Signor Galileo, under cover of his school for young men, should not impress on them troublesome and dangerous opinions, because he had heard I know not what, and that Your Highness, for goodness sake, should be careful . . . not to sow the seeds of error over the States.[64]

The ambassador must have understood that the pope was alluding to questions of serious heretical subversion, since he demanded the vigilance of the Florentine government over the Galilean school. Niccolini then tried to clarify this mysterious accusation once and for all. He replied that he could not even imagine that Galileo was a heretic, that though he "might dissent from true Catholic dogma in some part, each person in this world," the ambassador cleverly insinuated, "has people who are envious of him and are full of ill will."[65]

"Enough, enough," the pope immediately broke in—as if to say, "don't make me say more; it's just as well." They then went on to talk again about the Roman authorization and the technical aspects of the infractions imputed in the *Dialogue*.

Later on, during the audience of March 13, 1633, when the trial was imminent, the pope will repeat that a trial cannot be avoided because it is a mat-

[63] Ibid., p. 392. [64] Ibid., p. 392 ff. [65] Ibid., p. 393.

ter of "new and dangerous doctrines" and "of the Holy Scripture where the best thing is to go along with common opinion; and may God help Ciampoli, too, when it comes to the new opinions, for he is also a friend of the new philosophy,"[66] the ambassador will again try to insist. But the pope this time will react harshly: "he was here to obey, to cancel or retract everything that would be pointed out to him in the interest of religion. In short, it was an affair of state. Ambassador Niccolini decided not to insist and went on to "other business." On April 12, the trial against Galileo began.

Rather than in a cell, the defendant stayed in the apartment of the fiscal procurator, in a kind of guest house in the Palace of the Holy Office. This was a provision of exceptional consideration for an exceptional defendant.

The trial is very well known: dozens of studies have been devoted to it, and the proceedings have been published in the national edition of Galileo's *Works*. But Galileo's trial, at this point, does not have a history, it is only a judicial appendix, the execution of the incriminatory depositions sifted and gathered together in the preliminary-hearing phase.

Apparently, the pope's behavior is disconcerting: instead of having had the inquest proceed on its normal course in the Holy Office, he has arrogated the inquest to himself; he has concealed the denunciations, letting only the suspicion of their existence leak out to justify the trial. The pope has gathered together a commission where Galileo's denouncers are represented by a not-very-authoritative personage and, what is more, are in a minority with respect to two theologians suspected neither of enmity for the Galileans nor of friendship for the Jesuits. There is talk of "suspect materials" and of various books by Galileo which have been incriminated, and then it turns out that only the *Dialogue* is incriminated for the infraction of an injunction.

Confronted with all these apparent incoherences, one must not forget that the Galileo affair was an affair of state, a very serious affair of state. If it were to come out that the pope's official scientist was a suspected heretic against the faith, it would be scandalous. As can be seen, the situation was in many ways the same as that which had arisen when *The Assayer* had been denounced. The difference was that now the times had changed, aggravating all the consequences: the scandal was more serious, and more serious were the mounting pressures exerted by the intransigents. The room for maneuver was more restricted, and a trial of Galileo—his official condemnation—was the only way out.

The surprising procedural irregularity; the pope's anger at being implicated in such severe charges and at such a difficult moment, owing to his friendship with Galileo; the assurances of Cardinal Barberini concerning the intentions of the commission of preliminary investigation; the justificatory

[66] F. Niccolini's letter to A. Cioli, March 13, 1633, *Works*, XV, p. 67 ff.

allusions to the exceptional nature of the incriminations; the probable presence of several of Galileo's incriminated books—all these apparently incomprehensible and contradictory elements are in reality welded together by a coherence destined to remain secret for reasons of state.

The trial was inevitable for the same ideological and political reasons. Cardinal Barberini had clearly explained this to the ambassador of Florence: there was no way one could avoid trying Galileo, no way to put an end to the affair. A public demonstration of renewed firmness was required, as with De Dominis ten years before. Only this time Galileo would never cease benefiting, within the limits of the possible, from the interested protection of people in power.

That the official trial involved a different and complex matter, however, can be seen from a series of extrajudiciary measures, of settlings of accounts, of defenestrations and demotions that historians of the Galileo affair, polarized with respect to the trial's proceedings, could not consider as a whole. But the first of these exceptional extrajudicial procedures took place right in the Palace of the Holy Office at the beginning of the trial. The beginning of the trial revealed an obstacle to the steady progress of the procedure set on that secondary rail. At the first audience, as is very well known, Galileo presented an impeccable line of defense. He had with him the statement, released to him in 1616 by Cardinal Bellarmino, which proved that during the famous meeting at the Vatican, Bellarmino had informed Galileo of Copernicus's sentence and of his imminent inclusion in the Index, but which did not contain any further injunction as regards the exposition of Copernican ideas. Compared to that statement signed by Bellarmino, the list of charges in the hands of the judge and commissary general from Firenzuola, Father Vincenzo Maculano, had all the weight of confetti. The planned trial line was blocked; instead of collaborating, the defendant was making difficulties.

Argument is suspended. The judge must report to his master, Cardinal Barberini. An incredible extrajudicial procedure is decided upon. On the afternoon of April 27, 1633, the judge has a private colloquy with the defendent, without chancellors, lawyers, or witnesses. The colloquy is secret, always the case when dealing with questions that should best remain private.

Scholars who have inquired into Galileo's trial have entertained the hypothesis that Father Maculano had tried to persuade Galileo to submit to the judiciary ritual, threatening him with greater legitimate procedural rigor. That is possible, but then it is hard to understand why the threat required such a surprising infraction of the procedure, secrecy and the absence of witnesses.

And is it not perhaps more in keeping with the exceptional circumstance to hypothesize that during the confidential tête-à-tête between judge and defendant, the former explained with convincing arguments that to hinder the trial's line based on an infraction of the 1616 injunction was most counter-productive—*in primis* for the defendant himself? It is possible.

It is certain that the arguments used were convincing. Galileo, at the next session, in deference to the official Copernican incrimination, gave vent to a sensational Copernican self-accusation—indeed, too sensational. He renounced his line of defense and said that, having had a chance to reread the *Dialogue*, he realized that he had defended Copernicus's theory, but only because he was carried away by literary self-indulgence.

The father commissary heaved a sigh of relief. He no longer doubted that the trial would go back to the pre-established track. Immediately after his secret meeting with Galileo, he had hastened to reassure his master on the happy outcome of the initiative: "The court will maintain its reputation, and the defendant can be treated benignly. His Holiness and Your Excellency will be satisfied."[67] This satisfied Galileo's protectors, satisfied the formal decorum of the judicial ritual and silenced the devout, and satisfied also the defendant, who will receive the minimum penalty.

On June 22, 1633, in the clear early light of a Roman summer day, in the habitual decorum of the Church of Santa Maria sopra Minerva, the brief conclusive act of the great judicial drama is played out in that theater of shadows. After the reading of an official sentence (which focused on serious suspicions of heresy) for the infraction of important ecclesiastical decisions, Galileo abjures Copernicus's doctrine.

> We say, sentence, and declare that you, Galileo, by reason of the evidence arrived at in the trial, and by you confessed as above, have rendered yourself in the judgment of this Holy Office vehemently suspected of heresy: namely, of having believed and held the doctrine, false and contrary to sacred and divine Scripture, that the Sun is the center of the world and does not move from east to west and that the Earth moves and is not the center of the world; and that an opinion may be held and defended as probable after it has been declared and defined to be contrary to Holy Scripture.[68]

Neither persuaded nor completely dissatisfied with the official sentence of this farce of a trial, the Cardinal-Inquisitor Borgia polemically abstained

[67] See Father Vincenzo Maculano's letter to Cardinal Francesco Barberini on April 28, 1633. This letter (*Works*, XV, p. 106 ff.) was discovered in the Barberini archives in 1833. On this obscure extra-judicial episode of Galileo's trial, see S. Pieralisi, *Urbano VIII e Galileo Galilei*, Rome 1875, p. 197 ff.; G. Gebler, *Galileo Galilei e la Curia romana*, Florence 1879, II, p. 44; F. Costanzi, *La Chiesa e la dottrina copernicana*, Siena 1898; G. de Santillana, *The Crime of Galileo*, Chicago 1955; A. Koestler, *The Sleepwalkers*, London 1959; G. Morpurgo Tagliabue, *I processi di Galileo e l'epistemologia*, Milan 1963 (2nd ed., Rome 1981).

[68] *Works*, XIX, p. 405.

from signing the sentence.[69] Galileo was condemned to penance and prison for life; but, by order of the pope, instead of serving his sentence in a cell at the Holy Office's palace, he could immediately return to the residence of the Florentine ambassador and then serve his sentence under house arrest in his own home, "The Jewel," at Arcetri.

Extrajudicial Acts

While the judicial ritual was unfolding at the Holy Office, Rome followed with bated breath as the punitive measures of the real drama struck without scandal at various actors who would be singled out and sent away from Rome in order to stifle every possible jarring echo of the affair. The targets of these purges included the slandered and the slanderers. Gabriel Naudé, the French libertine and observer of this theater of shadows, wrote to Gassendi in April 1633, when the trial was imminent, that "Galileo has been cited owing to the intrigues (*menées*) of Father Scheiner and other Jesuits who want to ruin him, and who would certainly do so were he not powerfully protected by the Duke of Florence."[70]

In papal Rome, as in every dictatorship worthy of the name, persons displeasing to the regime (as well as those too compromising and compromised) were punished by being removed from the centers of power. Distances in the seventeenth century were greater than today, and the gravity of the defenestration was easily measured in kilometers.

Let us begin with the slanderers. One of the first to be rapidly removed from Rome in 1633, perhaps even before the official trial was over, was Father Orazio Grassi. There was not even the usual pretext of removal for reasons of promotion: removal and that is all, exile from Rome and silence. It was done with so much discretion that Galileo's historians never even noticed it.

Father Grassi was consultant to the rector of the Collegio Romano. He had been professor of mathematics for more than ten years. He was only fifty years old, and a brilliant future as a scientist lay ahead of him. He was, above all, the architect of the Collegio Romano's church, which was rising tall and beautiful.

But that building did not bring good luck, and so a year after its financier was exiled from Rome, its architect suffered the same fate. Father Grassi all of a sudden left the work site of that great, prestigious construction which had made him the Society of Jesus' official scientist. Substitute architects

[69] Ibid.

[70] See G. Naudé's letter to Gassendi, April 1633, *Works*, XV, p. 87 ff., especially p. 88.

were hastily summoned, but they were not equal to the task and the work suffered serious delays.

Perhaps Father Grassi did not have on his conscience the *Dialogue*'s incrimination for Copernicanism. In a letter to a friend, written from Savona three months after the conclusion of the trial, he will protest that he had absolutely nothing to do with the official condemnation. On the contrary, he wrote that his frank tolerance of the *Dialogue*'s Copernican ideas had amazed those who knew well his aversion for Galileo.

Nonetheless, even though Father Grassi had never thought to denounce the Copernican doctrine, he was sent away from Rome. Transferred suddenly, he was stripped of his posts. The prestigious scientific exponent of the Collegio Romano was brutally removed from the scene without even the excuse of another official position within the Society of Jesus. As long as Urban VIII lived, Father Grassi did not even hold a scientific or philosophical teaching position, or any position at all. Confined to his native city of Savona, very far from Rome, he lived in exile in that city's provincial college, with only his priestly duties.

And yet, he was still a great mathematician. He was approached by the Republic of Genoa as a professional consultant on naval engineering, and it was also in his capacity as a mathematician that the Society turned to him for a plan of the new College of Genoa (1633) or for the internal revision of difficult books of geometry. Revising was a very important task, given only to trustworthy people.

However, Father Grassi was no longer publishing anything. Never again will he publish a line, even on nonscientific subjects; the *Ratio* was his last work. But the most surprising thing is the fact that the *Ratio* and its author, "Sarsi," will not even be quoted any more, even by Father Grassi's numerous pupils and fellow Jesuits (who nevertheless will again use against the Galileans the dissuasive weapon of the Eucharistic heresy, spread by Sarsi in that celebrated and compromising book).

The uncertainties of politics had also made Galileo's illustrious adversary a victim of the drama. But colleagues at the Collegio Romano—Father Guldin, Father Zucchi, and the new scientific orthodox generation of Jesuits, Fathers Casati and Cabeo—considered the exiled prestigious scientist and architect their master, while hoping that sooner or later the situation might change.

Another actor in the affair, who had been mysteriously removed from Rome for some time now, was the Dominican priest Niccolò Ridolfi, Master of the Sacred Palace at the time of *The Assayer* case. Ridolfi had left Father Riccardi his post at the Holy Office in order to assume that of father general of the Dominican Order. On his return from an official tour of the order's

Parisian convents, which took place between late 1630 and 1631, Father Ridolfi was confined to the church of San Pietro in Vincoli. The pope will then dismiss him and send him to Naples—with this punitive measure disregarding the traditional respect for decision making autonomy accorded that great preaching order. The reasons for the punitive measure are secret to this day; officially it was said that they were owing to "personal motives."[71] In that punishment, moreover, many saw a reflection of the political crisis: the order rebelled against the new father general named by the pope, and this was a serious schism. Only after Urban VIII's death will Father Ridolfi be rehabilitated and exonerated of the alleged "excesses" that had been attributed to him.

While some important persons implicated in the Galileo affair were sent away from Rome, one person, already far from the Church's capital, no longer returned to it: the Bishop at the town of Teano, Father Giovanni di Guevara. The Galileo affair shattered his ecclesiastical career.

After all the diplomatic services Father Guevara had rendered the Barberinian pontificate, everyone, above all the Regular Minor Clerks, was expecting that his nomination to the Episcopate of Teano would be the first step toward the purple and toward high political posts in the Curia. It was not to be: Father Guevara remained at Teano, a failed cardinal *in pectore* until his death on August 23, 1641. The last letter from Guevara to his friend Galileo that has come down to us is from early 1628; in it he spoke of the old discussions on the problems of atoms and indivisibles. In the *Discourses*, Galileo will quote with respect his defender's comment on Aristotle's *Mechanics*. Another missed cardinal's cap, despite the insistences of the Warsaw court, was that of Father Valeriano Magni, he too a friend of the new philosophy. We will meet him again, lined up in its defense and sharing in the defeat.

Itinerant Libraries

What was happening meanwhile to Galileo's friends and protectors? In 1630 two great friends of Galileo had died: Kepler and Prince Cesi. Prince Cesi had died prematurely, without leaving any precise instructions as to the destination and safeguarding of the Academy of Lynceans' intellectual patrimony. The dean of the Academy, Francesco Stelluti, when telling Galileo the tragic news, expressed as early as August 2, 1630, his worries over the fate of that laboratory of ideas which was the precious and suspect library on the Via Maschera d'Oro.[72]

[71] See J. Quétif and J. Echard, *Scriptores ordinis praedicatorum . . . recensiti*, Lutetiae Parisiorum 1721, II, p. 503 ff.

[72] F. Stelluti's letter to Galileo, August 2, 1630, *Works*, XIV, p. 126 ff.

Meanwhile, in 1631, the political crisis brought the situation rapidly to a head, as we know, and even the very powerful protector and associate of the Academy, Cardinal Francesco Barberini, had his hands tied and could no longer compromise himself too openly with that institution, which had a widespread reputation for heterodoxy. A successor to Prince Cesi had to be found, another aristocrat tied to the regime and its new ideas. At first, Cardinal Barberini had suggested the name of Monsignor Pietro Sforza Pallavicino.[73] But the pressure was on, and the plan, as we shall now see, was abandoned.

Monsignor Sforza Pallavicino had also been directly implicated in the scandal, since in 1624 he had undertaken Galileo's defense against Father Grassi's famous denunciation and was besides very closely connected with Monsignor Ciampoli, the great artificer of "Operation Sarsi." Monsignor Pallavicino also saw his prestigious career in the Curia go up in smoke because of the Galileo affair. He was defenestrated and sent in 1632 to govern Jesi, Orvieto, and Camerino, following in the footsteps, as we shall see, of Monsignor Ciampoli.[74]

However, Pallavicino was, together with Father Grassi, the only actor and witness of the scandal to be able to return to Rome, in a dramatic reversal that, as contemporaries said, stunned all of Roman society, both fashionable and intellectual: "Rome was dumbfounded at such an incredible outcome."[75] In fact, Father Pallavicino returned to Rome in the same habit as Father Grassi: in the spring of 1637, he made up his mind to give to the flames a part of his poetic work—in poetry, too, he was Monsignor Ciampoli's pupil—and become a Jesuit. On June 21, a neophyte of high rank, Pallavicino entered the College of Novices at Sant'Andrea, on the Quirinale, under the direction of Father Oliva. Pallavicino had before him a rapid and illustrious academic career as professor of philosophy and theology at the Collegio Romano, the cardinal's purple, and the immortal fame of an apologetic work on the Council of Trent *contra* Paolo Sarpi's book.

The official chronicles of the time described the stupefying conversion of Sforza Pallavicino with these words: "Once the opinions of the innovators were exploded, he made peace again with Aristotle."[76] If one thinks of the previous scandal, this explanation seems coherent.

But Father Pallavicino did not cease to regard himself as an open-minded and modern intellectual, even though his cautious philosophical ventures

[73] See F. Stelluti's letter to Galileo, August 30, 1631, *Works*, XIV, p. 292.

[74] See ibid., p. 250.

[75] See Oldoini and Ciaconio, *Vitae et res gestae pontificum Romanorum et S.R.E. Cardinalium*, vol. IV, cols. 738-41, especially col. 739.

[76] Ibid. See P. Ireneo Affò, *Memorie della vita e degli scritti del card. Sforza Pallavicino*, Parma 1794, p. 20.

will cost him many criticisms, disavowals, and polemics within the Society. In his letters, he gives a picture of himself as a philosopher similar to Galileo in his last phase: Father Pallavicino, too, thought that Aristotle's logic was insuperable, even if in his physics and metaphysics there were confused obscurities and "perhaps also many errors." Father Pallavicino considered himself among the "Galileans in one sense, having great esteem for Galileo in mathematics and in the experiments and speculations on motion":[77] mathematical phenomenism. The old battles over the new philosphy were forgotten.

Thus it was that instead of becoming the new leader of the Academy of Lynceans, Sforza Pallavicino became a great theologian and an illustrious cardinal of the Society of Jesus. Even though the Academy of Lynceans no longer had a leader, it still disposed of the patrimony of the Cesi library. Negotiations with the heirs dragged on and on, as such things always do. To preserve the integrity of that scientific patrimony, the great Roman collector Cassiano dal Pozzo, a Lyncean friend of the family and a great virtuoso and patron, offered to purchase the entire library and put it under his protection in his palace–art gallery on Via dei Chiavari. Protection was assured, since Cavalier dal Pozzo was one of the principal collaborators of the cardinal-nephew. But now the cardinal-nephew also had to avoid further scandal.

In fact, when he prepared to sell, everything was done strictly according to the book. The library's catalogue, appraised by a Roman bookseller, was submitted to the Holy Office at the beginning of 1633, just before Galileo's trial. What for so long a time had been feared, now took place.

In the middle of February, in fact, the sale of the Cesi library was authorized, with a regular license signed by Father Riccardi, Master of the Sacred Palace. But, though the handing over of the library to Dal Pozzo was authorized, a series of books and manuscripts "which are listed on a separate sheet"[78] were sequestered. We do not know which ones they were. Perhaps the famous manuscripts of Abbot Antonio Persio today almost entirely lost, disappeared on that day from Federico Cesi's library. Another dangerous library, however, was still in free circulation: that of Monsignor Ciampoli. But when the Cesi library was censored, Monsignor Ciampoli's library had already been for some months far from Rome, together with its owner, he too defenestrated.

As the most visible and powerful Galilean in the regime's entourage, the orchestrator of *The Assayer*'s official success and of the Galilean "marvelous conjuncture" in Rome, Monsignor Ciampoli was obviously the most ex-

[77] Sforza Pallavicino, *Lettere*, Rome 1668, p. 80.

[78] See the document signed by Father Riccardi in Gabrieli, "La prima biblioteca lincea," p. 615.

posed. From the first sign of the serious political crisis of 1623, he was a predestined victim of the change in course. For years now, the calumnies of the intransigents about "orgies"[79] said to be given by Monsignor Ciampoli in the Vatican, as well as his friendships corruptive of orthodox convictions, had brought demands for his head; but this time his being sent away was inevitable.

Monsignor Ciampoli had been removed by the pope in April of 1632, immediately after the scandal of March 8 in the consistory. On November 24, already discharged from the secretariat, he left Rome. Because of his great intellectual and diplomatic prestige and the role he had played during the initial phase of the pontificate, however, the pope gave that enforced defenestration the appearance of a transfer-for-promotion and continued his sinecure (that is, the rich appanage of a canon at St. Peter's). Monsignor Ciampoli was exiled to the Apennines with the title of Governor of Montalto di Castro, an administrative post of secondary importance. Later, he will move on to the governorships of Norcia (1636), Sanseverino della Marca (1637), Fabriano (1640), and finally Jesi.

He will never stop sending recommendations and implorations to his protectors and friends of the good old Roman days, with whom he remained in epistolary contact during his exile: these included many important cardinals, such as the cardinal-nephew, Cardinal Mazarin, and Cardinal Maurizio of Savoy; and friends the likes of Cavalier Coneo, secretary of his old master Francesco Barberini and Sforza Pallavicino, who had been, as we know, rehabilitated. Ciampoli will never return to Rome. But his library will.

We know that on Ciampoli weighed a very serious and legitimate suspicion of being sympathetic "in spirit" to the new opinions, or—as the pope had said—a "friend of the new philosophy" that had infiltrated the discredited policy of the Holy See to its very summit. Besides the many other accusations and calumnies that historians were spared concerning Ciampoli, there also hung over him the suspicion of a double-cross carried on secretly with the Spanish faction in Rome. The truth of the accusation was supported only by echoes of the calumnies at that moment of the transfer of power. Monsignor Ciampoli, in his letters from exile, will consistently reject it as completely unfounded.

As political insurance, before handing over his Vatican post to his successor Monsignor Herrera, Monsignor Ciampoli had been able to have copies made of his diplomatic archives. He was ordered also to leave the copies in his office. But Monsignor Ciampoli, skilled in such matters, had wisely taken care to make two sets of copies and left Rome with the second set packed together with his copious personal archive.

[79] See A. Pozzobonelli, "Vita di Giovanni Ciampoli" (1644), in G. Ciampoli, *Lettere*, Macerata 1666, pp. 58-77.

Many mules were needed, trailing after Monsignor Ciampoli's carriage, to transport the archives and library: there was a large private correspondence (including the correspondence with Galileo), scientific papers, his books and Virginio Cesarini's. There were many incomplete manuscripts and "many books," as witnesses to his exile will relate, during which Monsignor Ciampoli "devoted himself rabidly to speculation,"[80] to redeem his defeat, reflect on his own correctness, and rehabilitate his intellectual memory.

A possibility for intellectual rehabilitation was offered him by King Ladislao IV of Poland, who had known and esteemed Ciampoli at the time of the Holy Year. The most tolerant Catholic monarch of the period, he too was a victim of extremist pressures and the Jesuits' political intrigues.[81] Ciampoli had remained in contact with Father Magni, theologian and philosopher of the Warsaw court. It was thanks to this father's intervention that Ladislao IV commissioned Monsignor Ciampoli to write a history of the Polish Kingdom, to which the former Vatican diplomat will devote the last years of his life.

When Monsignor Ciampoli died at Jesi on September 8, 1643, he left his illustrious patron four volumes on the history of Poland. He also left him in his will all his manuscripts, including those on natural philosophy which he had put in order, as well as the scientific, poetic, literary, and moral writings, some of which had been prepared for publication. Monsignor Giovanni Ciampoli obviously did not trust anyone but the King of Poland and wanted the archives of his library to be kept in safety far from Rome and outside Italy.

Perhaps also Virginio Cesarini's famous manuscripts had already been packed in boxes to be sent to Poland, before the snows blocked the mountain passes. And this must have been a voluminous shipment, for the inventory of the manuscripts destined for Poland has been conserved, together with the testament, in the Vatican Archive, sole witness to the "new opinions" on scientific and philosophical questions of that prelate with Galilean tendencies.

But the mules that were supposed to carry the Ciampoli archives to safety never took the road north. Instead, at the Palace of the Governor of Jesi there appeared functionaries of the Holy Office, with an armed escort and a sequestration order.[82] Monsignor Ciampoli's archives took the road to

[80] See Anonymous, "Vita di monsignor Giovanni Ciampoli," p. 111.

[81] On the relations maintained even after the crisis of 1631 between Ciampoli and Father Magni, see Ciampoli's letter to the Capuchin of Warsaw, in *Lettere*, 3rd ed., Venice 1661, p. 44. On Father Magni's declared philosophical Galileanism, we should remember that in 1648 he publicly celebrated Copernicus's memory in astronomy, and Galileo's "in many questions of physics"; see P. V. Magni, *Philosophia Virgini Deiparae dicatae, Pars prima*, Varsaviae 1648, p. 12.

[82] Anonymous, "Vita di monsignor Giovanni Ciampoli," p. 115; and Pozzobonelli, "Vita di Giovanni Ciampoli," p. 77.

Rome. At the very moment that the mules cross the threshold of the Holy Office's main portal, the scientific manuscripts, the writings on natural philosophy, the dialogues on heat and light written by Monsignor Ciampoli cease to exist. Not a trace of them is left. Were they lost? Allow me to imagine that they should still be where they were actually taken.

And what about the letters, the entire correspondence with Galileo, of which not a trace remains? We do not know. Later on, Ciampoli's secretary at Jesi will say that he has burned everything: "an excellent idea which freed me of all danger."[83]

And yet, the poetic, historical, and moral manuscripts of the Ciampoli archives at a given moment came out of the Palace of the Holy Office to be distributed among various libraries, above all to the Barberini and Casanatense libraries. From the Holy Office came the history of Poland, the letters, the poetry, and the prose writings which were printed between 1648 and 1667 through the good offices of friendly cardinals, such as Savelli and Colonna, but above all thanks to Father Pallavicino's insistence and censorious curatorial vigilance.

Later, an even more heavy-handed, corruptive censorship, aimed at toning down every possible trace of the new philosophy, would oversee the posthumous edition of *Dei fragmenti* in 1654. Among these were some sections of the "manual of correct philosophizing," which we have mentioned. Even published in this manner, Ciampoli's philosophical fragments revealed that their source of inspiration was the contemplative Augustinian spirit of Galileo's Copernican letters and *The Assayer*. The style, however, was different: the rational autonomy of the knowledge of nature is no longer affirmed polemically against Scholasticism, but is presented through mediated, prudent arguments.

One should remember in what state of mind Ciampoli had revised his notes during his exile. "I am offended, I am terrified, and the perfidy of the persecutors has taught me to fear even the benevolence of masters,"[84] he wrote in his letters; or, "we are navigating with reefed sails, and we speak in conformity with the pettiness of the present state of affairs."[85]

Nonetheless, the pedagogic intent of his fragmentary preface to *Filosofia naturale* [*Natural Philosophy*] was clear. Ciampoli had witnessed Virginio Cesarini's spiritual conflicts, and he knew the danger of the anti-Christian skepticism which, without a vigilant methodological knowledge enlightened by faith, threatened the study of nature's secrets. In order to avoid going astray in the immense Theater of Nature, whose dimensions and num-

[83] Ibid. For Giovanni Ciampoli's testament and the list of manuscripts left in bequest to Ladislao IV, see D. Ciampoli, "Monsignor Giovanni Ciampoli, un amico del Galilei."

[84] G. Ciampoli, *Lettere*, Bologna 1679, p. 130.

[85] Ibid., p. 129.

ber of worlds not even Galileo's telescope could estimate with precision, Ciampoli advised one not to be seduced by metaphysical pride—as, for example, Bruno was.

He urged his readers to find the criterion of "undoubted" certainty in the study of the only "authentic codex" thanks to which it is possible to decipher the book of the universe: namely, "the undoubted subtleties of mathematics" applied to the study of the "sensible world." His readers were also urged to renounce "metaphysical investigations," "philosophical fantasticating" on "impossible things" and "hopeless materials." Only "sensible experience is the teacher of philosophical certitude." The methodology recommended by Ciampoli consists in proceeding *ex notioribus* from the most obvious things to the vision of "occult truths" through the "numbered steps of necessary deductions."[86] The fragments from *Natural Philosophy* reconfirm the religious aspiration of the original Galilean Roman battles: the theory of parallelism between revelation and reason, each having its source in God, "the two Bibles in which the Lord God is master."

Galileo, too, in the Copernican letters of 1615, had used this religious idea above all in a defensive manner. In Ciampoli there was a greater religious conviction: the new philosophy was the ally of the new Christianity against Averroistic naturalism and against libertine and magical irreligiosity. One must abandon Aristotle, a pagan philosopher, inherited by Christian Europe owing to contingent events and then allied with Catholic theology in Scholasticism—a mixture that has made Aristotle a "fatal authority."

Ciampoli hoped that this hybrid of Catholic dogma and Aristotelian philosophy would be dissolved, so as not to "expose ecclesiastical authority to the dangers of ridicule." What dangers? In one of his speeches, entitled "On Novelty," Ciampoli again took up this problem of the tragic confusion between the arguments of faith and natural arguments. One of the examples he gave was the metaphorical explanation, in terms of traditional natural philosophy, of the Eucharistic dogma.[87]

In these fragments of Ciampoli's, there was an intuition of religious rationalism allied with the new science which was common to the religious preoccupations of a Mersenne or a Jean-Baptiste van Helmont. But did not Ciampoli know that the Catholic reform proposed by Van Helmont against the paganism of Aristotle had already been condemned in Brussels by the Inquisition?

To know what new scientific content lay within the horizon of Ciampoli's

[86] G. Ciampoli, *Frammenti delle opere postume*, Bologna 1654 (2nd ed., Venice 1655). The most strikingly significant variations and censures made by the editors of the posthumous editions of Ciampoli's *Prose* and *Frammenti* are listed in E. Raimondi, "Avventure del mercato editoriale," in *Anatomie secentesche*, Pisa 1966, p. 116 ff.

[87] See G. Ciampoli, "Della novità," in *Prose*, Rome 1649, discourse VI, p. 137 ff., especially p. 154.

ideals of reform, one must know his *Nuova Fisica* [*New Physics*] and his other scientific manuscripts. In 1640 the first book was already being edited, for Ciampoli wrote, "I am composing a *New Politics* and a *New Physics*: two vast works, each of which will spread over more than thirty volumes." We know the first, not the second.

Why did this work and other scientific manuscripts disappear? What might they contain that was dangerous? Ciampoli was not an astronomer; none of the catalogue titles of his manuscripts seems to suggest treatises in favor of Copernicus.

No, for Ciampoli too the issue was over another doctrine, and Copernicus had nothing at all to do with it. The reasons for the official condemnation of Monsignor Ciampoli's scientific works were revealed much later, in a new edition of his *Prose* [*Writings*] in Rome, but only in 1667. These reasons were revealed to us by the editor of that edition, Giacomo Antonio Celsi, who explained in his preface the "considerations" which had led to the fact that "nothing was published" of Ciampoli's philosophy. After saying that Ciampoli had used as the teacher of that philosophy a "Personage" who tried to avoid the role (Cardinal Pallavicino?), the editor clarified his statements:

> The second consideration is that Ciampoli in his early youth had encountered some very highly esteemed Philosopher of his city [Florence], who, hostile to Aristotle, had revived the ideas that attribute all change to the new coupling of small incorruptible corpuscles, without ever acquiring substance or quality, or losing its being.[88]

The posthumous denunciation of Ciampoli continues:

> And this philosophizing was therefore very much in keeping with the boldest and liveliest age, because of the arrogance of the human intellect, owing to which new things always find favor with young people. . . .

He then explains that Ciampoli, a poet, was a poor metaphysician and loved only "the Philosophy that never leaves imagination behind." He loved, in other words, philosophical fancies instead of Aristotle's rational entities:

> Since such a philosophy followed by Ciampoli is proved false in the most limpid and subtle light of nature—and was therefore opposed by that intellect which, by common estimation, was nature's most intimate secretary [Aristotle] and (although the followers are unaware of it) not going in accord with the teachings of religion is thus rejected unani-

[88] "A' lettori," in G. Ciampoli, *Prose*, Roman ed. 1667, dedicated to Clement X.

mously by the teachers of divinity, it is hence judged proper for such compositions not to earn Ciampoli praise.[89]

Again in 1667, at Rome, the atomism of the Galileans was condemned because of the impossibility of its "going in accord with the teachings of religion." We know very well what impossible translation was at stake.

Patience is the historian's prime virtue, because sooner or later the reasons and motivations behind events are revealed. One only has to know how to wait.

This belated, posthumous condemnation harked back to the success among young Roman intellectuals of the corpuscular ideas of *The Assayer*, brought up again in the *Dialogue*. It threw a bright, retrospective light on the pope's worried remarks in 1632, when he said that Ciampoli, too, was in that group.

Ciampoli, Father Ridolfi, Father Grassi, Father Guevara, and Sforza Pallovicino: while Galileo was tried for the Copernicanism of the *Dialogue* in order to cover up a more serious scandal, all the actors in another plot were kept away from Rome. Father Castelli was also kept away for the time necessary to silence the "rumors."

But Galileo too, so much later, was entitled to a posthumous condemnation in regard to that old, embarrassing episode. What mattered in 1632, however, was to keep up appearances.

[89] Ibid.

NINE. SAVING APPEARANCES

And your Galileo's lofty work,
marvel of the centuries, made known
the very atoms and brought closer the stars
. . . .
Those opaque, rough globes
which perhaps have seas and mountains
will never enslave or reign over us,
for we are the smiths of our own destiny.*

Theological Speculations

The Galilean scandal in Rome having subsided, the problem remained, and would remain for a long time. We shall follow it for only about fifty years until it reappears officially against Galileo.

How to "save [Eucharistic] appearances" from the ever more invasive success of corpuscular theories in physics? How to reconcile speculative theology with scientific and philosophical speculation when they spoke of the same things: color, taste, and smell? The Eucharistic controversy was a general line of apologetic conduct adopted in the seventeenth century by Jesuit scientists and philosophers against the advent of mechanism in physics. After Galileo, almost all the major Catholic scientists and philosophers, willingly or not, on their own initiative or forced by the polemical initiatives of others, had to come to terms with these great problems: the separation of substance from quantity, or extension; and the reality of accidents without subject. This problem had already coursed like a deep current through the entire intellectual history of Europe, and now it rose ever more toward the surface, clashing with the emergence of the new philosophy.

The fact was that the question had, from the beginning, constantly been the order of the day because of the Protestant challenge to Tridentine dogma and the echoes of those ultramontane heresies even in Italy, in the very hearts of zealous, heterodox (if not actually heretical) Eucharistic theologians.

* "e del tuo Galilei l'alta fatica, / meraviglia dei secoli, fè noti / gli atomi stessi e avvicinò le stelle . . . / . . . Che non giammai quei globi opachi e scabri / ch'an forse e mari e monti / destinar ponno a noi servaggio o regno / che noi di nostra sorte a noi siam fabri."—G. Michele Milani, *La luce, Canzone con annotazioni dello stesso* [The Light, Song with Notes by the Author], Stark, Amsterdam 1698, p. 44

We must say that the initial denunciation of *The Assayer* had been from this point of view very timely and practically inevitable, since just at the moment of publication of Galileo's book the Eucharistic question was the most dramatically present and urgent one. *The Assayer* had barely come out, and it was as though it had given a signal. We have said (Chapter Four) that in 1624 in Alsace, an illustrious Parisian pastor, Jean Mestrezat, a philosopher and worthy adversary of the Jesuits, had again mounted an attack against Cardinal Bellarmino's and Cardinal Du Perron's Eucharistic apologetics in the very well known book *De la communion à Jésus Christ au Sacrement de l'Eucharistie contre les cardinaux Bellarmin et Du Perron*. The arguments were those of the whole tradition of sacramental heresies: "the bread is the body of Christ truly and really, but not substantially and essentially, because if the bread no longer had substance, it would no longer be bread; it would be nothing and thus would not be sacrament."[1]

It was, as always, the old nominalist temptation served up in a new linguistic form. In fact, against Bellarmino's linguistic argument, which had affirmed the rigorous designative value of the sacramental formula, Mestrezat proposed a theory "of the signs that signify the thing."[2] This new enemy of the Church then also presented a theory of the subjective perceptions of sensible phenomena, thus applying his point of view to the problem of sensible Eucharistic phenomena. A real presence for Mestrezat signified a problem tied to sensory perception.

When this new anti-Eucharistic heretical provocation bounced back to Rome, everyone must have thought that this French Protestant had learned the lesson of *The Assayer* and had used it for his own perverse purposes. Indeed, under the pen of the Protestant polemicist reappeared the suggestive Galilean metaphor of the tickle to demonstrate the subjectivity of physiological sensations, and all this now served to justify the sacramental idea of the real presence without transubstantiation. "We say," Mestrezat wrote, that

> something is so close to our hand that we can touch it, because the hand acts by touch; we shall say that a sound is present to the ear (a roll of thunder, for example) even if it is produced very far from the ear. We say that the Sun is present to our eyes, even if locally it is quite far from our eyes, because the ear and eye do not require for their action the local contiguity between their own substance and their object. The same thing happens for the faith. It has its own present object without local contiguity between it and the Sacrament.[3]

Mestrezat's book was published in Sedan in 1624 and in that center of Protestant sedition which was the Academy of Sedan. Its lesson was imme-

[1] Mestrezat, *De la communion à Jésus Christ* . . . , p. 81. [2] Ibid., p. 159. [3] Ibid., p. 84.

diately assimilated by a young philosopher, a rebellious and restless student, David Derodon, who was soon to make a name for himself as one of the liveliest Protestant and atomist philosophers in France.

Derodon, terribly eager for experiences, had also gone through a direct and negative experience of Jesuit Catholicism at the College of Vienna; then, returned to his original faith, this roamer became a professor at the Academy of Die, as his father before him, and then at Orange. Fame for Derodon will come quite soon, thanks to a book on Christian mysteries immediately condemned to the stake by the Parliament of Toulouse in 1658, as the result of a denunciation and pressure exerted by the Jesuits. Derodon, however, was given the chair of philosophy at the Academy of Nîmes.

As a philosopher, Derodon is expert. He knows how to use the Aristotelian dialectic better than his old Jesuit teachers and is an Aristotelian full of interest in the new scientific achievements, which he reflects in his theology, so much so as to be considered tainted by heresy even by his coreligionists.

But worse was yet to come. Derodon, an Aristotelian in logic, was an atomist in physics. Before and after 1650, he was an adviser for theses on the atoms which attacked with great decisiveness the theory of substantial forms. For the moment, that was only an echo of the new philosophy, but it gave a foretaste of the inevitable direct confrontation with Tridentine dogma. Derodon launched his challenge in grand style in 1655 with his great *Dispute sur l'Eucharistie*, published in Geneva.

Derodon's book followed in Mestrezat's tracks against Bellarmino. Nothing new on the theological plane: these were the usual accusations concerning the Council of Trent's idolatry. But Derodon had infinitely greater skill at logical disputation than Mestrezat and was more convincing. For Derodon, the sacramental formula is endowed with productive, not transubstantial value; it is akin to God's having created light *ex novo* saying, "Let there be light." The sacramental formula also produces something entirely new, of a spiritual nature.[4] Of this conversion the bread and wine are the signs, the figurative symbols, not real accidents without subject. They are signs with the value of names—names, not reality; the old heresy.

Also, Derodon thought that accidents necessarily derived from the subject. Heat, for example, is knowable through a more-or-less agreeable subjective sensation of the sense organs: "the fire, or its subtle parts, insinuates into the pores of the body, and [those parts], by thus making room for themselves, separate the parts of our body . . . and since it separates them, it signifies necessarily that it is substance."

Fire is a corporeal substance, and so is light. Derodon expatiates at length

[4] D. Derodon, *Dispute sur l'Eucharistie*, Geneva 1655, p. 4 ff. On Derodon's intellectual character, see E. Haag, *La France protestante*, Geneva 1966, vol. IV, p. 229 ff.

on the phenomena of refraction, on thermoluminous phenomena, in order to explain that light and heat depend on the increase or diminution of their fiery parts. As a good atomist, he shares the theory of Galileo and Nicole d'Autrecourt on the noninstantaneous propagation of light, something that not even Descartes had done. But Derodon was more traditionalist on the theory of colors: he knew that they were appearances tied to the matter of light, but he held to the idea of a mingling of light and darkness. Smell, on the other hand, was explained by the local motion of odorous corpuscles; and, indeed, he makes a point of mentioning that Democritus drew nourishment for three days just from the smell of freshly baked bread.[5]

The Protestant Derodon was a Gassendi-influenced follower of Democritus, and he explained the fluid and solid state by the hook-shaped figures of the atoms. In the case of the Consecrated Bread and Wine, it is God who maintains the motions and shapes of the atoms, together with their substance.[6]

But a problem such as that of the new atomist theological speculation deserved another book, an entire book of physics in which to discuss and utilize polemically the most diverse sources: *Philosophia contracta.* We learn from it that Derodon knew Galileo quite well, but that he was a fervent anti-Copernican Galilean, probably out of fidelity to the Scriptures. He considered Galileo's argument about the tides, "sententias quasi fabulosas," not even worthy of refutation.[7] Derodon's great *maître à penser*, however, was the Provost of Digne, Gassendi, "the greatest philosopher of this century."

And just as Gassendi had converted Democritus's ideas to Christianity, so Derodon converted Gassendi's ideas to Calvinism. He proved, in fact, that atomism compels one to replace the concept of transubstantiation with that of transmutation: "a kind of movement and change . . . by which matter passes from one form to another,"[8] though maintaining its substance. A follower of Gassendi in physics, Derodon continued with paradoxical effect to argue like an Aristotelian in philosophy.

Derodon knew very well that the "teachers of Rome" had unofficially condemned the nominalist identification of substance with quantity; but he insisted that "quantity is nothing else but matter" and mocked the Jesuits, citing Father Arriaga and Father Honoré Fabri as philosophers who also have left the door open to a corpuscular physics.[9]

In 1658, Derodon published a very violent anti-Roman pamphlet focused on the Eucharistic controversy, the famous *Tombeau de la messe.* Denounced immediately by the Bishop of Nîmes, Derodon will be banished from France

[5] Derodon, *Dispute sur l'Eucharistie*, p. 208 ff. [6] Ibid., p. 223.

[7] D. Derodon, *Philosophia contracta, Secunda pars. Physicae*, 3 vols., Geneva 1654, p. 138.

[8] Derodon, *Dispute sur l'Eucharistie*, p. 222.

[9] Ibid., p. 122 for heat and p. 207 for the substantial theory of light.

in 1663. He took refuge in Geneva, where he continued to teach his philosophy of the Eucharist.

For the Society of Jesus, the most embarrassing thing in Derodon's sensational provocations had been the skillful stratagem of invoking some Jesuits in support of the atomist thesis. The Piazza del Gesú replied immediately, directing, after the attack of 1632, a drumfire of warnings to all colleges of the Society, advising them to abstain completely from atomist physics and from the geometry of indivisibles. The warnings and alarms followed each other in a progressive crescendo as atomism gradually forged ahead among French Catholics and Protestants: in 1641, in 1643, and a good three times in 1649.

Jesuit scientists and philosophers had thus been put in a state of high alert. By dint of their education and theological knowledge and vigilance, they were entrusted with the defense of Tridentine frontiers in the Catholic countries. And it was thanks to this vigilance that one could extinguish at birth certain hotbeds of Eucharistic heresy which had taken root among philosophers and theologians even in Italy. Here, however, the fire was easy to control: it had broken out in Sicily, the land of the Catholic king.

A Theological Galilean: Giuseppe Balli

A cultivated aristocrat of Palermo, his name was Giuseppe Balli, or Ballo. Had he not donned the religious habit, he would have become Baron of Callatuvi. He went to Spain to study theology and mathematics. But in Spain, besides Aristotle and Thomas Aquinas, he got to know from close contact the fascination of the young, pulsing heart of Catholic mysticism of St. John of the Cross and St. Theresa of Ávila. When he returned to Italy, appointed by Philip IV as the king's chaplain at Bari, his theological, poetic, and scientific vocation was set under the sign of a contemplative and mystical inspiration in the name of Dionysius the Areopagite.

Don Balli carried out a diligent program of theological and philosophical studies at Bari and was in contact with Cardinal Bellarmino. These were the years during which Galileo presented the new philosophy in the form of a Neo-Platonism which smacked of mysticism and which easily got the support of the most open-minded ecclesiastical circles. Reading the *Lettere sulle machie solari [Letters on Sunspots]*, in the second of which Galileo put forth a version of the principle of the persistence of motion, and then learning of Father Cavalieri's *Settioni coniche*, probably permitted him to join the restricted ranks of those who followed Galilean teachings: the generation of Cavalieri, Baliani, Gassendi, and Descartes. It is in this way that this theologian was able to participate equally—along with Cavalieri and Baliani, and before this last—in the research program for a generalization of the principle of inertia.

His contribution to the history of Galilean dynamics was contained in a few pages of a memorandum entitled *Demostratio de motu corporum naturalium*, which was published by him in an appendix to an exacting book on theology, *De foecunditate Dei circa productionem ad extra*, issued at Padua in 1635.[10] Don Balli's *De motu* is one of the most elegant seventeenth-century scientific texts in the history of the principle of inertia. It is written in a flowing style and in a language that reveal the author's most rigorous Aristotelian logical origins—searching for a purely rational demonstration of the persistence of motion and arguing against Pereira's physics, in the light of a metaphysical principle of conversation produced by the intellect of a divine agent.

In *De foecunditate*, Balli reveals how much the mystical ideas about light in Galileo's letter to Monsignor Pietro Dini must have attracted him, just as *The Assayer*'s atomist materialism must have offended his religious sentiments. He mentioned *The Assayer*'s theories, evoking again the atomist metaphor of characters in the book of the universe.[11] He distanced himself from a corpuscular theory of light by means of excellent physical arguments, which instead led him to embrace a theory of luminous expansion through the reproduction of luminous "spheres."

Nonetheless, Balli was not averse to a corpuscular physics, in which he recognized the rational and theological value of a motor principle in nature produced and maintained by God. Therefore, he understood Galilean corpuscularism as inspired by Ecphantus of Syracuse's mystical philosophy of the causal action of God.[12]

However, as a Scotist theologian, declaratively Augustianian, Don Balli knew what sort of problem was an obstacle to a dynamic natural theory of this kind. He thus decided to publish in 1640, again at Padua—where he ended his life in a local convent of Theatine fathers—his book on the problem of the Eucharist: *Resolutio de modo evidenter possibili transubstantiationis*.[13]

To tell the truth, Balli had been preparing that book for several years, but Cardinal Bellarmino had convinced him that it would not be at all prudent to publish it. Nevertheless, Balli did publish it later on, as a great act of faith in the light of reason, since in his Sicily the Jesuits had already begun denouncing him on the basis of indiscretions which circulated concerning the secret Eucharistic theory of this follower of Galilean science. But, instead of

[10] Balli's *De motu* is reproduced in R. Giacomelli, "Un contemporaneo di Galileo. Giuseppe Ballo," *Atti della Reale Accademia delle Scienze Fisiche e Matematiche di Napoli* 15, ser. 11, no. 10 (1914), pp. 1-35, especially pp. 24-35. Balli's scientific work is cited in Poggendorff, *Biographische literarische Handworterbuch*, vol. I, Leipzig 1863; and P. Riccati, *Bibliografia matematica*, Modena 1860. See, also, the note to p. 810 of the commentary on the *Discorsi sopra due nuove scienze*, eds. L. Geymonat and A. Carugo, Turin 1958.

[11] G. Balli, *De foecunditate Dei circa productionem ad extra*, Pataviae 1635, p. 122 ff.

[12] Ibid., p. 126.

[13] This is a very rare book, a copy of which I have studied in the Library of the University of Padua.

calming people, publication of the book (despite the author's death that same year) threw oil on the old, incandescent polemics of twenty years before.

And yet, Don Balli's *Resolutio* appeared with every possible guarantee: the approval of the Padua Inquisition (albeit with the opinion of a Theatine father); several warnings from the author; and even a long assurance from the printer, Sebastiano Sardo, concerning the book's Catholic orthodoxy so that it will not incur any official ecclesiastical prohibition. Don Balli's endeavor was to release the Eucharistic mystery from the bonds of Scholastic philosophy, to present it in a light in which the various "classes" of modern philosophy could recognize themselves without conflict. In fact, for his new presentation of the Eucharist, Balli invoked Ockham and the nominalists, Telesio—for whom he hoped a speedy rehabilitation—as well as Galileo, whom Balli did not quote but who was hidden among those modern philosophers who "think like the ancients that all substantial forms and all things, apart from matter, are accidents."[14]

Balli's *Resolutio* is not a speculative solution of the mode of the real presence, but a theory about how it is possible to conceive of such a mystery of faith in a simple manner by suppressing a useless accumulation of miracles and metaphysical entities. In fact, the solution of the problem is based on the reality of things: bread is an ensemble of sensible properties; of smells, tastes, and colors. These are at bottom the "substance" of bread. Hence, if the names must indicate what one senses, Consecrated Bread will be all of this: "I therefore explain transubstantiation not like those who improperly make a distinction between substance and Scholastic qualities, but in the sense that the bread is transubstantiated; for one cannot tell anything about bread without such entities, and therefore it is transubstantiated with all of them."[15]

Where, then, is the miracle? The miracle does not lie in the complex conceptual alchemy of the dialectic between quantity and qualities, but in the infinite freedom of God's creative work, which can transfer from bread to Christ's very body all the sensible effects, not just the substance. Don Balli's principle is precisely this transfer of sensible effects from thing to thing through the work of divine freedom. The mystery is contained wholly in the "efficient bond" that Christ maintains in an immediate fashion beneath the Eucharistic species. But Christ is not "hidden" beneath those appearances: He is visible. The sensible actions of the bread and wine manifest Him "as a second cause, declared and not hidden."

"It is a trick!" will be the objection brought against Balli. And Don Balli will reply that it is precisely a sleight of hand (*praestigiae, phantasticum*),[16] but unlike the tricks of magicians, who know them very well to be such,

[14] Balli, *Resolutio*, p. 71. [15] Ibid., p. 6.

[16] See G. Balli, *Responsio ad objectiones*, Pataviae 1640, p. 30.

these deceptive Eucharistic appearances are a religious mystery. It is Christ's body and blood which actively produces that emanation of sacramental species.

The problem was precisely that of the nature of the "species," the term used in the council's formulation of the dogma. Here Don Balli's *Resolutio* time and again explicitly takes a position against Father Suárez's Eucharistic philosophy, in order to vindicate a pregnant sense of the word, which Balli considers even more appropriate to St. Thomas's literal meaning. The species, therefore, are not real qualities without substance, but that which strikes the senses, like "that which is the matter of the species,"[17] as in the case of light.

In support, Balli cited the Neo-Platonism of Giovanfrancesco Pico, but probably he was also thinking of Telesio, Patrizzi, and Galileo in his letter on light to Monsignor Dini, when he said that the Eucharistic species subsist "without subject" like light, which God made subsist for a good three days before creating the sun and the stars.

One must be sustained by the mystics' great audacity to oppose official Scholastic theology or pour into the old formula of transubstantiation the afflatus of Christian Platonism and Telesio's and Galileo's substantialist ideas on light. And, for the moment, it was only an abbreviated foretaste. The printer Sardo let it be known that Balli had ready a generalized version of his theory, the true and proper *Enigma dissolutum*, but that they were waiting for the reactions of the Catholic public.

The *Enigma dissolutum* was never published. It remained among the manuscripts of mathematics and astronomy left in Balli's will to the convent of Paduan Theatines. Don Balli only had time left to publish some brief writings in defense against the criticisms provoked by his theory, which like all the natural philosophies encouraged reconciliation.[18]

For, examined closely, whatever the audacious "psychological" conception of Eucharistic appearances gained in economizing on miracles was lost in fidelity to Tridentine orthodoxy. After all, it was by no means owing to theological myopia that Cardinal Bellarmino, in his time, had advised Don Balli to abstain in this field from the French fashion of theological mysticism. It was not too difficult to see that Balli's thesis violated the famous canon of the thirteenth session of the Council of Trent. This canon affirmed that the Eucharistic species were maintained (*manentibus*), i.e. that they were exactly as they were before, whereas Don Balli proposed that they commenced to exist, as if by magic, at the very moment of the Consecration.

To block this intrusion of Neo-Platonic mysticism into the theology of the Eucharist and bring things back to the Tridentine letter of the law, a Jes-

[17] Ibid., p. 31.

[18] See G. Balli, *Assertiones apologeticae cum suis dilucidationibus*, Pataviae 1641 (post.)

uit mathematician from the College of Palermo—a person who enjoyed great renown in Sicily, Father Gerolamo La Chiana, archbishop and royal censor—promptly intervened with an anonymous infolio. Therein, Balli's theory of Eucharistic species was directly ascribed to the realm of the symbolism of sacramental heresy.[19]

Don Balli was dead. But his immediate follower, the Sicilian canon Don Chiavetta, who in 1643 had published at Monreale a *Trutina qua Josephi Balli sententia de modo existendi Christi Domini sub speciebus panis et vini expeditur*, suffered an inevitable censorship: his *Trutina* was put on the Index in the spring of 1655.

Did Balli's ideas, which perhaps appear to foreshadow Descartes' and above all Leibniz's later Eucharistic solutions, thus end in a provincial theological and judicial story, where it was much easier to stifle at birth the new philosophy's claims of Catholic legitimacy? Or did they survive and perhaps cross the Alps, where the conflagration lit by the new philosophy was impossible to control. The menace, as usual, came from France.

From the Book of the Universe to *The World*: Cartesian Heresies

The menace was Descartes' philosophy, it too entangled, like Galileo's, in the net of theological surveillance set up by the custodians of Tridentine faith. Descartes' acrobatics, in an attempt to reconcile his corpuscular physics with the Eucharistic dogma and to avert their being divulged publicly under his nose, are well known to Cartesian scholars. In general, the solution proposed by Descartes is regarded either with indulgent irony as an incident in his *esprit de système* or with an apologetic concern aimed at saving Descartes' irreproachable religious sentiments concerning this delicate question, too. Thus we know above all the Cartesian arguments and those of his most benevolent and respectful interlocutors: Antoine Arnauld and Father Mesland. On the other hand, it is completely ignored that the Eucharistic denunciation which Cartesian philosophy and physics incurred was inserted into the general strategy of theological discussion against the new philosophy that had been perfected against Galileo.[20] I will confine myself here to demonstrating the continuity of this problem from Galileo to Descartes.

We had in fact left Descartes in Rome, lost in a crowd of pilgrims during

[19] See Anonymous (Father G. La Chiana, S.J.), *Opusculum quo probat substantia corporis Christi quae sub speciebus panis continetur non possunt appellari imaginem Corporis Christi*, n.p., n.d. (Palermo 1642?).

[20] The great polemic on the Eucharist, the most serious theological and theoretical difficulty that Descartes encountered, is described in H. Gouhier, *La pensée religieuse de Descartes*, Paris 1924, 2nd ed. 1972; J. Laporte, *Le rationalisme de Descartes*, Paris 1945, bk. III, chap. 11, p. 405 ff.; more recently in H. Gouhier, *Cartésianisme et augustinisme au XVIIe siècle*, Paris 1978; Armogathe, *Theologia Cartesiana*, in which is described the role played in the Clerselier controversy by Father Poisson, Father Viogué, Maignan, and Rouhault.

the Holy Year of 1625, while *The Assayer*, which had been in the bookstores for a year, was being celebrated in pro-French circles. If we are to believe his friend and biographer Baillet, Descartes had stayed in Rome from the beginning of the Holy Year until the spring and, during those months, had become close to Cardinal Barberini, protector of the Academy of Lynceans.

It cannot be excluded that Descartes, who knew the books of Telesio, Campanella, Bruno, and Basson, had also bought in one of the bookshops on Piazza Navona Galileo's new book, about which there was so much talk in Rome because of its new theory of atoms.[21] In any event, it is certain that in 1638 Descartes was perfectly well acquainted with Galileo's atomistic theories. This is obvious, since for almost ten years Descartes had worked on a "discourse" about light and colors, which he now prepared to insert in the *Dioptrics*. Eight years before, however, the initial project had been much more audacious: he was thinking of discussing the phenomena of comets and light in a general exposition of physics; and, from this point of view, *The Assayer* could very well constitute a model.

At that time, November 1630, Descartes had written to Father Mersenne from Amsterdam precisely about this enterprise. The trial of Galileo would begin only two years later. But in Paris, Sarsi's *Ratio ponderum librae et simbellae* had already been published three years before. It cannot be proved that Descartes read that book, but to exclude it would be much more difficult, since Descartes was well informed about the debate between Sarsi and Galileo, knew who Sarsi was, and in his *Principles* (III, art. 128) will mention the *Libra*, "written against Galileo."[22]

In any case, when Descartes promised in 1629 to elaborate a "complete Physics," he was perfectly aware what the fundamental theoretical problem facing him was; and, being rigorous, he did not want to conceal it as Galileo had done, but rather wanted to confront it head on:

> I want to insert there a discourse in which I shall try to explain the nature of colors and light, which has held me up for six months and which is not yet half finished. But in fact it will be longer than I had thought and will contain almost a complete Physics; in such a manner, I assert that it could serve to discharge me from the promise I made to you of completing my *World* in three years, but it will in fact be a kind of summary. . . . I believe that I will send you this discourse on light just as soon as it is finished and before sending you the rest of *Dioptrics*. In fact, since I wished to describe colors in a certain fashion and, in conse-

[21] See Descartes' letter to Father Mersenne, February 9, 1639, apropos of Galileo's *Discourse on Floating Bodies*, in *Œuvres de Descartes publiées par C. Adam e P. Tanner*, 12 vols., Paris 1897-1913 (2nd ed., Paris 1969), hereafter *A-T*, vol. II, pp. 493-508 (and *Works*, XVIII, p. 25).

[22] For the *Principles*, see *A-T*[1], VIII, p. 178; and the letter to Dupuy (January 5, 1645), *A-T*, IV, p. 150 ff. It should be noted that Descartes had studied all the refutations of Galileo; see *A-T*, I, p. 579.

quence, *being obliged to give you an explanation of the way in which the whiteness of bread persists in the Holy Sacrament*, it would be a good idea to have it looked at by my friends before it is seen by everyone.[23]

Descartes will never cite *The Assayer*; nor will he relate to it the theoretical and theological problem of the reconciliation of corpuscular physics and the Eucharist. But unfortunately, not even this silence is proof for us that Descartes was unaware of that episode, given his well-known reticence about indicating sources, recognizing intellectual debts, and explicitly mentioning books read, even if parsimoniously and hurriedly. On the other hand, in *The Assayer*'s case, an omission of this kind could be justifiable, since that was a book which on the whole could not have pleased Descartes.

On comets it contained a discussion and a polemic without head or tail. It taught only that it was necessary and urgent to establish *ex novo* a rigorous theory of comets. As for philosophy, Galileo's book could only confirm in Descartes' eyes all the faults and limits of pure physicists when they set about philosophizing: multiple digressions, an absence of expository order "without having considered the prime causes of nature"; in short, the very same criticisms that later, in 1638, Descartes will feel he has to address to the author of the *Discourses*.[24]

Even with so many defects, *The Assayer* could be extremely rich in suggestions for the thirty-year-old Descartes, still immersed in occultist lucubrations in the manner of Della Porta. *The Assayer* could, in fact, direct Descartes toward the analysis of sensations and the fundamentals of physics, toward a radical criticism of Scholastic qualities, toward corpuscularism. *The Assayer* was full of stimulating ideas: the book of the world written in mathematical characters; the subjective nature of sensory impressions; the corpuscular structure of light. The book, however, must be rewritten completely.

The opportunity to rewrite *The Assayer* presented itself in 1629, when Descartes came across the observations made by Father Scheiner at the Collegio Romano on sunspots.[25] It will be recalled that in *The Assayer* Galileo had advanced a theory of comets as illusory luminous phenomena, exactly analogous to sunspots. Here there was an opportunity to return to comets and light through a "complete Physics":[26] *The World; or, An Essay on Light*.

The World; or, An Essay on Light was much different from *The Assayer*. It did not set out with a theory of comets, but arrived at a discussion of comets after having presented the general cognitive problems of sensations and physics. Also, the theory of comets was radically different. Descartes, how-

[23] Descartes' letter to Father Mersenne, November 25, 1630, *A-T*, I, pp. 177-82, especially p. 179.
[24] Letter to Mersenne, October 11, 1638, *A-T*, II, pp. 380-405.
[25] Letter to Mersenne, October 8, 1629, *A-T*, I, pp. 22-29.
[26] I cite again the letter to Mersenne of November 25, 1630.

ever, seemed to recall very well what Galileo had written in *The Assayer*—so well that *The World*, too, in order to criticize the Scholastic theory of sensible qualities, used *The Assayer*'s very suggestive example of the feather, which produces a tickling sensation in the sensible parts of the body without possessing any tickling properties in itself. "A sleep child," one reads in *The World*, "over whose lips a feather is softly passed, will feel a tickle: do you think that the idea of a tickle conceived by him is similar to something which is in the feather?"[27] "That titillation," one might go on to read in *The Assayer*, "is entirely in us and not in the feather."[28]

The reasoning, in reality, was the same: touch is the most certain of the senses; if a tactile sensation—tickling—is subjective, then too the objectivity of all other sensible phenomena must be illusory. The nature of these phenomena is affective, and we must reconstitute the material structures that, with their invisible actions, produce the sensibly experienced phenomena.

Like *The Assayer, The World* offered a corpuscular explanation of heat and light, proposing the identification of substance with quantity as the only way in which "to imagine" what substance is. And like *The Assayer*, the new *World* eliminated the qualities and real accidents of Scholastic philosophy, supplanting them with local motions of parts of matter: minimal parts of earth, air, and fire.

In 1633, *The World* was completed. We can now understand somewhat better than has been traditionally understood why Descartes, when he heard of Galileo's sentence, suspended publication of that book "on the nature of material things," as he will mention later in his *Discourse on Method. The World* will be published posthumously, in 1664; the caution of Descartes, much better informed than we are today on the possible profound reasons for Galileo's incrimination at Rome, was a justified exercise in the intellectual virtue of honest dissimulation.[29]

He again resorted to that virtue in the *Principles of Philosophy*, wherein he self-censored his Democritan materialism. Nevertheless, in the *Principles* he also continued the criticism of real qualities and the idea that color and the other sensations are subjective facts. He also continued to make the identification of matter with extension a presupposition of Cartesian physics. Three years before, in 1641, Descartes had published the *Méditations*, the great metaphysical work, prudently subjected to the preliminary revision of theologians.

In fact, the reservations and preoccupations of the young Sorbonne theologian Antoine Arnauld are known; that is, the *Quatres objections* to the *Méditations* which rightly intimated that, precisely on the Eucharistic ques-

[27] R. Descartes, *The World*, in *A-T*, XI, p. 5 ff. and p. 7 ff.

[28] *Works*, VI, p. 348, ll. 31-32.

[29] See Descartes' letter to Mersenne, November 1633, *A-T*, I, p. 270.

tion, grave difficulties would arise.[30] It is also known that in view of these loyal preoccupations, Descartes presented his opinion on the reconciliation of dogma and the new philosophy—first to Arnauld, then to the Benedictine of Chartres Dom Robert Desgabets,[31] and above all to Father Dionis Mesland, philosopher at the Jesuit College of La Flèche.

The objections were the same ones brought against *The Assayer*: the corpuscular concept and the identification of substance with extension or quantity, was tantamount, as everyone knew, to denying the objectivity of the sensible Eucharistic species and rendering transubstantiation contradictory. If, as Descartes postulated, matter is extended, then, since in the Consecrated Host the original extension remains, the original substance must remain as well. As one sees, to translate the second part of the dogma—that concerning sensible species—into Cartesian terms meant falsifying the first part, the transubstantiation.

At first, Descartes had replied with infinite caution, looking forward "to a time in which this opinion, which allows for real accidents, will be rejected by theologians as dubious for faith, repugnant to reason, and completely incomprehensible."[32] Arnauld, to whom this hope had been addressed (in the reply published in the appendix of the 1658 Latin edition of the *Méditations*) was, however, an obliging theologian. Father Mesland was also well disposed. Descartes, advising him not to circulate these confidential considerations, offered other hypotheses.

He hypothesized that the effect of the Consecration was to place Christ's body and blood within the sensible dimensions that belong naturally to bread and wine, "as if" those substances were truly in the Sacrament.[33]

The prudent Father Mersenne, in 1642, had sent Descartes information concerning Wyclif's condemnation at the Council of Constance, to put him on guard. "I thank you for what you sent me about the Council's condemnation of Wyclif, but I do not see that that can do anything to harm me," Descartes told him. "They have not established that such accidents are real, which is what I wrote that I did not see in the Councils."[34] Descartes was right: it was a question of viewpoint.

Descartes' point of view was that the idea of the permanence of real accidents was absurd and that the council's words must be subjected to a differ-

[30] R. Descartes, *Méditations*, Amsterdam 1658, p. 259 ff., *A-T*, IX, p. 177 ff.

[31] See P. Lemaire, *Dom Robert Desgabets, son système, son influence et son école*, Paris 1902, p. 100 ff.; and Armogathe, *Theologia Cartesiana*, pt. II, with bibliography and documents on Desgabets.

[32] See *Réponses aux quatrièmes objections*, in *A-T*, VII, p. 229 ff.

[33] See Descartes' letter to Father Mesland, February 9, 1645, *A-T*, IV, pp. 161-75, especially p. 164. For the other letters to Father Mesland, see ibid., pp. 215-17, 344-48, and 348-50. Add to them Descartes' letter to Clerselier on March 2, 1646, pp. 371-73; and the Cartesian Eucharistic fragment communicated by Clerselier (May 22, 1645), *A-T*², IV, pp. 741-47 with P. Costabel's note.

[34] See letter to Mersenne of March 1642, *A-T*, III, p. 545; and ibid., p. 349.

ent exegesis: Christ's body, instead of designating matter, must in all probability indicate Christ's shaping soul: "I do not see any difficulty," Descartes said to Father Mesland, "in thinking that the whole miracle of transubstantiation . . . consists in the fact that, by virtue of the consecratory words, Christ's soul informs the particles of bread and wine without these mixing with the blood of Jesus Christ, as it was supposed to. . . ."[35] Descartes did not see any difficulty in his version of purely subjective Eucharistic appearances, although it was foolhardy if not formally heretical according to official theology.

The letters to Mesland concerning the Cartesian synthesis of corpuscular physics and Eucharistic dogma, even if they were known and duly censored by the Jesuits, remained officially and prudently anonymous.[36] Father Mesland, whether by chance or under compulsion, went on a mission to Martinique. The *Principles*' corpuscular physics, however, circulated and was sufficient by itself to raise the problem, even without theological hypotheses. But in 1649 the *Principles* were struck by a denunciation of Eucharistic heresy brought by Father Thomas Compton Carleton, an English Jesuit mathematician and professor of theology at the College of Liège.

The form was according to protocol. In Father Compton Carleton's *Disputationes physicae* (in the eleventh dispute, to be exact), we read: "I have recently come across a modern writer who seems to eliminate from the universe every substantial form (apart perhaps from man and other living beings) and who claims that fire, earth, and water are nothing but raw material having various motions" (Descartes, *Principles*, pt. 4).[37] To the denouncer, Descartes seemed "more anxious for novelty than for truth, while pretending to continue to be a Catholic." It was as though Descartes were being denounced for Galileo's heresy. As for the Cartesian "libels" in reply to Arnauld, it was not difficult for Father Compton Carleton to say that the identification of matter with extension, despite those explanations, was in any event foolhardy.

Louvain's theological faculty, on September 7, 1662, condemned the two Cartesian theses: the rejection of real accidents without subject and the association of substance with extension. The year after, in Rome, Descartes'

[35] Letter to Father Mesland, *A-T*, IV, p. 168. On the Cartesian theory of Eucharistic concomitance, see the next letter, ibid., p. 347 ff.

[36] The letters to Father Mesland were imprudently sent to the Jesuit Father Honoré Fabri, an Aristotelian noted for his tolerance of Cartesianism. On April 15, 1660, Father Fabri sent a censure from Rome. See G. Sortais "Le cartésianisme chez les jésuites français du XVIIe et du XVIIIe siècle," *Archives de Philosophie* 6 (1924), III, p. 47 ff. and p. 51; Armogathe, *Theologia Cartesiana*, p. 89. Concordant contemporary testimony attributes to Father Fabri a role on the Jesuit side in the decision of the Index to suspend all of Descartes' works (1663). See F. Bouillier, *Histoire de la philosophie cartesienne*, Paris 1868, vol. I, p. 466 ff.

[37] Father Th. Compton Carleton, *Philosphia universa*, Antverpiae 1649, p. 238. Book approved by Father A. Gottifredi, sent to Belgium by Father General Vincenzo Carafa.

work was condemned by the Congregation of the Index, while awaiting correction. In 1672 Dom Desgabets, passionate follower of the Cartesian line in this field, will also be reprimanded.[38]

But the fire, contained on one side, flared up on another. In 1636, at the University of Wittenberg, Daniel Sennert, professor of medicine, dared to praise Democritus in his *Hypomnemata physica* in order to dispose of the Aristotelian theory of natural generation, proposing to explain that phenomenon with the motion of atoms, since every "corpus is per se a quantum."[39] Obviously, by 1639 the *Hypomnemata physica* had already been put on the Index; but vigilance could not be relaxed, because the heresy of the new philosophy spread everywhere, even among religious orders, those closest to the new philosophy.

Father Emmanuel Maignan of the Order of Minims, in his *Cursus philosophicus* (1652) and then in *Sacra philosophia* (1662), boldly defended the reconcilability of Cartesian physics with the Eucharist, leading Descartes into the temptation of denying any sort of objective reality to Eucharistic species: "pure names," Maignan called them; sensory impressions produced by God in our sense organs.

The Jesuits now had to intervene with a heavy hand from Paris and through the good offices of the illustrious and fashionable controversialist Father Théophile Raynaud, who in his great dispute on Eucharistic philosophy (*Exuviae panis et vini in Eucharistia qua ostenditure esse veras qualitates*, published in Paris in 1665) denounced once and for all the heresies or reckless acts of Descartes, Maignan, Balli, Ockham and all modern "nominalists," Pythagoreans, or Platonists.[40]

The denunciation was serious, giving no quarter: all the new philosophers were generically compared to Wyclif; even poor Don Balli was compared to the diabolical Berengarius. The Tridentine faith flatly demanded that one subscribe to the theory of Eucharistic accidents as real qualities without subject. The principle of authority of the theological tradition demanded that the identification of substance with quantity or extension be excluded. Let us be brief: Father Raynaud's book renewed the positions once expressed by Father Suárez and Father Grassi, now brought up-to-date to meet the Cartesian challenge, in compliance with the renewed Roman instructions of the Society of Jesus against the new physics. It is time to return to Italy.

[38] Dom Desgabets modified in a Scholastic sense his Eucharistic doctrines in *Explication familière de la théologie eucharistique*; see Dom Calmet, *Bibliothèque Lorraine*, Nancy 1751, cols. 396-403. But the anonymous Dutch pamphlet entitled *Considération sur l'État présent de la controverse* (1671), earned Dom Desgabets a censure from the Benedictine Order. For the report of the interrogation, see Armogathe, *Theologia Cartesiana*, pp. 133-35.

[39] D. Sennert, *Hypomnemata physica*, Frankfurt 1636, pp. 103 and 86 ff.

[40] Father T. Raynaud, "Exuviae panis et vini in Eucharistia qua ostenditur esse veras qualitates," in *Opera omnia*, vol. VI, Lugduni 1665, pp. 427, 450, 466; and id., "Theologia naturalis, sive entis increatis et creati . . . ex naturae lumine investigatio" (1622), ibid., vol. V, Lugduni 1665.

Theological Police

The advent of an intellectual transformation of great scope, of a revolution that historians will call scientific, around the nucleus of the tranforming ideas of physics, was by now evident in midcentury. It was not the development of scientific knowledge in itself that aroused fear. Astronomy, for example, did not cause fear; and the Jesuits had demonstrated better than anyone else that one could be Tridentine Catholic and anti-Ptolemaic in astronomy, with much more modern and refined theories with respect to the observations of the questionable Copernican theory. In physics, however, it was not possible to be Tridentine Catholic and anti-Aristotelian, at least from the point of view of those who considered the defense of the dogmas of Tridentine religion the prime purpose of their religious vocation and intellectual activity.

The Roman leaders of the Society of Jesus were mobilized by a series of internal circulars expressing alarm over atomistic philosophy. The threat was indeed serious and necessitated these conservative and restrictive measures even at the risk of sacrificing the excellence, modernity, and traditional boldness of Jesuit scientists in this field of scientific study.

For the rest, the Jesuits' religious greatness lay in their ability to adapt to changing religious needs. As at the time of the new astronomical discoveries, the Jesuits had tolerated the fact that their scientists were in the avant-garde, so as not to be second to anyone on the plane of observational and mathematical truth. Now, confronted by the inevitable theological contamination produced by the new philosophy, their function was one of prevention and repression: a vigilant task of theological policing.

Like every other police force that aims to prevent rebellious danger with a great deployment of means, the Jesuits too had traditionally tried to use the methods and language of their potential or immediate adversaries. Their colleges and libraries benefited from a freedom of discussion and research unknown to the other Catholic intellectual orders, since it was not an individual freedom, but an efficious (if not official) freedom and a common good.

That is also why the Eucharistic denunciation of the new Galilean corpuscular philosophy, perfected by the Society's scientific personality, Father Grassi, had become common property. But we have seen, too, that outside Rome, this freedom of inquiry allowed Jesuit professors to forget the traditional standards of conformity to commonly accepted doctrines. We have also seen what a source of embarrassment was created for the Society of Jesus by certain local infringements of doctrinal uniformity. Respect for that uniformity demanded that this theological police force also have an internal police force, as in any institution of a military character.

The Society's censors constituted this internal police force, subject to the direct and exclusive control of the father general, to whom reported the commission of five general censors at Rome. We know that this organism of information and control was entrusted with the delicate task of approving the fathers' books destined for publication. Therefore, the Society's censors (Father Grassi, as we know, was one of them) were fathers particularly well prepared on the moral, theological, and scientific planes; but, above all, they were reliable men, asserters of the order's religious orthodoxy and well aware of the need for a flawless doctrinal uniformity.

The *ratio studiorum* rigorously established the tasks of this work of control: "nobody shall introduce, in matters of a certain importance, new questions or any opinions not supported by a reputable author without consulting the superiors. . . . [E]veryone shall follow the most approved doctors and those opinions that, with the passage of time, are accepted in the Catholic Academies."[41] The use of the guidelines was, one sees, a delicate matter. Until the Galileo affair, right down to 1632, the deliberate tactics of the Vitelleschi leadership had made discreet use of them, and in the field of astronomy the criterion of progressive enrichment of the Society's intellectual patrimony had prevailed over the fear of novelty. With the Galileo affair, the new physics, and the new Father General Vincenzo Carafa, the use of guidelines changed in a discretionary sense.

The Galileo affair had been resolved, as we know. Yet the very serious threat of his philosophy had come through that official condemnation on the part of the Church unscathed. Thus, respect for truth fell entirely on the Jesuits' shoulders, on their preventive and repressive action. Above all, they were to be internally vigilant. Prudence was necessary. To bring such dangerous ideas and debates into the Society during a state of religious war was certainly inadvisable. St. Ignatius's great army must close ranks, concentrate its forces on the primary theological objectives of the moment.

This must have been the censors' perception of the state of affairs; and they decided, starting with the Galileo case, to sacrifice freedom of internal discussion to those objectives. The guidelines are now applied ever more liberally. An inquiry is begun in all the Society's colleges in order to prepare a list of "exotic"[42] or suspect propositions, to be forbidden in teaching and internal discussion.

Not everyone, obviously, was in agreement. In 1648 Father Pallavicino, an old Roman intellectual innovator, suffers censorship of the theological manuals compiled by him for the standard teaching of theology at the Collegio Romano, and he tries to oppose the indiscriminate censure and inter-

[41] See *La Ratio studiorum e la parte quarta delle costituzioni della Compagnia di Gesú*, ed. Father M. Barbera, S.J., Padua 1942, p. 144. On the cultural politics of the Jesuits in the second half of the century, see Costantini, *Baliani e i gesuiti*, p. 95 ff.

[42] See Historical Archive of the Superior General Curia of the Society of Jesus, *Fondo gesuitico* 657, p. 564, ibid., p. 97.

nal denunciation of the father general's superpolice. He is defeated in 1649 at the Society's Ninth General Congregation.

The hard line taken by the censors and the father general prevails. In March 1649, Father Pallavicino, secret admirer of the new philosophy, must publicly retract at the Collegio Romano some of his statements on the constitution of the continuum as well as the opinion, already condemned—as we know—that "quantity is composed of points."[43]

We today should not be shocked: this was not blind but, rather, infinitely perspicacious intransigence. Behind the transformation of physics it sensed a return to nominalist empiricism, and behind this the danger of seeing again the crumbling away of the Catholic theology that the Council of Trent had rigorously established. Respect for Thomist tradition in theology and Aristotelian tradition in philosophy was essential because on one matter they coincided.

So we should not be shocked if the Society of Jesus' top echelon and security apparatus wished to restrict the intellectual initiative of the order's most forward-looking and brilliant exponents. One was a Jesuit in order to defend the Council of Trent, not to make one's scientific career an end in itself. How could one imagine that the Society of Jesus might degenerate to the degrading tolerance of the Franciscans, the Dominicans, the Minims, and the other orders, with their sinister renown for harboring innovators and heretics within their ranks?

Father Pallavicino was imbued with the spirit of the Galilean period. His principled objections concerning that critical fidelity to Aristotle traditionally cultivated by the Jesuits were the sophisms of an intellectual. He mistook for intellectual pettiness and fear of the new a state of grave necessity, which did not escape the sense of responsibility of the chiefs at Piazza del Gesù.

The new father general appointed by the general congregation at the end of 1649, Father Piccolomini, reasserted the necessity for the hard line sponsored by the Society's security organs, but also approved by the majority of fathers directly involved in the struggle against intellectual heresy. There was issued an *Ordinatio pro studiis superioribus*, which sanctioned the strict observance of the bond between Scholastic theology and Aristotelian philosophy. The most qualifying part of this ordinance was the prohibition of propositions that had already been condemned officially and unofficially.[44] At the beginning of 1652, the Society's Tenth General Congregation confirmed this line and strengthened the censors' power.[45]

[43] See G. M. Pachtler, *Ratio studiorum et institutiones scholasticae S.I. per Germaniam olim vigentes*, Berlin 1890, vol. III, p. 76.

[44] On the *Ordinatio pro studiis superioribus* (1651), see *Institutum S.I., Regulae, Ratio studiorum*, Florence 1893, III, pp. 13-15.

[45] Ibid., II, p. 374 ff. It should be remembered that the requests on the part of scientific Jesuit groups

We should not forget that in these international congresses of the Jesuit party which were the congregations, the majority of the fathers from various provinces approved this line of conduct because, on the basis of what we have seen, they had no alternative. This line rewarded the apologetic devotion so passionately defended by Jesuits against outside threats.

Abhorrence of a Vacuum

To react to the outside was the other imperative of the hour. The Copernican Galileo was just as dangerous dead as the living philosopher of *The Assayer* was in the 1620s. But the sentence of 1633 forced the Society to align itself officially with that ecclesiastical decision, withdrawing its defenses to the anti-Copernican line.

The Society of Jesus' apologetic and scientific strategy against their great adversary substantially followed three directions. First, they were to discredit the enormous scientific authority acquired by Galileo in astronomy by extracting and exploiting every possible advantage from the condemnation of his Copernican convictions. Here they must again emphasize the theological falsity of Copernicanism with respect to the Scriptures. This was thankless work, certainly not gratifying and to be left to the professional polemicists and ultraconservatives in the Society, such as Father Melchior Inchofer and Father Grandami. But it was also necessary to demonstrate rationally, with the arguments of mathematics and experience, the falsity of the *Dialogue*; that is, to confute the astronomical, mechanical, and cinematic concepts Galileo had used as illustrations.

This line of controversy of post-Galilean Jesuitical science is the one best known to historians. From the studies of Alexander Koyré and the clarifications of Paolo Galluzzi, we know of Father Riccioli's efforts to offer physical proof of terrestrial immobility with the real experience of weights falling from the Tower of the Asinelli in Bologna. Riccioli was actually inspired by the paradox, presented by Galileo in the *Dialogue*, of the semicircular trajectory of a heavy body dropped from the top of a tower. To prove the falsity of Copernican theory, Father Riccioli tried effectively to derive from that theory consequences incompatible with experimental observations.[46]

for greater autonomy in research were inspired by a desire for greater apologetic effectiveness vis-à-vis the outside world and against the "perverse minds of heretics." H. Fabri, *Metaphysica demonstrativa*, Lugduni 1648, p. 579. See Father Sforza Pallavicino, *Vindicationes Societatis Iesu*, Romae 1649. On the Jesuit re-adoption of Galileo's mechanics in at least one case (the theory of machines), on the part of Father Casati, see U. Baldini, "L'attività scientifica nel primo Settecento," in *Storia d'Italia*. Einaudi Annali 3, Turin 1980, pp. 469-545.

[46] See A. Koyré, *Chute des corps et mouvement de la terre de Kepler à Newton*, trans. J. Tallec, Paris 1973; P. Galluzzi, "Galileo contro Copernico," *Annali dell'Istituto e Museo di Storia della Scienza di Firenze* 2 (1977), pp. 87-148.

The technical difficulties of the experiment, in any event, got in the way of the credibility of a proof which would instead demonstrate physically the earth's movement (and which only much later would be realized). That logical incompatibility between terrestrial motion and the Galilean law of falling bodies was certainly the weightiest of the arguments brought forward to confute the Copernican system, together with the accompanying reproduction of the official sentence against Galileo in Father Riccioli's famous book, the *Almagestum novum* (published at Bologna in 1651 and then republished in 1665, provoking lively protests on the part of Galileo's supporters and Florentine institutions).

Among other fathers involved in the scientific confutation of the *Dialogue* was Father Scheiner, an old Galilean adversary, with his *Prodromus pro sole mobili et terra stabili* of 1651—while the task of scientifically criticizing Galileo's mechanics was assumed by Father Pierre de Cazre, in his *Physica demonstrativa . . . adversus nuper excogitatur a Galileo . . . de eodem motu pseudoscientiam*, published at Paris in 1645; Father Niccolò Zucchi, with his *Nova de machinis philosophia*; and, at Flèche, Father Etienne Noël in his *Aphorismi physici* of 1646. This list, however, is neither exhaustive nor schematic.

Also very well known to historians is the second line of Jesuit scientific controversy in conformity with the official dispositions emanating from Piazza del Gesú; that is, the opposition to the geometry of indivisibles, the natural ally of atomism, which was taken on by Father Paul Guldin[47] (a friend of Father Orazio Grassi), by Mario Bettini,[48] and by André Tacquet,[49] author of the very famous Jesuit manual of mathematics. This is not to mention the constant torrent of polemics in defense of the traditional Aristotelian rejection of the mathematical infinite, found throughout the century's Jesuit scientific literature.

The third line of dissuasion against the new philosophy has until now been less apparent. It is also the most delicate because it could not avail itself of the advantage of an official condemnation by the Church, or of a great tradition of philosophical authority, but only of recent fundamentals of dogmatic theology applied to natural philosophy. In France, as in the Holy Roman Empire and Italy, the weapon of Eucharistic theology combatted the new philosophy. But if the new Cartesian physics at least maintained the old Aristotelian idea of fullness, it had to be impugned again when Italian Galilean studies introduced, in French circles hostile to the Jesuits, the idea of

[47] P. Guldin, *De centro gravitatio trium specierum quantitatum continuae libri quatuor*, Viennae 1635-1641. But, vice versa, Father Fabri, a more independent Jesuit scientist and one tied to the Florentine Accademia del Cimento, could take positions favorable to the mathematics of indivisibles, at least under the pseudonym of Antinus Farbius. See, on Fabri's mathematical work, E. A. Fellmann, "Die mathematischen Werke von H. Fabri," *Physis* 1-2 (1959), pp. 6-25 and 69-102.

[48] M. Bettini, *Apiariorum philosophia mathematica*, Bononiae 1656, p. 62.

[49] A. Tacquet, *Cylindricorum et annulorum libri IV*, Antverpiae 1651.

the void in Galilean physics. When the *Discourses* arrived in Rome at the end of 1638, the void was one of the most advanced theoretical frontiers left by Galileo to his Roman students in Father Castelli's school.

Obviously, after what had happened, Galileo's students could no longer speak publicly of the structure of matter, of material or mathematical atoms, or of natural philosophy, of "color, smell, and taste." The incrimination of Galileo's natural philosophy obliged them to imitate the dissimulation of their teacher after the official condemnation: the art of intellectual prudence. Publicly they could engage only in mathematics and experiments.

I have in mind Father Maignan's Roman experiments at the convent of the Minims on the Pincio; those of Gasparo Berti, Father Castelli's successor at the Sapienza; those of Raffaello Magiotti and, later, Evangelista Torricelli, also one of Castelli's pupils and subsequently Galileo's successor to the post of mathematician in the Grand Duchy of Tuscany (mathematician only, no longer philosopher, in light of the misfortune which that qualification brought on Galileo), who developed the Heronian concepts of the void presented in the *Discourses*. In 1644, in two letters to Ricci, obviously never published—though their contents were spread all over scientific Europe thanks to Father Mersenne—Torricelli announced that he had happily realized, by means of a mercury-filled tube in which a partial vacuum had been created, the astounding experience of atmospheric heaviness.[50]

The hypothesis of the air's weight, put forward by the Genoan scientist Giovanni Battista Baliani in 1630 and elaborated experimentally by Galileo thirty years before, was now an experimental reality that could benefit from a new theory. It was a decisive experience for Galilean physics. Galileo had also evoked the idea of an internal "vacuum force" in the *Discourses*. Now one could speak of external pressure. Even if it was no longer permissible to speak of atoms, the idea of a vacuum, indissolubly linked to that of atmospheric heaviness, was after all a way of being closer to Democritus than to Aristotle.

That vacuum was a new, frightening breach through which the heresy of Galilean physics could again pour in like a flood. Jesuit science, already tested by such a threat and for many years already on the alert, ran for cover. The theoretical effects of that experience were in fact fundamental—for physics and for Galileo's Archimedean conception—centered as they were on weight as the characteristic parameter of physics.

It was the great scientific discovery of the century inasmuch as it permitted the generalization of heaviness as a universal and characteristic factor in the

[50] See C. de Waard, *L'expérience barométrique*, Thouars 1936, for a reconstruction of the milieu and the roles of the various actors (Berti, Magiotti, Torricelli, Mersenne, Roberval, Pascal, Magni, Zucchi, Kircher) and for printed documents by Zucchi and Magni reproduced in the appendix. (See also "Carteggio 1642-48," in *Opere dei discepoli di Galileo*, eds. P. Galluzzi and M. Torrini, vol. I, Florence 1975.)

study of problems in mechanics and hydraulics. The barometric tube was for the birth of modern physics what the optical tube had been for astronomy, as Vincenzo Antinori will say many years hence in his *Notizie istoriche relative all'Accademia del Cimento* (1841).

Now the scientific revolution was taking place on earth. Did its importance for Galileo's physics prevent the author of this experiment and other Florentine Galileans from commenting on it and participating in the great discussion that it, too, will provoke?

By abstaining from that debate, Torricelli and the Galilean physicists gave, in my opinion, further proof of the great intellectual strength of their time: the art of prudence, of honest dissimulation. "When free living has already been corrupted, free speech must at times be shackled. Whoever does not do this in due time will accelerate, not prevent, tyranny,"[51] said Count Virgilio Malvezzi, a direct witness to the incrimination of Galilean physics in Rome, at that moment. It was a very valuable lesson.

Torricelli's experience was as important as it was technically and theoretically delicate. It was a matter of obtaining a vacuum in a mercury-filled tube; that is, achieving something which until that moment was only a theoretical hypothesis, something that one could not see (like atoms) and that could only be inferred theoretically from the concept of the air's heaviness. But is it really the invisible vacuum that is created in the tube in a basin full of mercury, or is it one of those usual Galilean tricks which serious experimental inquiry could expose, by demonstrating, for example, that vapors of mercury remained in that tube? In any case, everyone tried to repeat Torricelli's experiment before accepting the proposed results, but it was anything but easy. At Genoa, Baliani, pioneer of the hypothesis of the air's weight in the theory of the void, absolutely could not repeat the experiment, and in 1647 he was still impatiently trying to carry it out. However, Father Mersenne had repeated it with complete success the previous spring.[52]

In 1646 the experiment had also been carried out independently at the Court of Warsaw by Father Magni who, like the theologian and philosopher he was, generalized its meaning as a great proof against Aristotelian thought. That striking vindication, of which we shall soon speak, immediately provoked reactions: Father Zucchi from the Collegio Romano and Roberval from the Collège de France both reacted.[53]

It is in France, as is well known, that Torricelli's experiment underwent a series of sensational repetitions and was commented upon theoretically in

[51] See V. Malvezzi, "Pensieri," in *Politici e moralisti del Seicento*, eds. B. Croce and S. Caramella, Bari 1930, pp. 265-83.

[52] See Baliani's letter to Mersenne, July 14, 1647, in Moscovici, *L'expérience du mouvement*, p. 125.

[53] V. Magni, *Demonstratio ocularis loci sine locato, corporis successive moti in vacuo luminis nulli corpori inhaerentis*, Varsaviae 1647 (2nd ed., Bononiae 1648).

the light of its immense benefit for the new, ascendant philosophy. Cartesians, Pascalians, Gassendists, and Aristotelians compare results among themselves and try to develop experimental control of Torricelli's tube by a revealing gamut of empirical tests. A visceral experimenter, Father Mersenne enthusiastically directed operations.

Still in France, the most successful in repeating Torricelli's experiment are the Jesuits of the College of Lyon, under the guidance of Father Honoré Fabri, who has made Lyon into a center of scientific study which, in regard to physics, has no reason to envy the Collegio Romano and is so advanced as to appear heterodox in more conservative settings. Indeed, Father Fabri publicly asserts that he has already carried out the experiment as early as 1642 and therefore is particularly reliable when he denies that the vacuum's existence is involved.[54]

The claim to first discovery was only the initial step in the great controversy, engaged in by the Jesuits from the 1640s on, concerning the important subject of the void. This was a great battle, into which anti-Galilean strategy threw the best front-line troops of the Society's official science. In fact, the Jesuits will throw into that battle all the resources of their research, all the prestige of their scientific institutions, and—in the end, in a desperate attempt to rescue the results of a compromised conflict—the best arguments of their intellectual faith, as well.

Indeed, great problems were at stake. But, as always, the real, profound arguments were not brought forth immediately, officially, and publicly, but were left in the darkness of religious prudence, behind a veil of reticence and caution. Without this realization, it is impossible to understand anything about the scientific battle, and one is compelled to repeat the inveterate reproach of obtuseness, fanaticism, and illogicality brought against the polemics of Galileo's enemies.

Father Fabri, claiming as his own the experiment with the tube of mercury, rejected the idea that it could prove the existence of the vacuum. To our modern eyes, his refutation reveals a total incomprehension of the subject of atmospheric pressure. It was 1648 when Father Fabri's book, *Metaphysica demonstrativa*, was published under the name of Pierre Mousnier. A year later, the Collegio Romano gave new life to the polemic with the intervention of Father Zucchi,[55] Father Kircher, and Father Leone Santi. In his first letter against the vacuum, Father Zucchi emphasized again that the vacuum of the barometric tube is only apparent, and that the experiments confirm the idea that what one is dealing with are the dilated vapors of mercury.

We have, then, an identity of views between Lyon and Rome, as well as

[54] Fabri, *Metaphysica demonstrativa*, p. 570.

[55] Anonymous (Father N. Zucchi), *Magno amico* [Father Grandami, La Flèche] *nonnemo ex Collegio romano S.I. experimenta vulgata non plenum sed vacuum et antiperistasim stabilire*, Romae 1648 (superiorum permissu).

an identity of obfuscation and misunderstanding in regard to the capital argument of the innovators: atmospheric pressure. From the very start, the Jesuits had invested all their energies in an effort to reduce the theoretical significance of the Torricelli experiment to the category of phenomena normally explained by Aristotelian categories. This harked back to the clearest and most important of those motivations which inspired their studies of experimental philosophy.

These preoccupations were the same ones advanced by Father Grassi against *The Assayer*'s physics. Father Honoré Fabri, in the above-mentioned book of 1648, had immediately pointed out the necessity of maintaining the hylomorphic perspective because of the theological necessity of safeguarding the Scholastic interpretation of the Eucharistic dogma "from which sacrament one must establish necessarily that an accident is distinct from every possible type of substance."[56] An identical appeal to the dogma was made on the part of Father Cabeo.[57]

The same motivating preoccupation is also found in Rome, where in 1649 Father Zucchi, in his *Nova de machinis philosophia*, subscribed to the great apologetic reasons for that battle in experimental philosophy. It was an operation of containment in order that, now as in Galileo's time, "out of a love of novelty, while some attack with hostile spirit the philosophy accredited by time, others shall not succeed in tearing down what is most important."[58] It was then explained that what was most important was the hylomorphic theory, and that the serious matter was that attempt of the new physics' "insidious opinions" to affirm that sensible phenomena, or accidents, "can naturally present themselves without a [substantial] subject."[59]

Therefore, the abhorrence of the vacuum on the part of the most educated and representative scientists of the Society of Jesus should not be mistaken, by our modern epistemological eyes, for a grotesque comedy of errors produced by scientific obscurantism. It is instead a far-sighted decision which looked beyond experimental philosophy all the way to the final outcome of the conflict between hylomorphism and the science that we today call modern.

Return to the Stage

In fact, Father Grassi stood behind the Jesuit-conducted polemic on the vacuum. Indeed, to be precise, Father Grassi stood in front of it, in the very front ranks.

The strategic center of the Jesuits' scientific Lyon-Rome axis was actually found at a halfway point: in Genoa, where the most important roles of a per-

[56] Fabri, *Metaphysica demonstrativa*, prop. 52, p. 184 ff.

[57] See N. Cabeo, *Aristotelis Meteorologicorum*, tome I, Romae 1646, q. III, p. 112 ff.

[58] N. Zucchi, *Nova de machinis philosophia*, Romae 1649, p. 105.

[59] Ibid., p. 144.

sonal nature were being played. In Genoa, in fact, was the illustrious and unrecognized precursor of the vacuum theory, Senator Baliani. But above all in Genoa was Galileo's great official adversary, Father Grassi. Grassi continued not to publish even a single page, but the exile to his native Savona had ended. In the year following the death of Urban VIII, he is back, an unforgotten protagonist on the scientific stage, having been appointed rector of the College of Genoa.

Father Grassi cannot teach; he cannot repeat Torricelli's experiment. His scientific career had been shattered by the Galileo affair, but an intense informal activity begins for him—he is the great puppetmaster in the Jesuit battle against the void.

There is no need to point out how closely connected Father Zucchi always was to the professor of mathematics at the Collegio Romano during the 1620s and to the architect of St. Ignatius. In any case, Father Grassi, in the spring of 1645, as soon as Pope Barberini died, was able to make an authoritative reappearance in Rome, where the Collegio Romano, as soon as possible, had called him for urgent reasons. The construction of St. Ignatius left much to be desired, for the substitute called in to execute Father Grassi's project had not been equal to the task. Its author had a lot to complain about: too high a façade, its design altered, the roof's cornice incorrect. Father Grassi was authorized to present his objections officially in a detailed report in June 1645.[60] A commission of architects purposely gathered to deal with that series of regrettable defects agreed with him completely. One can well imagine that Father Zucchi and Father Grassi, at the Collegio Romano, also discussed physics during the eminent mathematician's sojourn in Rome. As for Father Fabri's Lyonese group, we may recall the visit to Lyon of the Collegio Romano mathematician just before the *Ratio*'s publication in France. Now, in 1646, Father Fabri paid a visit to Genoa.[61]

In reality, Father Grassi had all the qualifications and knowledge to pull everything together and carry out the most delicate task. That is, he could play on the prudent deference of the Genoan Senator Baliani for the Jesuits, his embarrassment at being unable to repeat Torricelli's experiment, and (not least) the vanity of a precursor overshadowed by Galileo's followers in order to win over this authoritative innovator to the new anti-Galilean cause against the vacuum.

Thirteen years of absence from scientific polemics, of silence, and of tasks of a purely religious nature had not changed the man. Father Grassi remained the same. Having returned to center stage, he played the part of which he had been master. In 1647, with his usual urbanity and effrontery,

[60] See Bricarelli, "Il padre Orazio Grassi," doc. 11, p. 22.

[61] See Costantini, *Baliani e i gesuiti*, p. 78 ff. Father Fabri stayed in Genoa in 1646, during the trip that took him from the College de La Trinité, Lyon, to the Apostolic Penitentiary of St. Peter's, in Rome.

which we know of him since Mario Guiducci's Roman summer, Father Grassi—through a scientific correspondence which has come down to us with the Baliani bequest to the Brera Library in Milan—made his own the experiments of Father Fabri and Father Zucchi, trying to insinuate himself into Senator Baliani's affections and enroll him in the ranks of the vacuum's adversaries (Fig. 8).[62]

This was asking too much, and the Genoan operation was unsuccessful; for Senator Baliani, following Father Mersenne's instructions, repeated Torricelli's experiment in November 1647 and immediately sent to Father Mersenne, in Paris, an open letter of thanks and unequivocal adherence to Parisian theoretical positions.[63] The month before, again in Paris, there had been published Pascal's *Expériences nouvelles touchant le vide*. The voice of the new philosophy was taking on a higher tone.

Father Grassi continued to bring pressure on Baliani, because in Lyon and Rome there arose new confutations, counterexperiments, and hypotheses to rescue the phenomenon of the barometric tube. It was all in vain. Baliani did not bow to those convincing arguments. On September 1, 1648, Father Mersenne's great voice was extinguished, but a few days later Florin Périer, on the Puy de Dôme, successfully performed the experiment imagined by Pascal, a crucial experiment.

On October 1, Pascal galvanized Parisian innovators with his *Récit de la grande expérience de l'équilibre des liqueurs*. For the Jesuits, mobilized against the vacuum, it was a very serious tactical defeat. The Collegio Romano's scientific prestige was being ridiculed as at the time of *The Assayer*, with all that such derision implied. All the more since, as always, the Jesuits were the only ones to expose themselves and to fight, amid the incapacity of the other orders, which as usual were tolerant (indeed, they permitted others to seize every delicate occasion to line up with the innovators and challenge the Collegio Romano's authority and authoritativeness).

As we have said, the last "plague spreader" was dressed as a monk. He was the spreader of the innovator's heresy of natural philosophy in Italy and, to tell the truth, he was an old friend of Galilean circles. This was the "tall monk," as Pascal called him; that is, Father Valeriano, the gaunt Capuchin theologian and philosopher at the Polish court of Ladislao IV.

Galilean Metaphysics: The "Tall Monk"

At the end of the 1640s, Father Valeriano Magni was one of the last direct witnesses of the moment and climate of the "marvelous conjuncture" expe-

[62] "I know, then, that [my thought] does not displease them" and that "he walks with me on the same road," Father Grassi insinuated in a letter to Baliani on April 17, 1648; see Moscovici, *L'expérience du mouvement*, p. 236.

[63] Baliani's letter to Mersenne, November 25, 1647, ibid., p. 126.

rienced by the Galileans in Rome. Now, after the deaths of Monsignor Ciampoli and Don Balli, he was the last to give voice to the echo of Augustinian spiritualism in that "new philosophy." He was against Aristotelianism, a pagan philosophy that could not cope with the study of nature. He infused his exhortations with the Bonaventurian, mystical inspiration and the Franciscan intellectual tradition that made him famous in the controversies against Protestants and Jesuits alike.

Claiming Torricelli's experiment so much as to publish it at first as his own, Father Magni had the theoretical audacity, thanks to the protection of the Warsaw court, to apply to it his philosophic-theological speculations on light and then publish those speculations in Italy. It should be said that thirty years after the memorable Roman experience of light from the Bologna rock, demonstrated by Galileo, Father Magni had published in Rome (in 1642) a book of Neo-Platonic philosophy pervaded with mystical accents: *De luce mentium et eius imagine*.

This was the moment in which the epistolary polemic between Galileo and Liceti concerning the moon's secondary light had again made the problem of cold light—separated from the environment by means of the luminescent rock exhibited so many years before by Galileo—the order of the day. Also, Father Magni included himself among the "many philosophers of nature who have striven to elucidate the occult light gleaming in the darkness."[64] But his was a mystical solution, based on a general principle of consciousness through the divine illumination of nature as of God. It was a "mystical theology," which invested the Christian "new philosophy" with "innumerable atoms," "luminous species," "vivifying light," and sizes and shapes—that is to say, with the stages of rational intelligibility of the world along an itinerary which, through experiments and demonstrations, passed from known things "all the way to those occult, more occult, and most occult."[65] It was a mystical itinerary of the mind, all the way to God, the efficient cause of the world.

But in 1648, when the experiment of the vacuum seemed to offer him a solid point of departure for abandoning the Aristotelian philosophy of light, Father Valeriano's adventurous metaphysical journey was halted by the first of "those occult things" which opposed corpuscular philosophy: the dogma

[64] V. Magni, *De luce mentium et eius imagine*, Romae 1642; reissued in 1646 in Vienna, under the title *De luce . . . ex SS. patribus Augustino et Bonaventura*.

[65] See *Opus philosophicum*, Lytonysl 1660, p. 7. On Father Magni's works, see *Lexicum Capuccinum*, Romae 1951, cols. 1776-1777. See in V. Magni, *Principia et specimen philosophiae*, Coloniae Agrippinae 1652, pp. 69-72, the defense of the substantialist physics of light against the denunciation of Eucharistic heresy. On Father Magni's original role in the history of the barometric experiment (1644-1646), in relation to Gasparo Berti's Roman experiments and Torricelli's Florentine and Parisian ones, see C. de Waard, *L'expérience barométrique*, pp. 123-28. At the end of 1648 the election of the new, pro-Jesuit Polish King will lead the "tall monk" to leave Warsaw for Vienna.

of the Eucharist. It will fall to Leibniz to resume that journey in the footsteps of the "tall monk's" works of metaphysics and logic.

In 1648, in fact, Father Magni had republished in Bologna his irreverent *Demonstratio ocularis loci sine locato corporis successive moti in vacuo luminis nulli corpori inhaerentis*, first issued the year before, in which the attempt was made not only to insinuate the existence of the vacuum in the tube of mercury, but also to demonstrate that that vacuum subverted Aristotelian hylomorphism and the Scholastic qualities.

Father Magni observed, however, that the empty barometric tube let light through: it was empty, but it contained light. Light, then, truly revealed itself capable of subsisting also in the absence of any substantial support, since there was no longer any transparent air in the tube capable of being illuminated. This observation was even more exciting than the idea of the vacuum. It confirmed the suggestive intuition both of the Mosaic physics of light created before the sun and of an all-pervasive element. It recalled the words written by Galileo to Monsignor Pietro Dini such a long time before and, more recently, those of Don Giuseppe Balli on the mystical and intellectual significance of a new philosophical conception.

Father Magni's fundamental point in the *Demonstratio* was precisely this: the experience of the vacuum permits us to see the light in the empty tube "without it being inherent in any body."[66] No matter how one looked at that circumstance, at that separation of light from its substantial environment, one inevitably reached consequences "very contrary to the Peripatetics."

In fact, if light is produced from nothing, then there is no longer any point in thinking of a raw material that sustains accidental phenomena. If one wishes to say, in the Aristotelian manner, that this is a question of a substantial form, then light would have to be produced by nothing. If light persists without being united to any substance, it is not an accident.

Light, in the last analysis, could only be a substance in itself. And yet, Father Magni also knew in advance what a delicate problem he would run into by thinking that light was substance: "Some people oppose to these consequences a proposition that is almost a matter of faith." The moving sincerity of the Bolognese Capuchin was as limpid as his convictions, but he at least had the good sense to use a minimum of dissimulation, not saying what it was imprudent even to pronounce. But, reasoning about that light in the void, about accidents and substance, Father Magni, in all Franciscan humility, remarked: "I say that Christian faith does not entail that assertion."[67]

Points of view. But if even a monk could not permit himself to give lessons in theology on the basis of the atomist slogan, "Nothing is born from nothing," it really meant that the Jesuits had lost the battle.

[66] Magni, *Demonstratio ocularis*, p. 9.

[67] Ibid., p. 14 ff.

Immediately in Bologna there was an orthodox reply, in the form of a pamphlet by the philosopher Giovanni Fantuzzi against the raising of that Lucretian and Gassendist slogan.[68] Who told the monk that there was a vacuum in that Torricellian tube?

Exactly, but only the Jesuit scientists were left to say that there was no vacuum, and by now Father Magni's philosophical initiative had again exposed the scandal. It was no longer a question of the vacuum, but of everything else: Aristotle, the Catholic faith.

A radical change of tactics was demanded. It was no longer possible to play with Torricelli's barometric tube, interpreting it in the light of Aristotelian categories. The conventional weapons of experimental polemics were by now useless; the strategic weapons of dissuasion must be wielded. Sarsi would have known how to reply. But now Father Grassi was not publishing. Someone else, however, did so for him.

"There are people who vigorously urge me to write *a more adequate and distinguished* reply than that of Father Zucchi; especially since *there is need of some other doctrine*."[69] Thus wrote, in July of 1648, the thirty-year-old Jesuit Father Paolo Casati to the Aristotelian Giannantonio Rocca, from Parma. Father Casati had been graduated a few years before from the Collegio Romano, and he taught philosophy and theology at the College of Parma.

In the large and unexplored Casati bequest to the Palatina Library, in Parma, I found traces of the official relationship between Father Casati and Father Orazio Grassi, but no private letters between the young theologian and the old teacher. It is nonetheless a legitimate hypothesis to suppose that it was Father Grassi who authoritatively directed Father Casati in his work on the (for him) new theme with which he had been insistently urged to deal in terms of a "new doctrine." As a matter of fact, so that this book of controversy written in great haste might be published quickly, Father Casati went to Genoa, where the book with the title *Vacuum proscriptum* came out in 1649. And, despite the inexperience which made the official author hesitate, that book seemed to come from the pen of a scientist perfectly up-to-date on all aspects of the polemic. As for the needed "new doctrine," it had certainly, without the shadow of a doubt, come from Father Grassi's pen, since it presented Sarsi's arguments from the *Ratio* of 1626.

The *Vacuum proscriptum* was an informed, critical book—aware of the futility of the reticences and counterhypotheses used until that moment by the Jesuits against the theory of the vacuum. More than rejecting the vacuum thesis, it set itself a more advanced target: namely, to avoid the conclusions of natural philosophy connected with it which Father Magni's writings had now made evident.

[68] G. Fantuzzi, *Eversio demonstrationis ocularis*, Bononiae 1648.

[69] Father Paolo Casati's letter to G. A. Rocca, July 20, 1648, in "Lettere a . . . Rocca," *Continuazione del Nuovo Giornale de' Letterati d'Italia* 35 (1786), p. 9 ff.

The book signed by Father Casati again stressed the Aristotelian thesis that light is an accident necessarily inherent in a subject. Otherwise, the miracle of Eucharistic accidents subsisting without their primitive substance would no longer be the miracle defined by the Tridentine dogma. *Vacuum proscriptum* therefore took up again the grand line of the Eucharistic objection in its original version, the one spread officially by Father Grassi in a suggestively rhetorical form. "What miracle," the Genoan book again recited, "would there ever be in the separation. . . . What miracle if the Priest who recites the holy formula finds the accidents of the bread already separated from the body through the work of nature? Perhaps that miracle which the Council of Trent has called transubstantiation?"[70] Here one senses quite clearly Sarsi's hand.

This criticism, more than that of the bold Father Magni, was a statement against the Galileans and a renewed denunciation of Pascal's supporters. To everyone was directed the warning that had already been sounded against Galileo: "once the ineluctable bond between substance and accidents is dissolved, who would be prevented, if an accident were not in some way sustained by some body [as Father Magni affirms in regard to the light which pervades a tube of empty glass], from entering the controversy and saying also that the whiteness, the taste, and the other accidents of the bread persist separate from their body by natural necessity."[71]

Stressed here was an ineluctable bond that connected the value of a miracle, the dictate of a dogma, and the physics of Aristotelian qualities. In the same year, 1649, Father Pallavicino, by now in full orthodoxy, also officially condemned (in his book *Vindicationes Societatis Iesu*) that theory of substantial accidents without an inherrent subject, listing it among those which the Jesuits must of one accord carefully protect against their detractors.[72]

Again in 1649, the same year in which from Genoa the incrimination of Galileo's natural philosophy was repeated, Father Compton Carleton denounced the Eucharistic heresy of the physics in Descartes' *Principles*. Descartes died on February 11, 1650. One year later, however, Pascal wrote the great preface to the *Treatise on the Vacuum*. The vindication of reason against the principle of authority which *The Assayer* had proclaimed now became a great religious hymn. Pascal's text was not published. In 1654, Otto von Guericke successfully performed new experiments with the vacuum; the polemics on the vacuum, light, and heat will also continue on the part of Father Zucchi, Fabri, and other Jesuits. We will not follow these polemics except for the one that relates to the reappearance of Galileo's incrimination—a fugitive but significant appearance, as if to celebrate the fiftieth anniversary of *The Assayer*.

[70] P. Casati, *Vacuum proscriptum*, Genuae 1649, p. 7. [71] Ibid., p. 5 ff.
[72] See Sforza Pallavicino, *Vindicationes Societatis Iesu*, p. 224.

First, however, we will follow the official return to Rome of the man who fifty years before, as official spokesman of the Collegio Romano, had incriminated *The Assayer*. In 1653, Father Grassi definitively returned to his Collegio Romano, to be reinstated with full honors in his old post as prefect in charge of the construction of the Collegio Romano's church. The most serious unauthorized changes having been corrected and the great church's completion entrusted again to its indefatigable author, construction had by now arrived at the arch of the great cupola, which was never built. The church was large and very beautiful. In 1650, on the occasion of the new Holy Year, St. Ignatius was opened to the public in the presence of Pope Innocent X.

Father Grassi at least had the satisfaction of attending the great festival of the Collegio Romano and the Society. It was August 7, 1650, the octave of St. Ignatius. Nothing could illustrate better than that church the fidelity of the men of St. Ignatius to the Church of Trent and of Rome. The pope certainly felt the universal emotion of that great work and that absolute dedication to the cause: he summoned the great architect of St. Ignatius and congratulated him.[73]

Father Grassi was almost seventy years old, but his health was still good despite the official justifications for his absence. He often traveled among Savona, Genoa, and Rome, and he continued to work as he did in the past. He had prepared a treatise on light and color, but even now he could not publish because, as he explained to Baliani in 1652, the "substance" of his new book indeed comprised the new ideas on the physics of light. Given the intrinsic ambiguity of Father Grassi's statements, it is not permissible to conjecture as to whether his book was in "substance" favorable to the new ideas of optical physics or opposed to them for the well-known reasons.[74] In the Society there was Father Francesco Maria Grimaldi, who presented many new ideas on the subject. But the Society had forbidden internal polemics on such controversial questions. Father Grassi, moreover, always obsequious, preferred to make an act of "Holy Obedience." When he resumed his post in the building of St. Ignatius, he also prepared a treatise on architecture. Perhaps, given the subject, he could have published that text without difficulties, but death prevented it. He had a heart attack in his Collegio Romano during the great heat of the Roman summer, on July 23, 1654.[75]

His was a discreet death—without pomp, in the shadows—befitting the sort of Jesuit life he had always lived. A succinct domestic obituary was, as was the custom, the only commemoration of Galileo's great adversary. All of

[73] On the inauguration of St. Ignatius, see the document published in Bricarelli, "Il padre Orazio Grassi."

[74] See Father Grassi's letter to Baliani, August 25, 1652, in Moscovici, *L'expérience du mouvement*, p. 250 ff.

[75] See Bricarelli, "Il padre Orazio Grassi," doc. 111, p. 24.

his unpublished works, manuscripts, and letters vanish into the darkness, so it would appear, the same darkness in which Father Grassi had written them.

Now this central figure in the Galilean drama had forever walked off-stage. To his Jesuit colleagues he left his church and his ideas concerning the theological incrimination of the new philosophy.

Water under the Arno's Bridges?

The old incrimination against Galileo's *The Assayer* was also dug up as a condemnation *ad memoriam*, whenever and wherever necessary: notably, in Florence. Galileo was one of that city's glories. Evangelista Torricelli, first Father Castelli's pupil at Rome and then Galileo's at Arcetri, had been appointed grand-ducal mathematician, as we have said, and this ensured Galileo's posthumous glory an effective institutional protection. In 1647, however, Torricelli died in Florence, and in Bologna Father Bonaventura Cavalieri also died. At Bologna in 1654, the Dozza publishing house reissued (apart from the *Dialogue*) Galileo's works, including *The Assayer*.

The second generation of Galileans—Vincenzo Viviani (Torricelli's successor in Florence besides being the chief figure in the Accademia del Cimento), Giovanni Alfonso Borelli at the University of Pisa, and the other Pisan Galileans such as Rinaldini and Oliva—benefited in the Medicean grand duchy from an enviable freedom in their personal and collaborative research. They could thank the political protection of the Florentine government which, at least until the death of Cardinal Leopoldo de' Medici in 1675, guaranteed a state of immunity to prudent Tuscan experimental research.

With respect to the official cultural institutions, however, there also existed in Galileo's homeland the traditional regulations safeguarding Aristotelianism in university teaching and the traditional theological preoccupations of Jesuit culture and education. Therefore, more than in other places, on the banks of the Arno reigned a situation of delicate compromises and power relationships between teaching and research, orthodox philosophy and latent memories of the new philosophy. It was not always an easygoing *gentleman's agreement* [English in original—TRANS.], and like all delicate equilibriums, this one too was precarious and could be maintained only so long as the grand duke was there to oversee it.

Also in Florence, as in Rome, the first signs of Galileo's condemnation had been greeted immediately by a salvo of Aristotelian criticism aimed at the atomism of Galileo's philosophy. It should be said right off, though, that these were routine academic skirmishes, bombastic and inoffensive; irrelevant university quarrels, at times even complaisant, and incapable of bringing greater problems into play. For example, Claude Bérigard, the brilliant

French Aristotelian at the University of Pisa, in his first reply to the *Dialogue*—the *Dubitationes in Dialogum Galilaei Galilae in Lincei*, published in 1632—had been among the first to mention the atomistic revival of Galileo's famous book. Bérigard could not help but connect the *Dialogue*'s atomist expressions with those of *The Assayer*, and to connect both to Lucretius's thought: the substance of Galileo's atoms and shapes is visibly "born from nothing and ends in nothing." Again in 1661, when Bérigard will have moved to the prestigious University of Padua, in the chair left free by Liceti, he will summon up again, along with a eulogy of the illustrious Florentine mathematician, his criticism of atomist physics (also apropos of the famous Bologna rock). But he will always resort to the academic courtesy of invoking only the classic Aristotelian arguments against Galileo's memory and have the adeptness of letting transpire an even too indulgent sympathy for corpuscularism.[76]

The new spokesman for Aristotelianism in Tuscany, Giovanni Nardi, a court physician, also respected the rules of this academic game among laymen. Giovanni Nardi was a very erudite Aristotelian and was very curious about the strangest physical-naturalistic oddities. In short, he was an academic on the order of Fortunio Liceti, with whom he engaged in memorable Aristotelian debates. Nardo wrote about almost everything, so great was his culture: he had published books on milk and its derivatives such as celestial manna, on dew, and on heat; then he had written a very famous and greatly esteemed book on subterranean fire, which must have impressed even Borelli.

One should not be too surprised, therefore, that in 1647 Giovanni Nardi published a commentary more cajoling than peremptory of the poem which had sung of Etna's eruptions; that is, *On the Nature of Things*. Lucretius was, as we know, an author condemned by the Church. The year before, however, at Pavia, there had been published a sort of biography of Democritus (Magnen's *Democritus reviviscens*, inspired by the new Gassendist fashion), where Lucretius's thought was attributed to Empedocles' fantasies more than to Democritan materialist atomism.

In any event, Nardi also stated in his critical comment on Lucretius that the erroneous poetic fantasies of Lucretian atomism deserved greater understanding than the presumptuous atomist opinions of modern innovators: "I prefer the talent of our Lucretius to that of the moderns."[77] For the latter, negators of Aristotelian qualities and those aspiring to explain "nature with mathematical demonstrations," the Florentine doctor reserved his con-

[76] C. Bérigard, *Dubitationes in Dialogum Galilaei Galilae in Lincei*, Florentiae 1632, p. 17; id. *Circulus Pisanus*, 2nd ed., Patavii 1661, pp. 419 and 428, on the corpuscular theories.

[77] G. Nardi (1585-1654), *T. Lucretii Cari, De rerum natura, Animadversio III De atomis*, p. 32; and *In Osores qualitatum*, p. 160 ff.

tempt, filling his commentary with *Animadversiones* and *Exercitationes* in the manner of the Jesuit professors at the Collegio of Florence, with whom he was "extremely friendly."[78] As a physician, Nardi criticized above all Sennert's atomist opinions.

As for Galileo, for obvious civic reasons (that is, for reasons of courtly decorum) Nardi never mentioned him explicitly. He confined himself to calling the author of *The Assayer* and the *Dialogue* a "standard-bearer" who had rejected "elements, qualities, matter, and generations of new things and openly denied them" and who thought that things "varied only according to their shapes, doubting the senses."[79] This could have been Galileo, or Democritus, or who knows how many others. Giovanni Nardi's Lucretian commentary is in any case a useful source for the history of the Tuscan university encounters between Aristotelians and Galileans in the 1640s, because in it are described the lively university disputes against the opinions of innovating students. Only in a book published outside Tuscany, the *Noctes geniales*, issued at Bologna in 1656, was Nardi more explicit: he recalled the apology for *The Assayer*, and its famous digression on the doctrine of motion as the cause of heat, which at the time "was admired by everyone."

In sum, *The Assayer* had never been forgot. Even though so much water had by now passed under the Arno's bridges, the serious question of Galilean atomism in physics—with all that it had given rise to—was like a posthumous menace constantly hanging over the Galileans' heads, like a sin gone unpunished.

There was, however, nothing to fear from those insinuations and routine polemics on the part of Aristotelians and lay university professors. None of them ever allowed himself to be carried away by the temptation, dangerous even for someone without qualifications, of bringing up the most serious argument, the argument *par excellence*, against the new philosophy and Galileo's memory. But the most attentive and concerned to keep the real terms of the problem from reappearing were obviously the Galileans, who certainly could not afford to repay Medicean protection by compromising it again in an ideological and religious scandal similar to the one that had compromised Galileo's protectors in Rome.

The Tuscan followers of Galileo could not be suspected of the "new philosophy." They, above all the direct pupils and witnesses of the years of the condemnation, had learned from their teacher to cultivate the art of prudence—even before mathematics and experimental research—dissimulating officially their opinions on the most burning problems of physics.

We said that Torricelli was Galileo's official successor, but only to the title of mathematician. The title of grand-ducal philosopher, of which Galileo

[78] G. Nardi, *De igne subterraneo*, Florentiae 1640, imprimatur of Father Antonelli, S.J.

[79] Ibid., n. 77, p. 160.

boasted and made use of officially, was now left opportunely in the dark. This was not modesty, it was necessity. We also said that Torricelli abstained from participating in the debate on the vacuum unleashed by his own brilliant experiments. This was not distraction, it was necessity: the necessity of not figuring among the new heretics on the blacklist of Jesuit Eucharistic apologetics. Experiment is the best teacher.

There were only the experiments which were being conducted by the Accademia del Cimento, officially protected by the Medicean grand duchy. And when, in 1666, it was decided to publish a careful selection of the best experiments performed at the Accademia del Cimento, its secretary, Filippo Magalotti, took care to state that officially the Academy's first "intention" was to abstain from "speculative matters."

There was never enough prudence. We already know of Vincenzo Viviani's reticence in telling us, while leaving in the dark, the reason why the polemic on comets had been at the origin of Galileo's downfall. Likewise Borelli, professor at Pisa since 1656, had to think in the same way as his Florentine rival—so much so that in 1664 he chose to dissimulate under a pseudonym, with a precaution biographers might consider excessive, a publication on comets that was inspired by *The Assayer* and that inevitably recalled the Galilean events of forty years before.[80]

And yet, dissimulated behind that veil of mathematical phenomenonism there still survived, invincible, the great Galilean suggestions on natural philosophy, the forbidden "speculations" on heat and light. But they were cultivated in private and hidden in public beneath the irreproachable mantle of legitimate experimental curiosity.

A curious, friendly visitor such as Balthasar de Monconys had not failed to notice Torricelli's persistent private cogitations on astronomy and physics, following the old programs of Galilean research: the speculations on the Bolognan rock and on natural generation were still alive. Francesco Redi, in his revelatory experiments with fly's eggs did nothing more than follow the famous atomist passage of the *Dialogue* on the generation of "flies" in must, but he nevertheless did not compromise the Florentine court by inconsiderately flaunting the new Galilean philosophy. Also Alfonso Borelli (who, having gone to Messina, will in 1670 write his report on the 1669 eruption of Etna, his *Historiam et metereologiam incendi Aetnei anni 1669*) correctly explained that the great thermal phenomenon was caused by the development of heat according to the doctrine of motion as the cause of heat in *The Assayer*. Though he spoke of the movement of the air in Etna's pores and entrails, he nevertheless did not make *The Assayer*'s atomism his own in

[80] P. M. Mutoli (G. A. Borelli), *Del movimento della cometa apparsa il mese di dicembre 1644*, letter to Father Stefano degli Angeli, Pisa 1665, p. 5, for the theory of the optical nature of the comet, as "Galileo very wisely pointed out" (p. 15).

Poisa. In fact, Galileo's incrimination had caused atomism to be abandoned, and so long as that condemnation was still nearby, prudence was maintained. With the passing of time, however, much water had passed, diluting that self-censorship. In truth, with time, Galileo's official condemnation had become ever more discredited, to the point that it was no longer illegitimate for Galileo's successors and followers in Tuscany to imagine the possibility of rehabilitation. It was perhaps because of that hope that in 1665—when Father Riccioli published *Astronomia reformata*, wherein he repeated the specious scientific legitimation of Galileo's official condemnation for Copernicanism—that an excess of zeal, instead of obtaining the intimidating effect already obtained and which it was still reasonable to expect, aroused the protest and open revolt of such Galileans as the Jesuit priest Stefano degli Angeli, pupil of Father Cavalieri and Alfonso Borelli. To bring up again, as Father Riccioli did, the official condemnation, repeating that campaign of pressure and dissuasion under cover of anti-Copernicanism, trying to discredit Galileo by setting him against Copernicus, was a prevarication whose belatedness by now made it inopportune and counterproductive. The protection and consent of Copernicanism by Tuscan civil authorities now authorized the Galileans to rebel. But even some scientific exponents of the Society of Jesus (Father Honoré Fabri, connected with the Accademia del Cimento, and Father André Tacquet) no longer found it in themselves to sanction that untenable *mise en scène* aimed at saving appearances.[81]

Perhaps the important polemic on the earth's motion made it possible to glimpse the future possibility that the condemnation mounted against Galileo would collapse like a house of cards. But the passage of time, instead of going in that direction, turned the clock back to the 1620s, all the way to the profound and dissimulated reasons for Galileo's incrimination.

In 1664 Descartes' *World*, laden with memories, was published posthumously. For some time now, the *Principles* had been circulating in Italy, too, and as Descartes' book had been on the Index since 1663, the Jesuits, who from the very first knew its dangerousness, were on the alert. And in fact, faced by the appearance of Cartesian corpuscular theories in Italy as well, spread through the chain of convents of Father Mersenne's order, the Holy Office sent out from Rome an informative note concerning the serious danger raised anew by Cartesian philosophy. This censure, which brought up to date the censure against Galileo, is documented by the summons to vigilance, repeatedly cited by historians, sent in 1671 to the Archbishop of Naples by Cardinal Carlo Barberini. He, we must add, knew better than anyone else the need for avoiding further scandals to religion because of

> the opinions of a certain Renato de Cartes, who in years past had pub-

[81] See Galluzzi, "Galileo contro Copernico."

> lished a new philosophical system revising the ancient opinions of the Greeks on atoms, on the basis of whose doctrine some theologians are trying to prove the manner in which the accidents of bread and wine persist after the Consecration has transformed the substance of said bread and wine into that of the body and blood of Our Lord Jesus Christ.[82]

But the illusion was nursed in Florence that the old denunciations had by now been proscribed and that, after eliminating the equivocal official condemnation, one could once more become a Galilean philosopher; and some pro-Galilean Tuscans plunged headfirst into relaunching atomist physics through an interlacing of Lucretian atomism, Christian Gassendism, and publicly displayed Galilean physics. More than by anything else, the delicate equilibrium maintained at the price of so much prudence was upset by the loud imprudence of a forty-year-old philosopher and theologian from Livorno, Borelli's pupil at Pisa who was protected by Francesco Redi and, through him, by Cardinal Leopoldo: namely, Donato Rossetti. Here was an innovator animated by Galilean philosophical pretentions and Pisan university frustrations, to which he gave vent as the protagonist of university polemics at the end of the 1660s.

It is known that Donato Rossetti, reader in logic, never got the philosophy chair at the University of Pisa from which he proposed to philosophize as "a Galilean and mathematical physicist."[83] This was owing to the Aristotelian university resistance to him and to another atomist of that time, Alessandro Marchetti.

Nonetheless, Marchetti finally got a chair in mathematics. But Rossetti did not; indeed, he will leave Tuscany after his protector's death. It should be added, however, that perhaps not even Donato Rossetti's teachers could have a clear conscience in defending the academic recognition of that restless Galilean who wanted to refashion Galileo's philosophy. "My mind is not completely at ease and serene," Borelli wrote to Alessandro Marchetti apropos of that new philosopher,

> because he is inexperienced in such speculations, and I greatly fear that he might incur what happened to me, who, having had time many, many years ago to think over the aforementioned experiences, have often found it necessary to change opinion. . . . I have already written my thoughts about this to him, but it will not reach him in time because

[82] See L. Amabile, *Il Santo Officio e l'Inquisizione in Napoli*, Città di Castello 1892, II, p. 53 ff.

[83] See P. Galluzzi, "Libertà scientifica, educazione e ragion di stato in una polemica universitaria pisana del 1670," in *SFI, Atti del XXIV Congresso Nazionale di Filosofia*, L'Aquila 1973, vol. II (Rome 1974), pp. 404-412. On Rossetti, see A. Fabroni, *Historiae Academiae Pisanae*, III, Pisiis 1785, p. 405. After a disciplinary procedure of the Office of Pisa, Rossetti will take refuge in Piedmont, where he will die in 1685.

he has been in too much of a hurry to do something, which I suspect he will repent.[84]

Borelli's concern was perfectly justified. In 1667, Rossetti published a pamphlet entitled *Antignome fisico-matematica*, which openly professed the atomism of *The Assayer* and, indeed, even without obviously quoting it explicitly, violated the self-censorship agreement which had been maintained until that moment regarding the book, quoting from it in a direct, transparent mention of the famous Democritan passage on mathematical characters in the book of nature. "Among our letters," Rossetti wrote, "some go with the others perfectly, such as the vowels . . . and such must be the case with Atoms, because if every one conformed with every other in the same way, new species would every day be seen in the World and those already seen would vanish."[85] As if that were not enough, Rossetti polemicized against Father Fabri in defense of the corpuscular emanation of light, "effluvium of corpuscles," and then, a year later, in polemic with a Bolognese Aristotelian, he referred to the atomism in Galileo's *Discourse on Floating Bodies*.[86] If at least he had been content to be a Gassendist, the matter would not have been so serious, but to dig up the Galileo affair so imprudently was an unjustifiable provocation at that moment.

I say this because even Donato Rossetti was aware of what had really happened (and was butting in with the most untimely imprudence) and because his reference to *The Assayer* had given rise to immediate censures, which we do not have but which Rossetti complained about. He challenged the censorship again in 1671, presenting all those delicate problems at the beginning of his novel on atomist physics entitled *Composizione e passioni dei vetri*, in which he studied the atomic composition of crystal.

Rossetti proposed an atomist "metaphysics" of nature based on the existence of atoms of light flung off by the sun, endowed with reciprocal appetence at a distance from a "sphere of energy." The luminous and the dark atoms, owing to the appetence of such energy, combined by means of their "poles" to form molecules. But Rossetti's physical system was much more ambitious: he announced a treatise on light, one on perpetual motion, a "physical-mathematical Corpus," and a great Gassendist metaphysics (the *Polista fedele*, a book of reconciliation between the new philosophy and Catholic theology, with digressions on the Resurrection, the universal flood, and dogma). All these ideas were crazily whirling about at the same time in the brain of this belated, fervent reader of *The Assayer*!

It would have been better had he followed Borelli's prudent advice instead of stirring up trouble, this embattled provincial plague spreader of the at-

[84] Fabroni, *Historiae Academiae Pisanae*, p. 406.

[85] D. Rossetti, *Antignome fisico-matematica*, Livorno 1667, p. 66.

[86] See D. Rossetti, *Insegnamenti fisico-matematici sopra la prostasi*, Livorno 1669.

omist heresy. The serious fact is that he had actually written his atomist theology, whose publication he announced, and by 1673 the manuscript was ready; we find it among the Redi papers in the Palatino bequest at the National Library in Florence.[87] Probably, however, Francesco Redi had enough good sense to dissuade him and persuade him not to publish anything, as shown by the correspondence between Rossetti and Redi during that period. But Rossetti was impatient to reply to Galileo's enemies; and, in the introduction to the 1671 book, he had described what was at stake.

At stake, the Aristotelians insisted, was the fact that without their physics "it is not possible to defend *any of the mysteries of the Faith*" and that consequently atomism, which was "wholly contrary to that principle and that metaphysics, must be abhorred and shunned by the faithful."[88] Rossetti complained because the students of the Jesuit college in Florence came to the University of Pisa convinced that the atomists were "Ethnics [pagans] and Publicans" and that a Florentine Jesuit had gone so far as to order one of his students, so that he should not fall into the sin of atomism, "to swallow every morning as a preventative, and before reciting Christian prayers, a certain ridiculous medicinal tirade of his against atoms."[89]

All these suspicions on the banks of the Arno concerned the usual philosophical problem: "that in the Sacrament of the Eucharist beneath the only sensible species of bread and wine is, with all its substance, the true and entire body of Christ." Donato Rossetti proclaimed that he was a Catholic—"a faithful polist"—that the Tuscan atomists "are not Democritans except in name," and that he would never have "preached atoms . . . if there were the danger with them of causing darkness, where there is the very clear light of the Faith."[90]

But by now it was too late to repent for having brought the old atomist scandal under the light of faith, compromising the position of advantage so patiently gained by the Galileans. When Cardinal Leopoldo died in 1675, matters ended badly; and what he had perhaps wished to be a rehabilitation of Galileo and a celebration of *The Assayer*'s fiftieth anniversary was transformed into a bitter defeat.

A New Document on Galileo, an Old Sin

We cannot know who the Jesuits were who administered the medicinal tirade against atomism to their Florentine students. Father Sebastiano Conti,

[87] D. Rossetti, "Due proposizioni di disinganno . . . ove vedesi come il Democritico possa convenire coll'Aristotelico nello spiegare il Sagramento dell'Eucaristia . . . all'ill.mo sig. Francesco Redi," in *Zibaldone di F. Redi*, ms. Palatino 1099, VII, cc. 41-48, National Library of Florence.

[88] D. Rossetti, *Composizione e passioni de' vetri*, Livorno 1671, preface.

[89] Ibid. [90] Ibid.

Father Vincenzo Glaria, and Father Giovanni Francesco Vanni were the three philosophically most combative in the Society's Tuscan colleges. The first was consultor of the Florentine Inquisition and was quite familiar with the background of the Galileo affair; and the third, professor of philosophy at the college in Florence, distinguished himself later as one of the most vehement continuators of the Jesuit battle against Galileo.

But perhaps our researches into the reactions against *The Assayer* have allowed us today—thanks to a new Galilean document discovered in the libraries of the antiquarian market and presented below in translation—to know what the medicine was; that is, the short morning prayer which a Jesuit student who went to Pisa had to recite so as not to fall into atomistic temptation. It is a Latin prayer to exorcise atoms, which cleverly repeated the Lucretian dictum that "nothing comes from nothing."

The antiatomist prayer can be found printed as a kind of epigraphic exhortation at the beginning of an *Exercitatio de formis substantialibus et de qualitatibus physicis* (see Documents), anonymous and without any indication of when and how it was printed. But its characteristic typographical ornament declares it to be an academic publication from the Florentine presses (probably those of the printer Ippolito Navesi; see Fig. 14) which appeared during the second half of the seventeenth century.

Traces of successive signatures on the frontispiece (a false frontispiece) of the discovered copy reveal that the anonymity is probably because of accidental circumstances of binding (or artificial circumstances of antiquarian greed). Also this document is very rare owing to the fact that, as such printings were very small in number, it was unusual for them to survive and show up in the miscellaneous sections of local collections.

At the National Library of Florence, however, among the miscellanea in the Magliabechiano bequest which suffered from the 1966 flood, I discovered another *Exercitatio*, a twin of the aforementioned, containing a polemic against vacuum and heat in the barometric vacuum.[91] This second *Exercitatio* permits us to think that the first must also have been accompanied by a list of academic *Conclusiones physicae*, almost all the same in this period, which served as a summary of the doctoral theses in the Florentine Jesuit College of St. Giovannino, on Via de' Gori.

The conclusions recapitulated the essential principles of hylomorphism and declared that atomist opinions ran counter to the Catholic faith. The *Exercitatio* which accompanied them was a work of controversy on a question, book, or author and was developed, if not actually written, by the professor who had directed the "defender" of the thesis. The name of the candidate

[91] See F. M. Abba de Orlandis, *Conclusiones physicae propugnandae in Collegio florentino S.I., addita est Exercitatio de transitu ignis per vitrum et de vacuo*, Navesi, Florentiae 1678 (Magliabechiano 1051.12, National Library of Florence).

figured on the actual frontispiece of the thesis, printed at the expense of the student's well-to-do family.

The similarity of the two *Exercitationes* thus permits us to say that they were prepared in the ambit of the Jesuits' Florentine college and published in Florence between 1677 and 1678. For the rest, these hypothetical elements are confirmed by information given by the publication soon after of a thesis whose writer, the student Lorenzo Gherardini, explained to the reader that he had put together the two *Exercitationes* cited above and previously published separately after having revised and corrected them. The *publisher's* warning to this second edition is very useful because it permits us to know the name of the student to whom our *Exercitatio* should be attributed, but, above all, the name of the Jesuit who had inspired and directed, if not actually written, the book.[92] The student was the Florentine nobleman Francesco Maria Arrighi, who was working for his degree in 1678, later chose an ecclesiastical career prestigiously concluded as the Bishop of Montepulciano, and was esteemed for his knowledge of philosophy.[93] The Jesuit inspirer was Father Giovanni Francesco Vanni.[94]

The historical interest of the original edition, until now unknown, is extremely pertinent to the history of this marginal Florentine episode in the great Eucharistic controversy over Galilean atomism. In fact, when that original edition appeared in Florence, it must have provoked many immediate official and unofficial protests. Indeed, the second edition tried to mitigate the text's harshness and respectfully cited the experimental researches of the Accademia del Cimento, whose members must have objected to that *Exercitatio*, which again brought up the inveterate accusations against *The Assayer*'s physics.

In point of fact, Arrighi, Abba de Orlandis—the other student, under whose name the second *Exercitatio* appeared—and Gherardini, who republished both of them (the first in a much revised and corrected form) were

[92] *Exercitationes physicae. I: De formis substantialibus et de qualitatibus physicis. II: De transitu ignis per vitrum et de vacuo. Altera editio aucta et recensita pro disputatione habenda in Collegio S.I., a Laurentio de Gherardini*, Gugliantini, Florentiae 1678. The exercises republished in this edition have been described in M. Torrini, *Dopo Galileo. Una polemica scientifica (1684-1711)*, Florence 1979, p. 20 ff., introduction to Father G. F. Vanni's polemics. The copy of this second edition preserved in the Vittorio Emanuele II National Library (Rome) is attached to Father Vanni's volume, anonymous, *Exegeses physico-mathematicae de motu graviorum*, Romae 1685 (National Library, Rome, 12.33.E.16). Another copy is in the National Library of Florence (Magliabechiano 1248.13): *Conclusiones physicae propugnandae a Laurentio de Gherardinis in Collegio florentino S.I.*, ivi. As for the Arrighi Codex, compared to the Gherardini Codex, it offers significant variants in the exordium, p. 20 ff. and p. 25. On Lorenzo Gherardini (1659-1714), see F. Inghirami, *Storia della Toscana*, vols. XII-XIV, *Biografia*, Fiesole 1843-1844, III, pp. 11 and 140.

[93] On F. M. Arrighi's personality, see Monsignor A. Baudrillart, *Dictionnaire d'histoire et de géographie ecclésiastique*, IV, Paris 1930, col. 723; or G. M. Mazzuchelli, *Gli scrittori d'Italia*, Brescia 1753-1763, II, p. 1127.

[94] On Father Vanni's scientific work, see Torrini, *Dopo Galileo*.

three students in Father Vanni's philosophy course at the college in Florence. They must have received their degrees during the same academic year of 1677/78, in the same program (the published *Conclusiones physicae*), and published a few months apart from each other first the two separate *Exercitationes* and then, after a few months—following on the immediately elicited reactions, and perhaps precisely because of them—a new edition which, as regards Galileo, was manifestly toned down in form.

Father Vanni, who was responsible for these publications which denounced both *The Assayer*'s atomism and that of the new Galileans, must have been reproached in the first instance for having attacked an adversary who could not defend himself. In the prefatory note to the aforesaid edition (Gerhardini) it seems that the Jesuit philosopher tried to justify himself and to make peace with a more balanced and correct critical evaluation. In reality, he summoned up with great dialectical ability (quoting a passage in *The Assayer* in which Galileo addressed the professors of the Collegio Romano) the fundamental "error" of Galilean philosophy perceived by Father Grassi in relation to the "very serious controversy":

> Nor do I think [Vanni wrote under his pupil's name] that Galileo would have suffered with greater aversion the criticisms of my teacher [Vanni], who is in the habit of using and praising many of his books and is concerned only to express the doubts which in a very serious controversy shook Galileo's spirit (he proposed nothing in regard to atoms, unless qualified with the benefit of doubt). On the contrary, page 19 of *The Assayer* seems to encourage the diffusion of these criticisms, where Galileo expresses with respect to his eulogists this opinion worthy of a philosopher: "I can thank them for their affectionate feelings; but I would have preferred that they had rid me of error and shown me the truth, because I esteem more highly the usefulness of true corrections than the pomp of vain ostentation."[95]

It must be recognized that, cunningly hidden behind these words of Father Vanni's, was a true story.

The *Exercitatio de formis substantialibus et qualitatibus physicis*—in the original edition under the name of Arrighi—which had provoked these reflections was not a chance text of controversy. The contents of the polemic launched by Father Grassi against *The Assayer* and circulated among the Jesuits for several decades were poured into the academic form of an exercise.

Exercises were much more lively academic rites than lessons. In them, as in a theological dispute, a supporter of a false or heterodox opinion, which was then rejected and shown to be false, was opposed by an opponent armed with Aristotelian principles.

[95] Gherardini, *Exercitationes physicae*: "ad lectorem."

The schema of our *Exercitatio* reproduced exactly this sequence. The opinion to be defeated was *The Assayer*'s atomism. In fact, the first part presented through long quotations, gathering them in seven sections, the opinions of *The Assayer* on shaped substance, on the geometric characters of the book of the universe, on the nominalist character of sensations, on the production of these last by means of local motions of substantial particles, on the production of heat by means of the movement of fiery corpuscles, on the communication of heat by the emission of these corpuscles, and on the atomic constitution of light. The initial sentence serves to sum up this entire first part: "Galileo was very inclined to think that all things were constituted by atoms diversely shaped and combined."

Now the opponent makes his entrance on the stage, and he declares at the start that all this theory has the stench of sulfur, that it "breathes [*spirat*] Democritus from every pore." This is a Scholastic text and therefore very well organized in its arguments. There are, in fact, three orders of argument which make it obligatory to reject atomism. The most important, the first, is of a theological nature, the second of a logical nature, and the third of a physical nature, as the Galilean opinion is false with respect to experiences.

The first argument (pp. 12-13) highlights the fact that atomism does not conform to the Catholic doctrine of the second canon of Session 13 of the Council of Trent, the dogma of the permanence of Eucharistic accidents. If the corporeal substance is composed of "sensible atoms" (that is, if the atoms of a substance produce sensible effects), then, since in the Eucharist the appearances of the bread and not those of the body of Christ are sensible, it must mean that in the Eucharist there are atoms of bread, the substance of bread—so, according to Galileo, there are atoms of fire—and not the substance of the body of Christ; and this idea is false.

It is important to note that at this point, going far beyond the very serious initial controversy over *The Assayer* on the Eucharistic problem, subsequent developments are taken into account, as well as the attempt—above all of Descartes and Balli—to resolve the paradox with the theory of intentional Eucharistic species: the theory of the miraculous "as if" it were bread, though it is not. The Jesuit author opposes this with a dialectical argument. If the accidents of the Eucharist could exist miraculously separate from a substance, then not even the substance of the unconsecrated bread could be sensible, like the body of Christ in the Consecrated Host. But in that case the theory of sensible atoms is false, destroyed by its own hands. In reality, the effort is to prove that those who "usually answer" that there is a compatibility between atomism and the dogma of the Eucharist are mistaken.

The dialectical argument is actually very good, very stringent, but to tell the truth it is directed more against the author of the letters to Mesland than against the author of *The Assayer*. You atomists, Father Vanni seems in fact to exclaim, you distinguish on one hand what is for you substance and on the other, appearances, in that case, we can do the same thing. Indeed, we are in a better position to do it, since according to us, substance and accidents are really distinct, and it is therefore legitimate to conceive that God, if He wishes, separates them miraculously.

It was Descartes, not Galileo, who had asked, "When a corporeal substance is changed into another, and all the accidents of the first remain, what is there that is changed?" Father Vanni's dialectic was a reply to the tactic of Cartesian Eucharistic theology.

The second argument, conducted in the "light of reason," also shows us that *The Assayer*, more than in the odor of Anaxagoras and Democritus as it was for Father Grassi, was now in the odor of Cartesianism, the new doctrine officially on the Index. If, in fact, one recognizes in man a rational, vegetative, and sensitive spirit, one cannot exclude the sensitive spirit in animals and the vegetative spirit in plants. The objection was honored by the Index, thanks to Descartes. However, we must not forget that in the *Dialogue* Galileo's atomism had reappeared precisely on the problem of natural generations and corruptions.

As to the third order of refutation, those of a physical nature, the reader found in them, literally copied, Father Orazio Grassi's arguments in the *Ratio* against *The Assayer*. This part might really have been written by the student Francesco Maria Arrighi, so faithfully did it reproduce the pages of that classic of Jesuit scientific literature, the manual of the anti-Galilean controversy.

The same should be said of the objection relating to the impossibility of applying atomist theory to natural generations, exemplified by vegetables; and the same for the communication of heat in the hand and between bodies. Naturally, since the *Exercitatio* is a school text and not a scientific book, all the arguments that were hinted at briefly with the swift strokes of Grassi's polemical pen are here pedantically detailed and organized with the skill of a professor of logic.

It is permissible to think that it was the second edition of Father Grassi's *Ratio*, in which the Aristotelian argument against the mathematical infinite had been omitted, that was used as a pattern by the author of the Florentine *Exercitatio*, since even here the problem of indivisibles does not appear.

Instead, the physical argument of the impenetrability of atoms is included, as it was in the *Ratio*'s second edition. As for criticism of the atomic constitution of light, the Florentines had slavishly copied word-for-word Father Grassi's fine reasoning on the paradox of the closed lantern that emanates heat, whereas for Galileo fiery atoms were less rapid and therefore less penetrating than luminous atoms. And with the rejection of the substantial and corpuscular theory of light, ideally connected with the initial objection to the Eucharist, this restatement of the falsity of *The Assayer*'s physics concluded.

As we see, the text must be confined to an opportune mention of Galileo's traditional incrimination at a moment when Galileo's re-evaluation in Tuscany was being considered. The progressive discrediting of the official reasons for Galileo's condemnation; the threatening collapse of that dike which until then had more or less been able to contain the spread of Galileo's philosophy in Italy; the resurgence of the specter of *The Assayer* among new generations of Galileans—these induced the Jesuits of Florence to discourage further scandals and dreams of rehabilitation, reviving fifty years later the still persistent theological preoccupations which Galileo's philosophy had aroused and which were still the order of the day.

But, precisely because it is not an original text but a text that has been rendered up-to-date, Vanni/Arrighi's *Exercitatio* is interesting. It reveals the spread and long duration of the directing ideas of anti-Galilean strategy, designed for the purpose by Father Grassi, propelled by the ground swell of a century-old controversy.

The effort to update should be fully recognized, since it is unquestionable that this *pamphlet* [English in original—TRANS.] reflected the successive revisions and developments of a very serious controversy, taking new points from the anti-Cartesian books of Father Compton Carleton and Father Raynaud. Also, the part on physics, although it is more of a compilation, must have taken into account the books by Father Zucchi, Father Fabri, and above all Father Casati, since knowledge of such works appears between the lines of the other *Exercitatio*, as well, destined to combat the theory of the vacuum and the transmission of heat in the barometric tube.

As always, the argument highlighted was once again the Eucharistic objection. In this case, too, its use against *The Assayer* develops the *Ratio*'s official incrimination, but allows us to think also that perhaps the anonymous denunciation presented to the Holy Office against Galileo's book must have circulated, or in any event become part of the patrimony of an oral tradition at the Collegio Romano. It was there that Father Vanni had learned philosophy and physics, had known Father Orazio Grassi personally, and from him and the old colleagues of the battle of the 1620s had drawn a lofty and authoritative inspiration for continuing the defense of the faith against Galilean philosophy.

Of course, between Father Grassi's original denunciation and his Florentine restatement of it, a half-century had passed; and this was quite clear. Authors and languages alike had changed. Father Vanni no longer had any need to resort to Father Suárez's manual, to make a meticulous exegesis of atomism, or even to appeal to the classic Scholastic theory of the separation between quantity and extension. Now Father Vanni spoke of extension, knowing very well what the atoms of the Gassendist and Cartesian innovators were. The controversy had by now developed so much and become so refined as to permit a synthetic and effective polemic against the well-known replies of the adversaries, while Father Grassi had to imagine what they might be. Besides, recent Tuscan polemics had taught Father Vanni that, among all the dangerous passages of *The Assayer*, one must insist on dealing with the book of nature.

Thus, the research that I had undertaken, starting with the problem of light in Galileo, came to a conclusion fifty years after *The Assayer* with that belated reminder in Florence as to the impossibility of a corpuscular theory of light such as that proposed by Galileo and then set aside for logical and theological reasons which were stronger than he, his logic, and his scientific and religious faith. I do not at all exclude the possibility that further interesting echoes of the original incrimination of *The Assayer*'s atomism may be found even later than the last events and books cited. One should also mention the editions of the *Magisterium naturae et opus* by the Jesuit Father Lana-Terzi and his perplexity concerning the legitimacy of disguising atomism as a mathematical theory so as to avoid the theoretical and theological problems of physical atomism.[96]

To present the European intellectual background, which continued to exert an influence for many years, it would be necessary to leaf through the pages of Pascal's *Letters* devoted to the epic Eucharistic controversies of Father Magni and Port Royal against the Jesuits and the pages of the *Journal de Trévoux*. We would have to summon up again the Eucharistic controversies among the Minims, Olivetans, Benedictines, Jesuits, and Dominicans. We would have to recall the Eucharistic denunciations of the *Recherche de la verité* by Father Malebranche, on the part of the Jesuit Le Valois, professor of philosophy at the college in Caen, and the great controversy that exploded when Malebranche opposed a theory of Eucharistic species which relied on Cartesian physiology.[97] Malebranche: the great Augustinian metaphysician, heir to the Oratorian spirituality of the beginning of the century, the result of the effort which lasted throughout that century to wed the new philosophy to a renewed Christianity. On November 21, 1689, Male-

[96] F. Lana-Terzi, *Magisterium naturae artis et opus*, Briziae 1784 (1st ed. 1670): "some recent philosophers have thought they could avoid difficulties by affirming that a quantum body (i.e. quantity) is composed of indivisible points that occupy a divisible space" (p. 27).

[97] See N. Malebranche, *Œuvres complètes*, ed. A. Robinet, Paris 1958-1967, I, p. 465, for the emphasis of the *Recherche de la verité* which set off the controversy. The controversy's dossier is in vol. XVII, from p. 445. Ibid., p. 498 for Malebranche's theory of the Eucharist. See H. Gouhier, "Philosophie chrétienne et théologie. À propos de la seconde polémique de Malebranche," *Revue philosophique* (1938), pp. 23-65.

branche's *Traité de la nature et de la grace* was put on the Index. The *Recherche de la verité* and the other metaphysical works of the pupil of Cardinal Bérulle and Descartes will meet the same fate.

One could thus continue the thread of this history; as regards Galileo, however, I decided to halt my research when I reached this point. At this point, the essential reasons for the great controversy were placated—and not because that theological crime had now fallen under proscription. On the contrary, it was always in the headlines. It suffices to leaf through the 1704 Index (the most up-to-date version, covering works up to 1681), wherein one finds in succession almost all our authors from Descartes to Magni, from Maignan to Don Chiavetta and Balli.

But at that moment it is official history which turns a new leaf. On October 18, 1685, in the Palace of Fontainebleau, amid the Raphaelesque colors of the paintings of Primaticcio and the by now autumnal hues of the great forest, Louis XIV revokes the Edict of Nantes (in fact, for years inoperative). The "general conversion" of French Protestants so brutally eliminates the crisis begun by the Reformation that the innovators are silenced, and the Jesuit institution can look upon the new philosophy with less anxiety.

Father Grassi's work, even though unfinished while the author was alive, was now crowned with triumph. At that same moment in 1685, the *Universal Triumph of the Society of Jesus*, frescoed by Brother Pozzo, marvelously crowned the Church of St. Ignatius. Here is why I could now halt my research at this at once historic and symbolic date: 1685, sixty years after the Roman celebrations for *The Assayer*.

At this point, in fact, Galileo's natural philosophy no longer demanded the continuation of philosophic vigilance. And for a good reason: the initial "scruple" of the Jesuits against those Galilean ideas was now satisfied; their good theological reasons were now officially recognized. From now on, it was no longer a bitter history of ideas that unfolded, but rather a judicial chronicle.

If reasons of state had made it possible for Galileo to escape the real, serious incriminations to which his challenge to institutional Catholic culture was liable, now Father Grassi's memory and honor had been redeemed. Finally, after so many new heresies and controversies, which amply proved the well-founded truth of Grassi's original denunciations, Galileo's physics was also officially incriminated in Rome, at the Holy Office, as the most dangerous—more than that of Descartes and Gassendi. Moreover, it was done openly, no longer dissimulated behind the fragile veil of official condemnation for Copernicianism.

The public consequences of that by now officially recognized condemnation were already evident. On December 2, 1676, in the fateful hall of the Convent of Santa Maria sopra Minerva—precisely where Galileo had publicly abjured his Copernican opinions—the Olivetan priest from Lucca, Father Andrea Pissini, atomist and supporter of the subjective character of heat, taste, and odor, abjured *his* heresy: "I correct myself and retract, and confirm that in the Eucharist, whether one sees it or not, touches it or not, whether in a pyx or elsewhere, there is always actually the Sacrament."[98]

From 1688 to 1697, an important trial took place in Naples against a nucleus of "atheists" which had been discovered there: innovators of the Academy of Investigators and atomists, some Cartesian and some not.[99] By October 1691, it was no longer possible, not even in Florence, to save appearances. The Grand Duke Cosimo III condemned the subterranean, private, and word-of-mouth survival of *The Assayer*'s ideas, which had continued in dissimulation and in secret. Also, private teaching "in writing or by voice" of the "Democritan or atomist philosophy" was prohibited.[100] The most recent intemperances and threats to the faith were stigmatized in the form of a collective denunciation, the *Lettere apologetiche in difesa della teologia scolastica e della filosofia peripatetica* (1694), by Father Giovanni Battista de Benedictis, prefect of the Jesuit schools, in which he summed up the significance of a half-century of theological and cultural action by the Collegio on the important front of resistance to the "new philosophy."

For some time now, behind these official prohibitions with a preventive purpose and behind the judicial repression on the part of local inquisitions in both Naples and Venice,[101] there had existed in Rome, from where the operation was directed, an official incrimination. It could now be communicated also to Vincenzo Viviani, who had been so convinced of seeing his teacher rehabilitated with the cancellation of the 1633 sentence that he had prepared for the occasion a *Life of Galileo* in the form of a hagiographic eulogy.

However, the Jesuit Father Antonio Baldigiani, Consultor of the Holy Office, informed Viviani that the incrimination of Galileo, for reasons that had nothing to do with Copernicus, had now been formalized by the Holy Office:

There have been held, and are being held, extraordinary congregations

[98] See *Ritrattazione del padre A. Pissini olivetano*, Rome 1677, reprinted in G. M. Pezzo, *Dissertatio physica. Theologica de Accidentibus Eucharistia*, Neapoli 1735; and in F.J.A. Ferrari de Modoetia, *Philosophia peripatetica*, Venice 1754, II, p. 170. See Jansen, "Eucharistiques (accidents)," col. 1433 ff. Father Pissini had published a *Discorso filosofico sopra le comete* . . . , Ascoli di Satriano 1665, of a physical and astronomical character and had initiated the publication of an atomist work, *Naturalium doctrina*, Augsburg 1675. See H. Hurter, *Nomenclator theologiae Catholicae*, 3rd ed., Oeniponte 1903-1913, vol. IV, p. 173.

[99] See L. Osbat, *L'Inquisizione a Napoli. Il processo agli ateisti 1688-98*, Rome 1974.

[100] R. Galluzzi, *Istoria del Granducato di Toscana sotto il governo di Casa Medici*, IV, Florence 1781, p. 409.

[101] See A. de Stefano, "Un processo dell'Inquisizione veneziana contro M. Fardella," in *Siculorum Gymnasium* I (1941), pp. 135-46.

> of the cardinals of the Holy Office and before the pope, and there is talk of a general prohibition of all the authors of modern physics; long lists of them are being made, and *these are headed by Galileo*, Gassendi, and Descartes as most pernicious to the literary republic and the sincerity of religion. The chief persons to form a judgment of them will be the religious, who *at other times have made efforts to issue these prohibitions.* . . .[102]

The efforts of these religious in safeguarding the Tridentine faith were crowned by the recognition of the truth, even if very belated, that "maneuvers went on for a long time" behind "this affair," as Marcello Malpighi also confirmed to Viviani, writing from Rome.[103] Now the Galileo affair was truly resolved.

[102] Cited in A. Favaro, *Miscellanea galileiana inedita. Studi e ricerche*, Venice 1887, p. 155 (ms. gal. 257, a.117, National Library of Florence). Emphasis added.

[103] Cited in Torrini, *Dopo Galileo*, p. 28 ff.

CONCLUSION

I am very sorry that you have discovered everything to be atoms . . . so that you have given your enemies the opportunity to deny all the heavenly things that you have shown us.*

The Staging of a Drama

Sunday, the seventh day of the month of Bichat—the month of modern science—was, according to the Positivist calendar proposed by Auguste Comte (between 1849 and 1854), the anniversary of the scientist Galileo as martyr of the religion of humanity and eponymous hero of a week that began on Monday with the name of Copernicus.[1] The progressive publication of the trial records (between 1850 and 1907)[2] fed the fervor of Galileo's Positivist cult, producing an oleographic hagiography destined to make a lasting impression on our eyes.

In the autumn of 1982, when the present book was already finished, I was able to see at close range an image of that period at the Palazzo Vecchio in Florence, in a show entitled "The Secret Museum" which assembled works from private collections. From a private Milanese collection had been sent a painting from the middle of the nineteenth century—an important work (awarded the silver medal at the 1857 Florentine Exposition of the Society for Promotion of the Fine Arts) by the Tuscan painter Cristiano Banti (1824-1904) and entitled *Galileo before the Inquisition*.[3] It was a very realistic painting, worthy of being the stage setting for an opera of the period.

It depicted Galileo, in a room at the Palace of the Holy Office, before his judges: Father Commissary Maculano and his two assistants. Light from an unseen window illuminated the inquisitorial intentions of the father com-

* "Assai mi duole c'ha scoverto tutto atomi . . . Sicche had dato manica a'nemici di negar tutte le verita celesti che Vostra Signoria ci addita."—T. Campanella, letter to Galileo, March 8, 1614, *Works*, XII, p. 32

[1] See A. Comte, *Système de politique positive*, vol. IV, Paris 1854.

[2] Starting with Monsignor M. Marini, *Galileo Galilei e l'Inquisizione*, Rome 1850 (where the fateful transcript of Bellarmino's 1616 injunction was first published), down to the most exhaustive critical edition of the proceedings: A. Favaro, *Galileo e l'Inquisizione*, Florence 1907. There were decades of furious polemics between apologetic historians and anticlerical scholars. One recalls E. Wohlwill, *Der Inquisitions Process des Galileo Galilei*, Berlin 1870; the edition by Berti, *Il processo originale di Galileo*; and Moritz Cantor, who, like Alberi, saw his request for access to the original documents rejected. Only Gebler satisfied the need for a rigorous edition (though still partial) of those trial proceedings: *Galileo und di römische Curie*, Stuttgart 1876-1877. On this period, see A. Carli and A. Favaro, *Bibliografica galileiana (1568-1895)*, Rome 1896.

[3] Reproduction in catalogue, *La città degli Uffizi*, Florence 1982, p. 302. On the Florentine exhibition of 1857 and the figure C. Banti, see G. Matteucci's catalogue and monograph, *Cristiano Banti*, Florence 1982.

missary, standing next to a large crucifix, his finger threateningly pointed at the page of a volume lying open on the table (the *Dialogue*, most likely). There is also a page of protocol: perhaps the famous record of Cardinal Bellarmino's controversial injunction. Of the two Dominicans seated at each side, the one on the left, visibly alarmed, clutches the collar of his cape in anguish while the other, on the right, with a hood down to his eyes, seems instead to be overcome by a sudden attack of sleepiness (Fig. 18). It is not excluded that the artist intended this to represent the three deadly vices of ignorance: fanaticism, fear, and sloth. In addition, as if to defy them, Galileo is represented erect with features and stance that are extremely youthful for a seventy-year-old man suffering from arthritis, and he is circumfused by truth's luminous halo. A hand resting on the folds of a mantle confers on his posture the elegance of a bullfighter in the arena. Galileo looks disdainfully past his judges, toward the future. One might say that he is preparing to intone the aria of the truth that will be his downfall.

I would rather not express an opinion on the aesthetic qualities of this suggestive hagiographic image. As for its affected historical realism, I would say that it is a moving fake with all the necessary appearances of historical reliability—neither more nor less than any of the two thousand Galilean autographs in the famous Chasles collection fabricated during those same years (1867-1870) by the talented hand of the autodidactic counterfeiter Vrain Lucas.[4] Even though the paper and ink defied the scientific evaluation of the Academy of Sciences, they were fakes nonetheless.

At the beginning of our century, the great French scientist and philosopher Henri Poincaré declared in regard to modern scientists that "the faith of the scientist resembles the anxious faith of the heretic, a faith that is always searching and is never satisfied."[5] Was Galileo a heretic? It was precisely the unsatisfied disquiet of Galileo's physics that made us pose certain questions. Why such reticent dissatisfaction on the part of the author of the *Discourses* in regard to the theory and natural philosophy of *The Assayer*? Why so sharp a gap between the ambitious philosophy of the initial materialism, consecrated by intellectual success, and the succeeding purely mathematical solution, accompanied by the final, prudent methodological reservations?

Many plausible patterns of epistemological response were available. All were rationally adequate. But Galileo's mysterious and fascinating reticence was enough to prevent us from subscribing to them without further questions, and to prompt us to follow the line of those questions along the traces of the shadow left on the intellectual episode by certain moments of remorse, certain premonitions and judgments. Persistent and revelatory shad-

[4] See H. Bordier, ed. E. Mabille, *Une fabrique de faux autographes ou Récit de l'affaire Vrain Lucas*, Paris 1870.

[5] H. Poincaré, *Savants et écrivains*, Paris 1910, p. viii.

ows, as in a seventeenth-century painting, massed all around the luminous official support won by *The Assayer* in Rome.

It was the disquietude of Galileo's studies in physics—a great, dissatisfied disquietude—which obliged us to examine Galileo's human and intellectual parabola without falling victim to the effects of deforming retrospective illusion. Since we are in the seventeenth century, I will call this distortion the "Don Quixote" effect; that is, the image of a scientist who does not know dissimulation and wants to live in the declared truth at all costs, always. But perhaps in Galileo's experience there is another reality—reticent, elusive. We searched for the reality left in the dark by official history, asking ourselves whether it might be determined by the value of the extralinguistic and extrascientific truth of Galileo's substitution of theories in physics.

Insensitive to the old Enlightenment and Positivist schemas, this book has tried to be sensitive to the civilization of the seventeenth century: the century of the revelatory shadow in painting and of honest dissimulation, the art of prudence in civil and intellectual life. Galileo's time was the inexhaustible cradle of astuteness and stupefaction. It has reserved for us the emotion of new answers—of new documents hidden in the dark areas of Galileo's story, and in the vicissitudes of Galileo's story and of the scientific revolution in which he was a pioneer and, if one wishes, a martyr.

Retrospective Illusions

One might well ask me ironically whether the newly discovered documents explain everything. No, not all. They only teach us how to look at the same appearances in a different way, and that is what counts. They teach us some of that art of the gaze which our modern epistemological eyes have irremediably lost.

Now, if we try to evaluate the new documents and those re-evaluated by this book in the light of their logic and their history, one can no longer avoid seeing Galileo's heresy in a new light: that is, his scientific and philosophic disquietude, but also the disquietude which cost him the loss of the pope's protection and the condemnation of the Church. Therefore, this book does not explain everything: it tries to understand, and to understand from within. It was the old schemas that explained everything, so as to save appearances and impede comprehension. This book proposes to change the evidence, preserving the truth but, above all, the complexity of the past.

Was Galileo a heretic? This book has tried to show that Galileo's heresy did not stand at the center of a dramatic scene in which two truths confronted each other: the Copernican truth, affirmed by Galileo with a faith that seems to verge on scientific dogmatism; and the Scriptural truth, affirmed by the Church with a faith that seems to verge on irrational fanaticism. I have tried to show that the problem of Galileo's heresy was an interweaving of two disquietudes: the reticent disquietude of the turns and twists in Galileo's physics and the disquietude of the period's controversial and polemical theological speculation.

In place of the traditional stage setting for the affair, with two changes of locale, one in the apartments of Cardinal Bellarmino and the other in the trial hall of the Holy Office, we have constructed and installed new scenery. Galileo's drama has here, in fact, been put on the stage of that "theater of marvels" which was the Rome of the first phase of the Barberini pontificate, when Rome was young and liberal because young and liberal was its new pope, who renewed it on rediscovered Christian foundations, and when Galileo was at the height of his parabola as the official Catholic scientist.

I would like to think that a contemporary of Galileo would have appreciated this book's stage setting: a play of reflections that spring from muted sources and invisible mirrors, whose purpose is to make fictitious appearances seem real, to make work of astuteness and compromise appear rigorous and irreproachable, and to make that which was political and religious expediency find its reflection in official documents. Lights and shadows. The effort of this book has been to avoid anachronism at all costs, including the cost of eschewing our rational, clear, and evident comprehension for the comprehension "in conformity with reason" of Galileo's time. That reason conformed to brightness and shadow: "to proceed openly and hidden," as Torquato Accetto teaches, and as the natural reason of the seventeenth century imposes, since nature wishes that "there be day and night in the order of the universe."[6]

Thus we have gone to the academies and libraries of the "literati," those moralistic and noisy innovators, and to the Galilean "virtuosi," withdrawn and austere like new Christian ascetics. These were settings deserted by traditional Galilean historiography, since the positive scientific effects of that ephemeral philosophic and civil mobilization in the shadow of the new regime, and of its political liberality, were of no consequence for the history of science. And it could not be otherwise.

The effects were inconsequential, but also miraculous, when those literary men in Rome, without mathematics or astronomy, those courtiers of the new pope, swore on the book of nature a new alliance among literature, Galilean philosophy, power, and faith. Many of them were mystics, devout Christians, more than Counter-Reformation Catholics. In the political alliance with France they looked toward the Christian king and the new mysticism of the French theological school, the new cardinal-theologian of the new papacy, Bérulle, forgetting Cardinal Bellarmino.

[6] See T. Accetto, *Della dissimulazione onesta* (1641), ed. B. Croce, Bari 1928, pp. 25-26.

They are Copernican and Galilean Catholics, like the pope who protects them. Their philosophy is that of *The Assayer*, the great Trojan horse of a Roman cultural change. Theirs is a Christian philosophy which cannot be, and does not wish to be, a speculative theology like that of the Jesuits. It cannot and does not want to enter into the area of religion, because the latter goes beyond reason. Within these limits, "pure natural terms," the rights of reason are saved and must be protected. And what about the rights of faith?

Tridentine faith—the great, impassioned reawakening of Scholasticism—penetrates into the area of reason and philosophy. It does not have the right to do this in astronomy, but it has the right and duty to do so in physics and philosophy; for a great dogma, the great Tridentine dogma, demands the protection to the letter of those words of St. Thomas in which it is formulated. The spectacular De Dominis trial is the great revelatory warning that presents the consequences of any scandal of deviation from the Curia's Tridentine line. That trial's display, magnificent to the point of fanaticism, exorcizes the suspicion of such a scandal.

The profound voice of Scholasticism has its loudspeaker in the Collegio Romano, with the paroxysmal ostentation of its Tridentine faith: a university-church; a total and modern rational abnegation; a fidelity to Trent which finds it exalting to completely channel philosophy and science, renewing a century's research and culture on the basis of Tridentine dogmas and the safeguarding of them. If the Galileans, to our modern eyes, are an ideological movement tied to a new pope, the Jesuits are the universal party of the papacy, with a thankless task and exposed to all the personal and official frustrations of a theological police, often secret, always militarily organized. They are the party of loyalty to Trent and to the Counter-Reformation Church's reason of state, which binds them indissolubly to fidelity to the Catholic king and dedication to the Hapsburg bloc. These are the actors and the roles. The Roman followers of Galileo and the Jesuits of the Collegio Romano both feel they are the promoters of renewal, the interpreters of new times, the Catholic aristocrats of thought. There is a great charismatic afflatus at the religious and intellectual root of their innovative effort, and yet both claim to act within the Church. But their efforts were at loggerheads, and a clash was inevitable.

Thus, this book has paused in order to consider the tensions which each confronted vis-à-vis the Roman success of *The Assayer*. Certain possibilities sprang up, others were doused. It was as though both sides had been weighed in the delicate scales of time, in order to accept or reject the traditional formulas. For centuries, however, and above all in the seventeenth century, the Church did not entrust its fate solely to faith. It relied on its capacity of resistance to external adversaries and on a government that, along with other governments, played the game of politics and war.

Politics and war were flung, with their enormous weight, on the pans of that scale, shifting them. A turn to political liberalism and Catholic military success had brought Galileo triumphantly to Rome to accept recognition as "a devoted son of the Church." A political turn to conservative rigidification brought him back to Rome as a compromising culprit.

Was Galileo a heretic? Galileo, on orders of the pope, who had arrogated to himself the secret inquiry of the famous trial of 1633, was officially condemned by the Holy Office because vehemently suspected of having believed in the Copernican doctrine, despite his repeated formal denials before the ecclesiastic authority (in 1616 and in the *Dialogue*). This heresy was inquisitorial—that is, disciplinary, not theological or doctrinal—both according to the words of the manuals of criminal heresiology of the period and as reported by the most serious juridical scholars of the affair.[7]

Inquisitorial, disciplinary heresy—that is, in modern but exact words, Galileo was officially condemned in 1633 for high treason. That official condemnation was the resolution of the drama, not the drama itself. Galileo's trial was a liberating exorcism directly related to a very serious political crisis of the preceding year, just as a liberating exorcism was directly related to that same threat in those very same days, when the solemn Papal Mass of December 11, 1632, in Santa Maria dell'Anima (the church of the Germanic nation) gave thanks for the death of the King of the North, the "terror of the universe," at the Battle of Lützen.

Also, Galileo's trial was orchestrated with the maximum propagandistic effort. By order of the pope, Galileo's condemnation for high treason was sent to all offices of the nuncios; it was publicized in Vienna, Madrid, Prague, Paris, and Brussels. This was an affair of state, not a matter of conscience. It was "a manifest proceeding"—too manifest not to be apparent.

But this book has not been written to show that the Galileo case was an affair of this kind, but to show that, beneath its appearances as an affair of state, it was truly a very serious matter. To the "manifest proceeding" of political motivation corresponded inevitably—we should always remember that we are in the seventeenth century—a "hidden proceeding." Thus, we have seen a trial before the trial, a staged production in a theater of shadows and silent mimes: the secret silences which surround the extremely serious incrimination. This brought the situation to a head and compelled the pope to oversee it personally, through a special commission's secret inquiry, so as to put an end to the affair and the suspicions in an exemplary condemnation for high treason.

It has not been the purpose of this book to furnish Father Maculano with a complete trial dossier, to provide all the evidence he lacked in the spring of

[7] See, on this subject, the juridical point of view on the difference between doctrinal heresy and disciplinary heresy as explained by L. Garzend in *L'Inquisition et l'hérésie*, Paris 1913.

1633. That would have been a superfluous effort. How many books have tried to do so, from the most diverse viewpoints, all focused on that sentence as on the mirage of a retrospective illusion, owing to the great, oppressive ideological significance of the condemnation. The present-day Church, too, announcing that it is giving up its old viewpoints, proposes the publication, the republication, of all the documents in that court's file, ready to recognize possible judicial errors (as if Galileo's trial were susceptible of a classic analysis). The initiative is meritorious on the plane of erudition; on the plane of historical reality, however, it does not change the events.

One does not change the reality of a trial which is not susceptible of classic analysis because it was called for and directed by a threatened dictator, subjected to the mounting pressures of the gravest crisis of his regime. The recognition of judicial errors, the rehabilitation of the victims of political trials—these are of convincing effect for his successors, but of scarce consolation to historians.

Galileo's condemnation was also officially presented by this power as high treason; it was neither the first nor the last such case. Are we willing to remain entangled in the contradictory game within the mechanism of such a deceptive scenic apparatus as a political trial? No, as historians we are not. Was Galileo a Copernican? Certainly, at least as much a one as was needed for his official condemnation. A "monster," some shocked judicial analysts will call the sentence. A monster, yes, but with the significance that word had in the seventeenth century: a prodigy; indeed, an "inscrutable oracle," for that is what an official sentence of the Holy Office was.

We should not be dazzled by that oracle, with our modern eyes unaccustomed to the shadows. With the gaze of the men of the seventeenth century, we should learn to disassociate manifest appearances from the hidden reality: the appearance of a veil of innocence and authority mercifully thrown by the reality of a condemnation over a scandal in the Curia. It was a severe, official punishment (albeit alleviated by exceptional privileges), but an indirect punishment. Since Galileo had been proclaimed by the pope, in 1624, to be his devout son, the son's all-too-serious sin, even suspicion of it, would fall too heavily on the father's shoulders.

In the seventeenth century, whose complexity of political calculation and psychological phenomenology today escapes us, reasons of state and reasons of faith constantly have recourse to dissimulated, masked punishments, in order to avoid scandal and nourish the consolation of the people of God. This art of dissimulation, this art of prudence—prime political and religious virtue of the powers that be—does not leave behind clues. Almost never, for we actually did find some. From the light of that cruel piece of stage management—Galileo's spectacular abjuration—in the shadow of calculation and interest, precious as a nocturnal flower, the secret is born.

"Secret"—that word is the one most often repeated in this book, as it follows the traces of things that are done but not said, or problems that must not be pronounced because merely by speaking one can become an accomplice. In Galileo's time, the secret has a great ally: suspicion. And together these two are a universal presence which animates men's life and thought.

The denunciation of *The Assayer*, which remained in suspension and then was recalled by an inevitable logic, was a suspicion as well as a secret. It was a suspicion that could not be evoked without arousing, at the time of De Dominis's trial, the image of betrayals, troubles, and heresies.

It was the nominalist and atomist heresy against the dogma of the Eucharist. We have been able to appreciate how much it was the order of the day at the moment of *The Assayer* and how, in the hands of the Jesuits, it was kept up-to-date for fifty years as a formidable war machine and strategic weapon of dissuasion against the "architecture" of the new philosophy and the new physics.

We have also seen what were the qualifications of its modern rediscoverer and perfecter, Father Orazio Grassi, moral and scientific authority of the Collegio Romano, a great architect. He knew very well that the dogma of the Eucharist was the touchstone in judgments between heresy and orthodoxy, because that is what the great protagonists of the Tridentine tradition had taught. Father Grassi, high personality in the Society of Jesus, also bore in his heart the reflection of that great will to prevent another disintegration of Catholic theology, a new ruination of the Church.

Was Galileo a heretic? The accusation was immediately shelved. It was officially spread about, and the threat of denouncing the scandal became a possible source of pressure when the *Dialogue* set forth again *The Assayer*'s atomism. So far as one can know, and it is logical to think so, Galileo never pronounced himself on the problem of the Eucharist. On principle, physics and the new philosophy never entered the sphere of faith. It was faith, however, that could not avoid going into physics and philosophy.

Translated into Galileo's atomist and nominalist language, the sensible phenomena of the Eucharist could no longer be reconciled with Eucharistic transubstantiation. Hence, insofar as they were irreconcilable with a dogma, these were doctrines both false and dangerous, even though not raising directly the doctrinal point in question: in such a case "a false proposition [is] repugnant to Catholic faith insofar as irreconcilable with it," the *De haeresi* (1616), that reliable treatise of criminal heresiology compiled under the direction of Cardinal Bellarmino, had specified.

Only now, in a full political crisis of the papacy, which despite itself was involved in the affair, it certainly was not possible to upgrade the accusation of complaisant tolerance to the very serious suspicion of heresy. One could not, for reasons of state and of faith, bring up and punish publicly so compromising a suspicion. It must be rendered inoffensive.

Repudiate Galileo with great severity, obviously. But also get rid of the denouncers and witnesses, the living proof of that suspicion. And that is what was done.

Of this whole grand affair, until now unknown, we have taken apart the documents, we have found and put back in their proper places some missing pieces, we have reconstructed a history and have arrived at this conclusion: Galileo's condemnation was not a personal matter. Because of its unique aspects, Galileo's condemnation was a historic event. We have said how. But it was also a philosophic event because of what it contained that is exemplary for the history of the seventeenth century's scientific revolution.

A Common Good

Scientific revolution: a revolution of ideas. In physics this revolution of ideas is a stirring one, because that discipline, which is not supported by any classical mathematical tradition, wants to leave traditional metaphysical philosophy behind. It claims now to have the enormous privilege of walking solely with the legs of rational experience and mathematics and of traveling far and everywhere, into the heavens and the invisible structures of terrestrial matter.

It is a revolution greater than that which takes place in the astronomer's sky. It brings the heaven of ideas down to earth, and causes disarray in common and philosophical knowledge. Astronomy is always classical, always mathematical. But matter changes, passes from the continuous to the discontinuous, from the visible to the invisible.

Believe in order to see, proclaim the authors of *The Assayer* and *The World*: let us distrust sensible appearances; let us investigate first our possibilities of having objective knowledge, and then we shall see that beneath the sensible phenomena are hidden corpuscular structures, objectively knowable by reason. See in order to believe, retort the Jesuit scientists and philosophers: we see the miracle of the permanence of sensible Eucharistic appearances even after the substance of bread and wine is totally changed. For the miracle not to offend man's reason, the distinction between quality and substance must be defended; and the idea that the phenomena really do exist, that they are qualities independent of us, must be maintained.

Here we are at the heart of Counter-Reformation faith and, at the same time, of the scientific revolution. The Copernican question, with respect to these, is an incident along the way, a misunderstanding. Here we do not have a question of some Scriptural passage interpreted with archaic exegetical instruments, even taking into account the possibilities of the age; rather, it is a question of a new, rigorous dogma, one of the capital articles of Catholicism. This is a question of the great theological problem of the century, the great knot of faith and controversy.

We, historians of science, shrug our shoulders. What has Eucharistic theology to do with Galileo's and Descartes' physics and mathematics? It is a pseudoargument of some Jesuit poisonous as a scorpion, a purely dialectical argument to injure Galileo, a "Machiavellian maneuver" loaded with malice. It is an extrascientific argument—and therefore not serious, uninteresting—an infamy, unworthy of historical consideration, at the most to be included in the inventory of theological fanaticism of some cultivated Jesuit scientists. I think that it was arguments of this kind which led to the complete neglect of the problem, even when it was quite visible, against which Galileo's physics collided. And what about the Eucharistic theories of Descartes, Varignon,[8] and Leibniz?[9] We, historians of science, smile indulgently.

Indeed, how can one explain the bitter passion, incomprehensible to us, with which so many men of culture—scientists, theologians, and philosophers (often one and the same person)—gave the best of their intellectual concerns and also their lives to arguments which seem so academic to us today? How can one explain to those Catholic men of the seventeenth century that we today do not understand their speculative passions, and that we prefer to look elsewhere because they embarrass us, insofar as we are all Galileo's descendants and have after so much time become, Catholics and laymen alike, modern scientific bigots?

We smile aloofly. But in the seventeenth century, squads of theologians and intellectuals fought on the various fronts of that dogma. For many of them, that Tridentine formula was more than a principle of authority; it was a principle of identity, a reason for being.

Trent proclaims with philosophical certainty the real presence of Christ in the Host. It is the immense privilege of knowing and comprehending that God comes to earth by virtue of a sacramental formula, and that He is visible to men, reincarnated as Christ under the Eucharistic species. To see is to believe.

The formula of the dogma summed up a secular alliance between faith and

[8] See P. Varignon, "Demonstration de la présence réelle du corps de Jésus Christ dans l'Eucharistie," in Anonymous, *Pièces fugitives sur l'Eucharistie*, Geneva 1730, p. 8 ff.

[9] See "Systema theologicum Leibnitii," in A. J. Emery, *Exposition de la doctrine de Leibniz sur la religion*, Paris 1819, pp. 220-68. Mass and *vix motrix* were for Leibniz the only two permanent Eucharistic accidents, distinct from matter (p. 236). On the "irenic" nature of these mss. in the Hanover Library, see J. Baruzi, *Leibniz et l'organisation religieuse de la terre*, Paris 1907, p. 242 ff. See then Leibnitz's letters to Arnauld and Duke Federico, in C. I. Gerhardt, *Philosophische Schriften*, 7 vols., Berlin 1875-1890, vol. I, pp. 75 and 61; "De vera methodo philosophiae et theologiae" (1673-1675), in *Philosophische Schriften*, vol. III, Berlin 1890, pp. 154-59; *Demonstratio . . . Eucharistiae, Samtliche Schriften*, VI, I.

reason. Galileo appeared to challenge that rational pact sanctioned by speculative theology. Was this a "Machiavellian" trick? Or an *ad hoc* argument, the fruit of a second-rate ideology, purely tactical, a spur-of-the-moment objection? I do not think so: it was a great theoretical problem.

Do we want an example? Let us open an important scientific book, one which contains undoubted contributions to so-called positive science: the *Physico-mathesis de lumine, coloribus et iride*, the great posthumous work of 1665, from the Jesuit physicist Father Francesco Maria Grimaldi. In that large volume, we find the discovery of the diffraction of light and the theory of a luminous fluid (not necessarily substantial) whose vibratory state makes it possible to hypothesize the appearances of colors in bodies as owed to undulatory motions of the luminous fluid. And when the Consecrated Host is in the darkness of the ciborium, what then will become of the permanence of its whiteness? We read these long pages devoted by Father Grimaldi to the delicate Eucharistic problem with a concern neither to falsify his theory in regard to the dogma nor to put the Eucharist in contradiction to the proposed new theory of colors. How can "we save the eucharistic appearances?"[10] this physicist asks explicitly. And the answer is not easy. It was in fact a serious problem, not a facile "Machiavellian" maneuver.

Father Grimaldi, one will object, was a Jesuit, and his great book on the physics of light had fallen victim to its author's submission to the scientific persecutions and prohibitions set in motion by the Society of Jesus during those years, even within its own organization. So, then, let us withdraw from the orbit of those persecutions and threats. Let us get out of Italy and open the dialogue on *Astronomia physica, seu de luce*, published at Paris in 1660 by such a liberal and eclectic spirit as Jean Baptiste Du Hamel, a prudent Cartesian who had nothing to fear from the Jesuits, protected as he was by Cardinal Antonio Barberini and by Colbert. Here, too, we have a new scientist, a consecrated experimenter, secretary (from 1669 to 1697) of the new and already prestigious Académie Royale des Sciences. Modern science has already begun, and yet the characters in Du Hamel's dialogue cannot avoid discussing, before all else, the serious problem that a theory of light poses with respect to the dogma of the Eucharist.[11]

The endeavor to issue from the past was arduous, even for the men of the new scientific culture. It was a fundamental problem, a philosophical prob-

[10] See F. M. Grimaldi, S. J., *Physico-mathesis de lumine, coloribus et iride libri duo*, Bononiae 1665 (post.), p. 533 ff., for criticisms of atomism; p. 396 ff. and p. 406 for the Eucharistic problem of light as substance. On p. 407, see the adaptation of Grimaldi's hypothesis to the phenomenon of Eucharistic appearances. Obviously, the book's publication provoked censures and Roman revisions from the Society of Jesus.

[11] See J. B. Du Hamel, *Astronomia physica, seu de luce, natura et motibus coelestium libri duo*, Parisiis 1660, p. 2 ff.

lem and one of a mindset rooted for centuries in knowledge and ideas, not an argument of the moment.[12]

It is from this point of view, against this background, that Galileo's condemnation should be considered. Let us not be myopic. Let us not examine it starting with 1616 (when Copernicus was put on the Index). Rather, let us examine it from a distance, from a great distance.

A profound wave rises up against Galileo, from the patrimony of medieval culture all the way to the history of philosophic and scientific ideas. It reaches Galileo and continues to roll throughout his century. It is the history of speculative Eucharistic theology, of the joining of hylomorphism and Catholic religion at a point of enormous attraction.

It is a history that still remains to be written, one that this book has evoked only approximately. One must make the scientists speak, but also the theologians and the heretics—as well as the artists and the popular devotions and even the witnesses of gold and stone of this capital event in European thought and knowledge.

What is the advantage, one will ask at this point, of again digging up a controversy that was a dialogue among the deaf, having no beneficent (but only a delaying) effect on the history of modern science? Certainly the theories of physics, unlike theology, did not obtain results from the century-old struggle, only obstacles.

But there was an effect, and that history will make us appreciate it. It was the effect of making us conquer the autonomy of research and reason from which we benefit today. And one might appreciate the fact that it did not descend to earth from the heaven of Plato's ideas, but was conquered at great cost in the seventeenth century, like every other human freedom. It is a common good, which must be safeguarded.

I say this without partisanship, with that same freedom of comprehension which today is everyone's, asking forgiveness both from those lay scientists who would gladly prefer to expunge such negligible and thorny material as the philosophic and scientific problem of the Eucharist and from those devout scholars who would prefer to monopolize problems such as this. I want to conclude my book precisely on this aspect. Problems such as this do not belong exclusively to anyone, just as Raphael's fresco *The Dispute concerning the Holy Sacrament* is not anyone's particular property. At any rate, whatever the dramatic consequences might be, it is a question of problems which today belong to the heirs of that controversial culture—to all modern men. And like every intellectual problem, this one too is a precious common good.

[12] For the full critical and historical scope of intellectual tangles such as the one discussed here, see L. Febvre, *Au cœur religieux du XVIe siècle*, Paris 1957.

DOCUMENTS

G3*

Having in past days perused Signor Galileo Galilei's book entitled *The Assayer*, I have come to consider a doctrine already taught by certain ancient philosophers and effectively rejected by Aristotle, but renewed by the same Signor Galilei. And having decided to compare it with the true and undoubted Rule of revealed doctrines, I have found that in the Light of that Lantern which by the exercise and merit of our faith shines out indeed in murky places, and which more securely and more certainly than any natural evidence illuminates us, this doctrine appears false, or even (which I do not judge) very difficult and dangerous. So that he who receives the Rule as true must not falter in speech and in the judgment of more serious matters, I have therefore thought to propose it to you, Very Reverend Father, and beg you, as I am doing, to tell me its meaning, which will serve as my warning. [fol. 292r]

Therefore, the aforesaid Author, in the book cited (on page 196, line 29), wishing to explain that proposition proffered by Aristotle in so many places—that motion is the cause of heat—and to adjust it to his intention, sets out to prove that these accidents which are commonly called colors, odors, tastes, etc., on the part of the subject, in which it is commonly believed that they are found, are nothing but pure words and are only in the sensitive body of the animal that feels them. He explains this with the example of the Tickle, or let us say Titillation, caused by touching a body in certain parts, concluding that like the tickle, as far as the action goes, once having removed the animal's sensitivity, it is no different from the touch and movement that one makes on a marble statue, for everything is our subjective experience; thus, these accidents which are apprehended by our senses and are called tastes, smells, colors, etc. are not, he says, subjects as one holds them generally to be, but only our senses, since the titillation is not in the hand or in the feather, which touches, for example, the sole of the foot, but solely in the animal's sensitive organ.

But this discourse seems to me to be at fault in taking as proved that which it must prove, i.e. that in all cases the object which we feel is in us, because the act that is involved is in us. It is the same as saying: the sight with which I see the light of the sun is in me; therefore, the light of the sun is in me. What might be the meaning of such reasoning, however, I shall not pause to examine.

* Ms., Archive of the Sacred Congregation for the Doctrine of the Faith, Rome, Series AD EE, fols. 292r, 292v, and 293r. See Fig. 7.

The author then goes on to explain his Doctrine, and does his best to demonstrate what these accidents are in relation to the object and the end of our actions; and as one can see on page 198, line 12, he begins to explain them with the atoms of Anaxagoras or of Democritus, which he calls minims or minimal particles; and in these, he says continually, are resolved the bodies, which, however, applied to our senses penetrate our substance, and according to the || diversity of the touches, and the diverse shapes of those minims, smooth or rough, hard or yielding, and according to whether they are few or many, prick us differently, and piercing with greater or lesser division, or by making it easier for us to breathe, and hence our irritation or pleasure. To the more material or corporeal sense of touch, he says, the minims of earth are most appropriate. To the taste, those of water and he calls them fluids; to the smell, those of fire and he calls them fiery particles; to the hearing, those of the air; and to the sight he then attributes the light, about which he says he has little to say. And on page 199, line 25, he concludes that in order to arouse in us tastes, smells, etc., all that is needed in bodies which commonly are tasteful, odorous, etc. are sizes, many varied shapes; and that the smells, tastes, colors, etc. are nowhere but in the eyes, tongues, noses, etc., so that once having taken away those organs, the aforesaid accidents are not distinguished from atoms except in name.

[fol. 292v]

Now if one admits this philosophy of accidents as true, it seems to me, that makes greatly difficult the existence of the accidents of the bread and wine which in the Most Holy Sacrament are separated from their substance; since finding again therein the terms, and the objects of touch, sight, taste, etc., one will also have to say according to this doctrine that there are the very tiny particles with which the substance of the bread first moved our senses, which if they were substantial (as Anaxagoras said, and this author seems to allow on page 200, line 28), it follows that in the Sacrament there are substantial parts of bread or wine, which is the error condemned by the Sacred Tridentine Council, Session 13, Canon 2.

Or actually, if they were only sizes, shapes, numbers, etc., as he also seems clearly to admit, agreeing with Democritus, it follows that all these are accidental modes, or, as others say, shapes of quantity. While the Sacred Councils, and especially the Trident Council in the passage cited, determine that after the Consecration there remain in the Sacrament only the Accidents of the bread and wine, he instead says that there only remains the quantity with triangular shapes, acute or obtuse, etc., and that with these accidents alone is saved the existence of accidents or sensible species—which consequence seems to me not only in conflict with the entire communion of Theologians who teach us that in the Sacrament remain all the sensible accidents of bread, wine, color, smell, and taste, and not mere words, but also, as is known, with the good *judgment* that the quantity of the substance does not remain. Again, this is inevitably repugnant to the truth of the Sacred Councils; for, whether these minims are explained with Anaxagoras or Democritus, if they remain after the Consecration there will not be less substance of the bread in a consecrated host than in an unconsecrated host, since to be corporeal substance, in their opinion, consists, in an aggregation of atoms in this or || that fashion, with this or that shape, etc. But if these particles do not remain, it follows that no accident of bread remains in the consecrated Host; since other accidents do not emerge, this Author says on page 197, line 1, that shapes, sizes, movements, etc. do so, and (these being the effects of a quantity or quantum substance) it is not possible, as all philosophers and Theologians teach, to separate them in such a way that they would exist without the† substance or quantity of which they are accidents.

After Suarez t. 2 Met.: disp. 40 f. 2 n. 2

[fol. 293r]

And this is what seems to me difficult in this Doctrine; and I propose and submit it, as regards my already expressed judgment, to what you, Most Reverend Father, will be pleased to tell and to which I make obeisance.

Examination XLVIII*

. . . .

I must now reply to the digression on heat in which Galileo openly declares himself a follower of the school of Democritus and Epicurus. But since he has dealt here in a few lines, without any development, with a problem that deserves an entire book, and since it is difficult for me to discuss it with him, whose principles I do not know, for these reasons I will not make any statement on this opinion. Let him defend it uncontested.

[p. 486]

On this matter, judgment will fall to those who, teachers of a thought in conformity with truth and of scrupulous language, watch over the safety of the faith in its integrity.

Yet, I cannot avoid giving vent to certain scruples that preoccupy me. They come from what we have regarded as incontestable on the basis of the precepts of the Fathers, the Councils, and the entire Church.

They are the qualities by virtue of which, although the substance of the bread and wine disappear, thanks to omnipotent words, nonetheless their sensible species persist; that is, their color, taste, warmth, or coldness. Only by the divine will are these species maintained, and in a miraculous fashion, as they tell me.

This is what they affirm.

† *lo [ro]*, "their," canceled.

* In L. Sarsi (Father Orazio Grassi), *Ratio ponderum Librae et Simbellae*, Lutetiae Parisiorum 1626 (see Figs. 5 and 6), in *Works*, VI, pp. 485-90. In the original text, passages of *The Assayer* were quoted in Latin translation (often in summary). Page numbers in the margin correspond to vol. VI of the *Works*.

Instead, Galileo expressly declares that heat, color, taste, and everything else of this kind are outside of him who feels them, and therefore in the bread and wine, just simple names. Hence, when the substance of the bread and wine disappears, only the names of the qualities will remain.

But would a perpetual miracle then be necessary to preserve some simple names? He should have realized how far he departs from those who with so much study have endeavored to stipulate the truth and permanence of these species, in such a way as to involve the divine power in this effect.

I know very well that certain sly and deft minds, by taking the adversary unawares, would manage to excogitate a loophole to escape this and also make it appear viable, if it were permissible to interpret as one pleases the precepts of faith's sacrosanct formulas and distort their authentic and common interpretation.

[p. 487] But, to tell the truth, what has not been granted for the opinion on the earth's motion, although its immobility is not considered among the fundamental points of our Faith, will be even less permissibile, if I am not mistaken, for that which constitutes the essential point of faith or contains all other essential points.

In the host, it is commonly affirmed, the sensible species (heat, taste, and so on) persist. Galileo, on the contrary, says that heat and taste, outside of him who perceives them, and hence also in the host, are simple names; that is, they are nothing. One must therefore infer, from what Galileo says, that heat and taste do not subsist in the host. The soul experiences horror at the very thought.

So as not to discuss the entire question, we desire in any event to subject to our examination those aspects which are, so to speak, tied to the question as such.

Before all else, I must at least discuss the argument which more than any other Galileo has advanced in support of his opinion; i.e. to maintain that heat and the other sensible qualities are nothing else but that which is felt.

He says in fact: "A bit of paper or a feather . . . touching between the eyes and the nose, and beneath the nostrils, excites an almost intolerable titillation. . . . Now that titillation is entirely in us, and not in the feather (or other light material which impresses it on us by contact). . . . Now, just as evidently, I believe there can be many qualities which are attributed to natural bodies such as tastes, odors, colors, and others . . . and [many of this sort] . . . which outside the living animal I think are nothing but [pure] names."

But, if the truth be known, since it proceeds from the particular this argument proves nothing.

May I, too, in fact be allowed to go against Galileo. The tickle exists in him who feels it thanks to that light attrition of the feather close to the nostrils, since no particle of the feather can penetrate our skin or our flesh. It is, therefore, precisely for this reason that I am led to believe that heat is produced in us by contact owing to the attrition of the body, and that there is no corpuscle released by this body and capable of penetrating, by virtue of its motion, the skin or flesh.

Instead, as regards the object of the discussion, I hold that nobody could cast doubt on the fact that the sensation of heat exists independently from him who feels it.

In fact, this seems to me a completely ridiculous question, because a sensation cannot exist outside him who feels it. One asks, therefore, whether he is something beyond the sensation of heat as such, and whether this heat pre-exists in the heating body or only in him who feels it.

Now Galileo denies that heat is something different from its sensation, that it resides in that which heats. On the contrary, I deny very firmly that heat, considered in this way, resides (to use his words) both in the heated body and in him who is being heated.

....

[p. 488] In the second place, the expression *Motion is the cause of heat*, whose significance was not in the past sufficiently examined by Aristotle, is now expressed in its authentic and true significance: "Here," he declares," is the motion that produces heat; that is to say, the motion of corpuscles which pass through living bodies."

But I say to you, Galileo, this motion is also the cause of cold, of tastes and of all the odors, and you know it well. There will in fact be no reason why what has been accurately stated for the cause of heat should not also apply to taste and smell. But, for my part, I would not consider those ancients, from whom you think Aristotle took this expression, to be so empty as to think that they attribute this cause to heat alone, as if it were essential to heating; for, on the contrary, the ancients considered it the cause of all sensations.

Undoubtedly one can imagine that, when they affirmed that motion is the cause of heat, they recognized a certain special connection between movement and heat, exactly as if someone were to say, pronouncing an almost common phrase, that the sun is the cause of beets and so saying would not be wronging cabbages, beans, and all other vegetables, since all things glory in having been equally generated by so illustrious a parent.

Galileo should not believe, therefore, that the ancients were so reckless as to assign this cause to heat as if it were the principal factor of heat alone. This same cause—that is to say, the motion of corpuscles in which the larger bodies are resolved—is set forth by the ancients for all qualities.

Moreover: "The diversity of sensations," he maintains, "comes from the different shapes of the respective corpuscles"—it can, in fact, happen that round-and-smooth corpuscles produce a sweet taste and angular-and-rough corpuscles a bitter taste.

A given sensation therefore requires corpuscles of a corresponding shape: heat would then be invariably produced by corpuscles of an identical shape. And when light warms, according to Galileo, it will be nothing else but a luminous substance dissolved in indivisible atoms. Indivisibles, then, while they pass through the flesh, divide it and by this means heat it.

But indivisibles do not have shape. Fiery particles are not of this kind, because (being divisible) they allow for shape. Besides, that which is indivisible, according to the meaning shared by philosophers, cannot divide since it does not occupy a place. Light cannot separate flesh, as it consists of distinct and separate indivisibles. Thus, it could not even heat, for heating would consist in a separation of our flesh.

But what is stupefying is precisely the fact that these corpuscles preserve their form, so that no clash or attrition could force them to change it.

....

[p. 489] Galileo, on the contrary, affirms, as I said, that it is at the moment in which something is transformed into indivisible atoms that light is created. He claims that such indivisibles are not physical, like the other aforementioned corpuscles, but mathematical and really bereft of parts: *in really indivisible atoms.*

From here on, I encounter even greater difficulties.

The first of these is the difficulty I mentioned before. In fact, if light exists through the dispersion of indivisibles, one must say, from the moment that indivisibles cannot separate a continuous body, that light cannot provoke any sensation, since, according to Galileo, there is no sensation which does not come from the separation of the continuum.

Light, however, also generates heat. Consequently, either light is not produced by dispersed indivisibles, or heat is not produced by separation of the continuum.

But since in this new manner of philosophizing it seems that one must dare beyond all measure almost anything, provided it be against the ancient and religious philosophy, what will become of him who will refute this thesis of Galileo's on which is founded the entire edifice of his dissertation?

"The tickle," he says, "is nothing outside the sensing body." Let me now become more desirous of novelty than a searcher for truth. I formally deny this proposition, without the slightest hesitation. But, you then come and say to me, to deny that would be to go against everybody's common sense. It does not matter: we are searching for the new, with all the zeal of which we are capable, even if it seems incredible, or almost so.

I declare, then, that the tickle, as well as heat, is a given quality, outside of him who feels it, and that any body might experience such a sensation, provided that its sense organs are naturally formed around the cheeks, above all in the nostrils, around the armpits, and on the soles of the feet. In fact, when one lightly approaches these parts (that is, in an appropriate manner), one excites by virtue of this quality a sensation partly irritating and partly pleasurable. But, you could ask me, Does the tickle reside in the paper or feather which one passes lightly over the cheeks? Precisely. If one means by tickle not the sensation, of which only the animal is capable, but the quality in itself, well, then this is all that makes the sensation exist. It is precisely on the basis of this criterion that I have said that there would not be heat in the fire if one assumed the sensation as the definition of heat. The heat is in the fire, if this sentence has meant anything, by means of that which creates the sensation of heat. But I have said these things only to please Galileo.

Besides, I do not have the words to say how uncertain what he claims appears to me: the sense of touch certainly resides in the entire body, but above all in the soles of the feet and the tips of the fingers. In fact, since these parts are always naturally more calloused and harder than the others, they seem least suited to sensitivity. Nevetheless, we know by experience that the heat [p. 490] or cold of other bodies is more easily felt when we bring close to them the back of our hand rather than the palm or fingertips.

But again, we would see nothing in the light if light were a very rapid and almost instananeous movement, according to Galileo, it will be the most active thing since all energy and activity derives from movement.

Moreover, since it is constituted of indivisibles, it passes through any other thing, for only the size of a body could form an obstacle to penetration.

Instead, fiery particles, which are corpuscles and therefore possess a slower and temporary movement, will be less active and less apt to spread through bodies.

But this is exactly contrary to what we observe: light, in fact, once a lantern has been closed, does not shine outside it. Instead, corpuscles of heat manage to come out of it. These corpuscles would therefore be more active and more prone to penetrate: thus, one must consider them endowed with a more rapid movement and smaller dimensions compared to particles of light.

Yet, from all that has been said one can conclude that everything consists naturally of indivisibles: anything can resolve itself into that of which it is constituted. If, then, all things are composed of indivisible atoms, all equally consist of indivisible atoms.

But may it now be permitted to ask whether indivisibles of this kind are finite or infinite.

They cannot be finite. From them, in fact, would descend innumerable absurdities, impossible in terms of mathematical demonstrations, as Galileo well knows. They are not even infinite in number: indeed, either they are arranged one in respect to the other in such a way that each taken singly is external to the place of the other, or it is certain that many of them would occupy the same space. This last idea, by unanimous judgment, is shown to be false. In fact, in a line a hand's breadth long, the indivisibles situated at the extremities are not found in the middle; nor in the middle are there are extreme indivisibles. Every indivisible will therefore be situated in its respective place, and each will be external to the other.

Therefore, in any line, the first indivisible is followed by another after it, then by a third, by a fourth, and so on. Then one cuts this line at the fifth indivisible. The line thus cut will be composed of five indivisibles. Hence, it will be impossible to divide it into two parts, even though Euclid teaches us that every line can be divided in two.

If someone affirms that indivisibles do not all behave in the same manner, but that some are outside in respect to others and that others occupy the same place together, one must first of all say what criterion makes it possible to affirm that some are in the same place and others in a different place. In fact, if certain corpuscles compentrate each other in the same place, they have this property by virtue of the same criterion of indivisibility, since an indivisible added to an indivisible does not produce anything larger: two indivisibles, given that they overlap completely, in reality occupy as much space as a single indivisible. But, on the basis of the principle by which all indivisibles touch one another, infinite indivisibles, where they touch, will also not occupy more space than a single indivisible. Therefore, they will not produce any extension: in fact, the infinite does not change the nature of things. This may be said in passing and very quickly on a subject which is still not sufficiently clear.

Exercise on Substantial Forms and Physical Qualities*

Nothing comes from atoms

All the bodies of the world shine with the beauty of their forms.
Without these the globe would only be an immense chaos.
In the beginning God made all things, so that they might generate something.
Consider to be nothing that from which nothing can come.

* *Exercitatio de formis substantialibus et de qualitatibus physicis*, anonymous, without printing information, in 4° (22 × 17 cm), 24 pp. (from *Conclusiones physicae propugnandae a Francisco Maria Arrighio in Collegio Floretino Societatis Iesu. Addita est Exercitatio* . . ., H. Navesio [?], Florentiae ca. 1677).

You, O Democritus, form nothing different starting from atoms.
Atoms produce nothing: therefore, atoms are nothing.

Galileo was very inclined to believe that all bodies were constituted of differently shaped and combined atoms and that there are no other qualities or accidents except the local motion of atoms.

This is what he expressly declares on the subject in his book entitled *The Assayer*, in which a digression on the subjects begins with the following words, on page 196, last line:

"I say that whenever I conceive of any matter or corporeal substance, I immediately feel compelled by the need to conceive of it as finite and formed in this or that shape; and that in relation to other things it is large or small, in this or that place or time. Nor by any imagination can I separate it from these conditions; but, be it white or red, bitter or sweet, sonorous or mute, of pleasing or displeasing odor, I have to make no mental effort whatsoever to perceive it as necessarily accompanied by such conditions. Therefore, I do think that these tastes, colors, etc., as regards the object in which they seem to reside, are nothing but pure names, and reside only in the sensitive body: so that, the animal being removed, all these qualities are taken away and annihilated."

These words permit us to formulate a first argument. It is necessary for us to conceive of a corporeal substance insofar as it has shape and insofar as it exists in a place and time, and insofar as it moves or remains quiet. However, it is not necessary to know that it is colored, tasteful, or sonorous: therefore, the shape is substance, but the colors tastes, and sounds are only pure names. The preceding term here implied can moreover be proved on the basis of the doctrine on page 25.

"Philosophy is written in this grand book which stands continually open to our gaze (I say the universe), but it cannot be understood unless one first learns to comprehend the language and to read the characters in which it is composed. It is written in mathematical language, and its characters are triangles, circles, and other geometric figures." Thus from this derives a

Second argument. The entire world is a book written in geometric characters; i.e. triangles, circles. Therefore creatures, which are the characters of this book, are constituted by shaped atoms.

....

Third argument. Tastes reside in a tasty body, odor resides in an odorous body, just as titillation resides in a feather. In the latter, titillation is a pure name. Therefore, in the species of taste and odor, taste and odor are pure names.

. . . .

The fourth argument is of this kind: the corpuscles of the earth produce touch; the watery corpuscles produce taste; the fiery corpuscles, smell; the corpuscles of the air, hearing; the luminous corpuscles, sight. Since they strike the senses by means of a local movement, those which are called qualities are therefore nothing but local movements.

. . . .

From these words results a

Fifth argument. In limestone there are fiery corpuscles which do not emit heat so long as they are quiet, whereas they emit heat when they are moved in a local motion of a greater weight of water: the heating and the heat are only the movement of fiery corpuscles.

It is starting from this doctrine that Galileo deduces in what sense he considers true the common axiom according to which motion is the cause of heat; that is to say, that the presence of fiery corpuscles is not sufficient but that their motion is also required to produce heat.

. . . .

Sixth argument. A greater weight or friction between two hard bodies pushes together the fiery corpuscles in a local motion, and this is what heating consists of. The cause and means of heating is therefore rigorously assigned.

. . . .

Seventh argument. Light is to the other qualities in the same proportion in which the indivisible stands to the divisible. Therefore, just as heat, taste, and odor are divisible atoms moved successively, so in the same way is light constituted by indivisible atoms moved instantaneously.

This opinion of Galileo's exudes Democritus from all sides. But we cannot admit it, above all for three reasons. The first is that it does not seem to be in conformity with Catholic doctrine. That is clear from the Council of Trent, Session 13, Canon 2, when it teaches that in the Eucharist the substance of bread and wine does not subsist.

Let us instead reason in this fashion, first of all: the corporeal substance either consists essentially in sensible atoms, shaped or combined in such a manner, or it does not.

In the first case, God cannot act so that where one finds sensible atoms, shaped and combined in such a manner, there is no corporeal substance. Nor can God act in such a way that where there are not the aforesaid atoms, there is substance.

In the Eucharist there exist sensible atoms of bread (in the same way in which Galileo says that sensible fiery atoms exist in fire), but in the Eucharist there are no sensible atoms of Christ's body.

In the Eucharist, therefore, there is the substance of the bread, but there is not the substance of Christ's body. Both of these two propositions are false.

If Galileo affirms the second, then in dealing with the problem of the essential constituents of the corporeal substance, he involves nonessential constituents.

Usually one responds that God cannot act in such a way that where there are atoms of bread, there cannot be the substance of the bread. In the Eucharist, however, it happens that they can appear, without really existing. Similarly, the body of Christ cannot exist where there are no atoms constituting the body of Christ; and yet it happens that in the Eucharist, although there exist atoms of this kind, they cannot appear as such.

This reply marvelously supports our opinion. In fact, since in the consecrated host there is the appearance of bread, without that which constitutes bread, one deduces that the species (that is, the accidents) are present without the substance. Hence, instead of the substance of bread, which exists in a conatural fashion beneath the species, beneath these species, without any perceptible change, the body of Christ begins to exist in a miraculous fashion. From this it follows that the substance of the bread, which exists naturally, is imperceptible, just as the body of Christ is imperceptible in the host.

The second reason pertains to the fact that the opinion of Democritus seems contrary to the light of reason. It is absurd to admit that there is a soul in man, a rational, sensitive and vegetative soul, and not to admit that there is a sensitive and vegetative soul in animals, or a vegetative soul in plants, or to admit only the one and not the other.

Once again, if in living beings one admits form different from a combination of atoms (that is, the soul), in relation to which one attributes to each being certain properties and operations that are proper to it, why should one not admit a form for nonliving things owing to which different properties can be attributed to different composites?

The same holds proportionately for that which regards physical qualities. One admits in man such vital actions as sight, hearing, and so on. One admits the existence of imaginary beings and intelligible species, all things which are not constituted by atoms but of a quality much more perfect than light, heat, odors, and tastes. Why, then, not admit these qualities, also?

The third reason derives from the fact that the aforesaid opinion seems contrary to experiments in physics. It is known that lettuce is generated from lettuce seed, and that nutriment is transformed into an animal's substance. Or, therefore, that in the seeds of the lettuce exist, as in the earth, all the constituent parts and that in the nutriment something transforms itself into flesh and bone; and only given these conditions can reciprocal choice and aggregation take place, not being possible otherwise.

However, if we admit the first, we see that innumerable other vegetables could be planted and grow in the same ground, in the place of the lettuce. Thus, it would be necessary to admit the existence in the ground of particles of each of these vegetables, which is almost impossible.

. . . .

I respond to the second problem. I distinguish here the antecedent; that is, that we conceive of a corporeal substance inasmuch as it is shaped, because nothing exists in the intellect which did not first exist in the senses. I concede the antecedent, that shape is the essence of bodies. I deny the premise and the consequence.

Let others decide whether Galileo also includes the shape of atoms under the name of prime and real accidents on the basis of what he says on page 197: that tastes and smells do not differ. Certainly he contends, on page 200, that fire consists of shaped corpuscles. Therefore, either fire is not substantially different from nonfire, or the shape in which alone it is claimed to be different must be the substance.

As for the second, to this day mathematicians have not accepted any proposition that is not known in its terms, or that is not demonstrated. So, when Galileo will have demonstrated that the universe is a book written in triangles and circles, we will admit the premise. But until that moment, we reject it.

As for the third, the legitimate syllogism is the following. Just as tastes exist in a body that tastes, so does the tickle exist in the feather. In the latter, the tickle is not produced by corpuscles that penetrate the flesh (just as taste is not produced by corpuscles that penetrate the tongue). Just as the act of bringing a feather close to the nostrils gives rise to an irritating tickle, so also does the application of a sapid body to the tongue give rise to taste. Nevertheless, the comparison is not absolute, since for the tickle the proximity of the feather to the nostrils is sufficient, while for taste it is necessary that the body applied to the tongue be sapid.

Formally, I deny the lesser term of the comparison. A tickle of the species that provoke irritation is more than a pure word. And we are not dealing with a word in a species of formal irritation: similarly, a sapid body is more than a word and is not even a word in a species of formal taste.

If, however, in order to say that colors are pure names, the fact were sufficient that some blind people are not able to recognize colors but at the most recognize sensible qualities, the combination of which results in colors, it might be deduced that colors do not consist in the movement of atoms. Because if that were the case, a blind person would have the perception of this movement, in the same way in which, when he touches an instrument as it is being played, he feels the local movement without which no sound would result.

Against the fourth argument: either sensible things permit one to recognize that they are individually a pure and simple local movement of atoms, or one cannot recognize them. Galileo affirms that they can thus be recognized, but we are dealing instead with the contrary because it is our experience that [the movement] is propagated only by the rippling and the local movement of air or water. But we should also experience the same thing as regards other sensibles, if each of them taken singly were a local movement of atoms.

. . . .

From these words [of Galileo's] we can deduce the following: the wealth of nature is inscrutable in the production of its effects. If, therefore, in regard to the production of sound we do not limit ourselves to a single and exclusive mode, why should we limit ourselves solely to local motion when it comes to the production of qualities? To a pure and simple combination of atoms in the generation of substance?

Those who affirm this last argument remove from nature every generative virtue; they want nature to produce new works starting from old atoms, in the same way that the most miserable Jews sew clothes from old rags and sell them as new.

. . . .

We come now to the fifth argument. There exists an affinity between the conversion of stone into lime, that of wood into charcoal, and that of clay into brick. In fact, after a very intense heating, they are extinguished in a furnace, or in any cavity perfectly closed to the entrance of air.

Hence, either there are also fiery particles in charcoal and bricks, or there are not. If one has fiery particles in these, too, then why when one puts charcoal or bricks in water do they not make it boil? If there are no fiery particles, then why should they be found in lime? And if they really are in lime, why then, by introducing lime into oil does it not burn, even though oil is amenable to fire and, far from extinguishing it, feeds it?

. . . .

As for the sixth: from what has been said concerning the fifth argument, it seems clear that the first part of the antecedent is false. I go on now to respond to the second part. It is undeniable that from a slow but very strong attrition of the pulleys and ropes in the windlass, when very heavy weights are being lifted, that so much heat can be released that it sets fire to the ropes. In what way, then, will such heat be produced? By the decomposition of the ropes and pulleys (which generally is very meager and, in any case, not in the form of fiery particles, but of dust) or by means of fiery particles that are presumed (without any basis) to exist in the ropes and pulleys? What degradation do our hands undergo when we rub them vigorously against each other, even when they are hard and calloused? And how could a pure-and-simple rubbing release so many fiery particles capable of producing so intense a heat?

Hence, the true cause is the ignition of the air owing to very violent friction. In the case of the friction of hands, the cause seems to be the ignition of sweat, which emanates from the hands by compression and rarefaction. From this it follows obviously that 1) Galileo has falsely affirmed, on page 192, that the air and water do not possess attrition and do not catch fire, and that 2) he says falsely, on page 175, that bodies which are not consumed by attrition do not burn.

As for the seventh argument, the assertions in regard to light destroy the doctrine on heat. In fact, Galileo states that light comes from the disaggregation of fiery particles into indivisible atoms and that light has instanteous motion.

It is, however, quite obvious on the basis of experience that light is the cause of heat. We therefore reason as follows: 1) Some indivisibles do not possess any shape whatsoever. The indivisibles of light burn. Therefore, that which burns cannot possess any shape. 2) Combustion is the separation of flesh on the part of fiery atoms. Atoms cannot divide the flesh, since they do not occupy divisible spaces. Consequently, either they cannot burn or the combustion is not a separation of the flesh. 3) Light is constituted by indivisible fiery particles, but heat is constituted by divisible fiery particles. Consequently, these can issue out of a closed lantern (traversing it, propagating the light in the air), which they can do more easily than not coming out from the pores of the lantern to diffuse heat. But that is contrary to experience. Consequently, light is not constituted by fiery particles but is a quality, whose propagation is not [*sic*] impeded by the opacity of the lantern. 4) The light of the sun is spread instantaneously, as Galileo seems to admit. Therefore, either the luminous atoms are in different places at the same instant, which is impossible, or light is not constituted by atoms.

In proper form, I respond that light is not comparable to other qualities from the point of view of the relationship between indivisible and divisible atoms, but that it is comparable to heat inasmuch as it is something which does not have its opposite with respect to that which has its opposite. And with respect to tastes and smells as a simple quality with respect to the qualities which are not simple.

God be praised.

INDEX OF NAMES

Library of Congress Cataloging-in-Publication Data

Redondi, Pietro.
Galileo: heretic (Galileo eretico).

Translation of: Galileo eretico.
Includes index.
1. Science, Renaissance. 2. Science—Philosophy—History.
3. Galilei, Galileo, 1564-1642. I. Title.

Q125.2.R4313 1987 509'.031 86-30581
ISBN 0-691-08451-3